Advances in Remote Sensing and GIS Analysis

Advances in Remote Sensing and GIS Analysis

Edited by

PETER M. ATKINSON

Department of Geography, University of Southampton, UK

and

NICHOLAS J. TATE

Department of Geography, University of Leicester, UK

JOHN WILEY & SONS

Chichester • New York • Weinheim • Brisbane • Singapore • Toronto

Copyright © 1999 by John Wiley & Sons Ltd,
                Baffins Lane, Chichester,
                West Sussex PO19 1UD, England

National   01243 779777
International (+ 44) 1243 779777
e-mail (for orders and customer service enquiries): cs-books@wiley.co.uk
Visit our Home Page on http://www.wiley.co.uk
        or http://www.wiley.com

All rights reserved. No part of this publication may be reproduced, stored in a retrieval system, or transmitted, in any form or by any means, electronic, mechanical, photocopying, recording, scanning or otherwise, except under the terms of the Copyright, Designs and Patents Act 1988 or under the terms of a licence issued by the Copyright Licensing Agency, 90 Tottenham Court Road, London, UK W1P 9HE, without the permission in writing of John Wiley and Sons Ltd., Baffins Lane, Chichester, West Sussex, UK PO19 1UD.

Other Wiley Editorial Offices

John Wiley & Sons, Inc., 605 Third Avenue,
New York, NY 10158–0012, USA

WILEY-VCH Verlag GmbH, Pappelallee 3,
D-69469 Weinheim, Germany

Jacaranda Wiley Ltd, 33 Park Road, Milton,
Queensland 4064, Australia

John Wiley & Sons (Asia) Pte Ltd, 2 Clementi Loop #02–01,
Jin Xing Distripark, Singapore 129809

John Wiley & Sons (Canada) Ltd, 22 Worcester Road,
Rexdale, Ontario M9W 1L1, Canada

Library of Congress Cataloging-in-Publication Data

Advances in remote sensing and GIS analysis / edited by Peter M.
  Atkinson and Nicholas J. Tate.
     p.   cm.
  Selected papers from a meeting held at the University of
Southampton, July 25, 1996, supplemented by invited contributions.
  Includes bibliographical references and index.
  ISBN 0-471-98577-5
  1. Remote sensing—Congresses.   2. Geographic information systems—
Congresses.   I. Atkinson, Peter M.   II. Tate, Nicholas J.
G70.39.A395   1999   99-10017
621.384′196—dc21   CIP

British Library Cataloguing in Publication Data

A catalogue record for this book is available from the British Library

ISBN 0 471 98577 5

Typeset in 10/12 pt Times by Vision Typesetting, Manchester
Printed and bound in Great Britain by Biddles Ltd, Guildford and King's Lynn
This book is printed on acid-free paper responsibly manufactured from sustainable forestry, in which at least two trees are planted for each one used for paper production.

Contents

List of Contributors vii

Foreword xi
David Rhind, Vice-Chancellor, City University, London

Preface xiii

1 **Techniques for the Analysis of Spatial Data** 1
  Peter M. Atkinson and Nicholas J. Tate

2 **Land Cover Classification Revisited** 7
  Paul M. Mather

3 **Image Classification with a Neural Network: From Completely-Crisp to Fully-Fuzzy Situations** 17
  Giles M. Foody

4 **Cloud Motion Analysis** 39
  Hugh G. Lewis, Franz T. Newland, Stéphane Côté and Adrian R.L. Tatnall

5 **Methods for Estimating Image Signal-to-Noise Ratio (SNR)** 61
  Geoffrey M. Smith and Paul J. Curran

6 **Modelling and Efficient Mapping of Snow Cover in the UK for Remote Sensing Validation** 75
  Richard E.J. Kelly and Peter M. Atkinson

7 **Using Variograms to Evaluate a Model for the Spatial Prediction of Minimum Air Temperature** 97
  Dan Cornford

8	Modelling the Distribution of Cover Fraction of a Geophysical Field John B. Collins and Curtis E. Woodcock	119
9	Classification of Digital Image Texture using Variograms James R. Carr	135
10	Geostatistical Approaches for Image Classification and Assessment of Uncertainty in Geologic Processing Freek van der Meer	147
11	A Syntactic Pattern-Recognition Paradigm for the Derivation of Second-Order Thematic Information from Remotely Sensed Images Stuart L. Barr and Michael J. Barnsley	167
12	The Rôle of Classified Imagery in Urban Spatial Analysis Victor Mesev and Paul A. Longley	185
13	Image Classification and Analysis using Integrated GIS Jackie C. Hinton	207
14	Per-Field Classification of Land Use using the Forthcoming Very Fine Spatial Resolution Satellite Sensors: Problems and Potential Solutions Paul Aplin, Peter M. Atkinson and Paul J. Curran	219
15	Modelling Soil Erosion at Global and Regional Scales using Remote Sensing and GIS Techniques Nick A. Drake, Xiaoyang Zhang, Eva Berkhout, Rogario Bonifacio, David I.F. Grimes, John Wainwright and Mark Mulligan	241
16	Extracting Information from Remotely Sensed and GIS Data Peter M. Atkinson and Nicholas J. Tate	263
Index		269

List of Contributors

Paul Aplin Department of Geography, University of Southampton, Highfield, Southampton, SO17 1BJ, UK

Peter M. Atkinson Department of Geography, University of Southampton, Highfield, Southampton, SO17 1BJ, UK (much of the work on this book was done while on study leave at the school of Mathematics, University of Wales, Cardiff, UK)

Michael J. Barnsley Department of Geography, University of Wales Swansea, Singleton Park, Swansea, SA2 8PP, UK

Stuart L. Barr School of Geography, University of Leeds, Woodhouse Lane, Leeds, LS2 9JT, UK

Eva Berkhout Kleine Gracht, 6221CA, Maastricht, The Netherlands

Rogario Bonifacio Department of Meteorology, University of Reading, Whiteknights, Reading, Berkshire, UK

James R. Carr Department of Geological Sciences, Mackay School of Mines, University of Nevada, Reno, Nevada 89557, USA

John B. Collins Department of Geography, University of Boston, 675 Commonwealth Avenue, Boston, Massachusetts 02215, USA

Dan Cornford Department of Computer Science, Aston University, Aston Triangle, Birmingham, B4 7ET, UK

Stéphane Côté HMR Inc., Beauport (Québec), Canada

Paul J. Curran Department of Geography, University of Southampton, Highfield, Southampton, SO17 1BJ, UK

Nick A. Drake Department of Geography, King's College, Strand, London, WC2R 2LS, UK

Giles M. Foody Department of Geography, University of Southampton, Highfield, Southampton, SO17 1BJ, UK

David I.F. Grimes Department of Meteorology, University of Reading, Whiteknights, Reading, Berkshire, UK

Jackie C. Hinton Section for Earth Observation, Institute of Terrestrial Ecology, Monks Wood, Abbots Ripton, Huntingdon, Cambridgeshire, PE17 2LS, UK

Richard E.J. Kelly Department of Geography, Birkbeck College, University of London, 7–15 Gresse Street, London, W1P 2LL, UK

Hugh G. Lewis Department of Electronics and Computer Science, University of Southampton, Highfield, Southampton, SO17 1BJ, UK

Paul A. Longley Department of Geography, University of Bristol, University Road, Bristol, BS8 1SS, UK

Paul M. Mather Department of Geography, University of Nottingham, University Park, Nottingham, NG7 2RD, UK

Victor Mesev School of Environmental Studies, University of Ulster, Cromore Road, Coleraine, BT52 7SA, Northern Ireland

Mark Mulligan Department of Geography, King's College, Strand, London, WC2R 2LS, UK

Franz T. Newland Remote Sensing Group, Department of Aeronautics and Astronautics, University of Southampton, Highfield, Southampton, SO17 1BJ, UK

Geoffrey M. Smith Section for Earth Observation, NERC Institute for Terrestrial Ecology, Monks Wood, Abbots Ripton, Huntingdon, Cambridgeshire, PE17 2LS, UK

Nicholas J. Tate Department of Geography, University of Leicester, Leicester, LE1 7RH, UK (much of the work on this book was done while at the School of Geosciences, Queen's University, Belfast, UK)

Adrian R.L. Tatnall Remote Sensing Group, Department of Aeronautics and Astronautics, University of Southampton, Highfield, Southampton, SO17 1BJ, UK

Freek van der Meer International Institute for Aerospace Survey and Earth Sciences (ITC), Department of Earth Resources Surveys, Geological Survey Division, Hengelosestraat 99, PO Box 6, 7500 AA Enschede, The Netherlands

John Wainwright Department of Geography, King's College, Strand, London, WC2R 2LS, UK

Curtis E. Woodcock Department of Geography, University of Boston, 675 Commonwealth Avenue, Boston, Massachusetts 02215, USA

Xiaoyang Zhang Department of Geography, King's College, Strand, London, WC2R 2LS, UK

Foreword

Remote sensing and Geographical Information Systems have come a long way in the last 25 years since ERTS (later Landsat) was first launched. In those days, the available information consisted of a few digitised maps and a small number of low resolution images. Moreover, these were typically available only to a few hundred (at most) people world-wide. The equipment was primitive by contemporary standards, in 1970, the organisation in which I worked added 32K to the memory of its PDP 9 computer at a cost equal to about £30 000 and, for the same price, took delivery of a very early hard disk of 3Mb capacity! Interfaces between system components were consistently troublesome and all software was home-grown to the extent that it usually needed the author to make it work.

We now have the capacity on our desktops to handle hundreds or even thousands of images and maps. Software tools are legion. The number of serious users of these facilities is counted in tens or hundreds of thousands and numerous courses and standard textbooks swell the number daily. The number of casual users of digital maping can be counted in millions since facilities such as 'show me the nearest 3 ATMs to where I am' are now freely available on the Internet. Moreover, we can reach out via the World Wide Web to obtain and download images and information from NASA, ESA and Terraserver sources.

Is this then the end of history so far as remote sensing and GIS are concerned? Has everything other than local application of these tools been rendered passé? Is remote sensing now a trivial pursuit lacking challenge for the university researcher and has GIS simply become a set of routine tools? Is everything in these fields now predictable? The answer is emphatically no! I know this for certain because I have just gone through a tortured process to produce, with three colleagues, the second edition of a large edited book on GIS. Published eight years after the first edition, it contains a review of the changes in this field in the interregnum. We set out to include the best people in the world, and ony 10% of the 80 authors were the same as the first time. Few of the chapters were on the same themes. Our review of the changes contains our confession that, in 1990/91, we completely failed to anticipate the advent of the World Wide Web and grossly underestimated the importance of the Internet. In addition to wholly new analytical techniques and legions of new applications, the public policy side of GIS grew enormously in significance in the 1990s.

It may be that we were simply dumb (but I'd like to think not!). On the basis of this experience, it seems to me that change in all the areas associated with remote sensing and GIS will be on-going, even if the emphasis will shift as time goes on. Moreover, the advent of wholly new sensors – notably high resolution optical ones and hyperspectral ones with data from both being routinely made available – may well transform what is possible and what needs exploring. The reality of commercial remote sensing and private sector information brokers has enormous implications for service levels, public policy and for education establishments. The inevitability of data combinations drawn from different sources will raise questions of accuracy and reliability as well as beneficial ownership. An ability to generate revenue from this information or knowledge derived from it may add significantly to the tension between the free exchange of scientific information and preservation of Intellectual Property Rights – which is becoming ever more important in some areas of science.

No vestige of a beginning, no prospect of an end, was the phrase used at the end of the eighteenth century by Hutton and Playfair to describe earth history. It fits just as well to advances in remote sensing and GIS. This book is to be welcomed therefore on two fronts, generally as part of a never-ending process and specifically because it addresses research problems of considerable theoretical and practical significance. I congratulate the authors and editors on putting together such a collection of valuable papers. And I specifically thank Nick Tate, Peter Atkinson and their colleagues for their valued collaboration between Ordnance Survey (a government department) and universities in experiments designed to tease out the future benefits of remote sensing for operational and policy-related purposes. Such collaboration is what makes technology useful and it also ensures that research gets funded!

David Rhind
Director General
Ordnance Survey
August 1998

Preface

The impetus for this book was a meeting held by the Remote Sensing Society Geographical Information Systems Special Interest Group (RSS GIS SIG) on the subject of 'Spatial Analysis for Remote Sensing and GIS' at the University of Southampton on 25 July 1996. This particular SIG of the RSS holds regular meetings where members of the RSS and other researchers can meet to present research related to GIS.

The meeting in Southampton in 1996, convened by Peter Atkinson, was the third in a series of meetings held by the RSS GIS SIG. It provided a forum for the presentation of up-to-date research in spatial analysis for remote sensing and GIS and also an open environment for the discussion and cross-fertilization of ideas between researchers. The meeting was open to researchers employing *all* aspects of spatial analysis in the context of remote sensing and GIS, and included a mix of research that is science-driven and techniques-driven. The meeting comprised 12 oral presentations, seven of which were selected to be included as chapters in this book. These chapters have been supplemented by invited contributions in similar subject areas from researchers with a more international base. The resulting collection of chapters presents examples of both the use being made of GIS and spatial analysis in remote sensing, and also more technique-oriented research which is currently being conducted in the remote sensing community.

We hope that the book will prove useful and stimulating to students and researchers both with a direct interest in remote sensing and GIS analysis and also to those from outside these fields who might wish to learn something of the 'flavour' of research being conducted. Since this book was first conceived, many new threads of research have appeared and are being pursued vigorously in what is, and continues to be, a thriving international research community.

We are grateful to all those who assisted with the original meeting and the production of this book. In relation to the RSS GIS SIG meeting we would like to thank the Ordnance Survey, Erdas UK, the Department of Geography, University of Southampton, and the Remote Sensing Society for sponsoring the meeting. We would also like to thank Paul Pan for his contribution and the local organizing committee (Daisy Nolan, Paul Aplin, Mohamed Embashi and Mark Cutler) for their unstinting assistance. In relation to the production of this book, we thank Sally Wilkinson, Sarah Read, Emma Bottomley and Isabelle Strafford at John Wiley & Sons, and Karin Fancett for their support and patience. We would also like to thank the following list of people who gave their time to review the chapters (and in some cases more than one chapter) of the book:

Stephen Anthony, Stuart Barr, James Carr, John Collins, Jennifer Dungan, Peter Fisher, Giles Foody, Jackie Hinton, Lucas Janssen, Ioannis Kanellopoulos, Richard Kelly, Nina Lam, Mitch Langford, Rick Lathrop, Paul Mather, David Miller, Martien Molenaar, Donald Myers, Margaret Oliver, Stephen Plummer, Andrew Rogowski, Chris Skelly, Geoffrey Smith, Neil Stuart, Adrian Tatnall, Freek van der Meer and Graeme Wilkinson.

Peter Atkinson and Nick Tate
January 1999

1
Techniques for the Analysis of Spatial Data

Peter M. Atkinson and Nicholas J. Tate

1.1 Introduction

The objective of this book is to bring together a collection of chapters which describe in a logical sequence some of the most recent techniques being applied in research using Remote Sensing (RS) and Geographical Information Science (GIS). Traditionally, remote sensing and GIS researchers have worked separately, making use of techniques which we might loosely describe as 'spectral' and 'spatial' respectively. However, there has been an increasing trend towards the integration of the data analysed by these two groups, particularly the increased recognition of the rôle that spatial information has to play in many contexts of remote sensing analysis. This is perhaps not at all surprising given that the image analysed in the RS context is effectively the raster in a GIS context, with both being similar operationalizations of a field view of geographical phenomena. Increased collaboration between RS and GIS researchers has led to an increased cross-fertilization of research ideas, and the formulation of new research agendas for both the GIS and RS communities (e.g., Star, 1991; ASPRS, 1994; Legg, 1994; Sample, 1994; Star *et al.*, 1997).

The coverage of the book is biased towards remote sensing, and there is much in the growing field of GIS which is not covered. In particular, the focus of the book is on remotely sensed and GIS data defined in terms of the raster data model, and we have excluded much analysis – GIS or spatial analysis *per se*, conducted on data defined in terms of the vector data model. There is also much remote sensing which is not covered in this book. The techniques described are focused in the main on analysis of the *spatial* aspects of remotely sensed images to the exclusion of research focused on understanding the interaction of light with properties of the Earth's surface. Thus, hyperspectral remote sensing and bidirectional reflectance distribution function (BRDF) modelling, for example, are not included explicitly.

We can conceive of three stages in a remote sensing or GIS analysis at the simplest level. These are (i) measurement and sampling, (ii) the application of models and techniques to achieve some objective and, finally, (iii) validation of the results achieved. This book is primarily about remote sensing and GIS analysis as defined within this framework, with an emphasis on the second and third stages: *modelling* or *techniques* and *validation*. The difference between modelling and the application of techniques, while they are not entirely exclusive, has to do with the objective of the analysis. In some cases the objective is to characterize (or model) mathematically the spatial process, or spatial form (variation) of interest. In such a case, the analysis is conducted in order to better understand that process or form and involves modelling. In other cases, the objective is to achieve some technological goal, that is, to 'do' something useful. Most commonly this objective is to estimate some unknown value from the sample data. In this book, the emphasis is on both modelling and techniques.

The objective of estimation can be divided conveniently into the estimation of *continuous* (e.g., biomass, leaf area index) *variables* and the estimation of *categorical variables* (or more commonly classification). The techniques may also be divided into those which operate on a single variable (e.g., techniques for interpolation), referred to as univariate, and those which operate on two or more variables (e.g., regression), referred to as bivariate or multivariate. The chapters of this book include the estimation of continuous *and* categorical variables using univariate *and* multivariate techniques.

There are many ways to classify the 14 main chapters of the book including the techniques used, the context of the analysis, the data sets used, the spatial resolution or scale of those data and whether or not the analysis involves univariate or multivariate analysis and so on. This made the grouping of chapters difficult and it is for this reason that there are no explicit section headings. That said, we have tried to organize this book around the use of several core research areas: artificial neural networks (ANNs), geostatistics and object-based analysis, for both estimation and classification tasks. ANNs are gaining prominence in remote sensing (Atkinson and Tatnall, 1997) because they consistently seem to achieve greater accuracies than alternative statistical approaches. Similarly, geostatistical techniques have been applied increasingly in recent years in remote sensing and GIS as researchers realize the wealth of spatial information in images and raster data can be utilized beneficially, for example, to increase estimation accuracies (Curran and Atkinson, 1998). In particular, the variogram provides a useful description of spatial dependence and can be used in techniques such as kriging (for estimation) and conditional simulation (for simulation). Object-based analysis, while at the core of vector-based GIS, provides a useful approach which can result in greater classification accuracies.

From Table 1.1, which summarizes the chapters of this book, it is clear that a wide range of techniques for the estimation of continuous variables are covered, including regression, geostatistical techniques such as kriging and cokriging, regularization to change the scale of the data, and ANNs. An even greater range of techniques is included for the classification of data including ANNs (for fuzzy classification), texture

Table 1.1 Summary of the substantive chapters of this book

	Author(s)	Objective	Subject	Technique or approach used
2	Mather	Fuzzy classification	Land cover	ANN
3	Foody	Fuzzy classification	Land cover	ANN
4	Lewis et al.	Classification	Cloud vectors	ANN
5	Smith and Curran	Noise estimation	Remotely sensed images	Variogram and others
6	Kelly and Atkinson	Estimation and mapping	Snow depth	Regression, IDW, kriging, cokriging
7	Cornford	Estimation and mapping	Air temperature	TSA, regression, kriging
8	Collins and Woodcock	Estimation and mapping	Snow cover	Regularization
9	Carr	Classification	Topography	Variogram, texture
10	van der Meer	Classification	Geology	Indicator kriging, cokriging, stochastic imaging
11	Barr and Barnsley	Classification	Land use	Graph theory
12	Mesev and Longley	Characterization	Settlements	Fractals
13	Hinton	Classification	Land cover	Object-based analysis
14	Aplin et al.	Classification	Land cover	Object-based analysis
15	Drake et al.	Estimation and mapping	Global soil erosion	Scaling of models

classification, geostatistical classification, object-based classification and classification based on graph theory. In addition, fractals are described and statistical techniques for changing the scale of models are presented. The chapters may also be classified according to the substantive context of the analysis. From Table 1.1 it is clear that several different subject areas are covered including meteorology (cloud vectors and air temperatures), hydrology (snow cover and snow depth), geology, geomorphology (topography and soil erosion), land cover, land use and the description of urban settlements. In the next section, we briefly summarize the content of the remaining 15 chapters.

1.2 Summary of Book Chapters

One of the most recent advances made in remote sensing has been the adoption of ANNs in the place of maximum likelihood (ML) classification, which is standard in most remote sensing software. ANNs have been found in many studies to produce higher classification accuracies than ML, although this is not guaranteed to be the case (e.g., if the Gaussian model assumed by the ML classifier is appropriate, then ML represents an efficient algorithm for classification). Another important development in remote sensing has been the realization that nearly all remotely sensed images contain some mixed pixels (single pixels representing more than one class), so that fuzzy classification techniques represent an appropriate alternative to the more common hard (e.g., ML) classification. In the first two chapters by **Paul Mather** and **Giles Foody** respectively, the application of ANNs for fuzzy classification of land cover is described, while the concern of the next chapter by **Hugh Lewis, Franz Newland, Stéphane Côté** and **Adrian Tatnall** is the application of ANNs to feature tracking in the context of clouds.

One of the problems which remote sensing investigators often face is that they do not know the amount of noise in the images which are being analysed. This problem is compounded by the fact that repeated measurement (a common way of estimating measurement error) is not possible for remotely sensed images which are often only available singly. This shortcoming is addressed in the chapter by **Geoffrey Smith** and **Paul Curran** who describe and compare several different techniques for estimating the signal-to-noise ratios of remotely sensed images. One of these techniques involves the use of the variogram, a function which appears repeatedly in this book. The chapter by **Richard Kelly** and **Peter Atkinson** is concerned with a geostatistical analysis of snow depth data for the whole of the UK. The authors compare kriging to inverse distance weighted (IDW) interpolation for mapping snow depth in the UK from sparse point observations, and cokriging to regression for mapping from point observations supplemented with a digital elevation model (DEM). Generally, the geostatistical approaches were more accurate, but some interesting results were obtained with regard to non-stationarity of variation across the UK. The next chapter by **Dan Cornford** continues the geostatistical theme with an examination of the scales at which minimum air temperature varies across the UK and the adoption of models and techniques which are appropriate to those scales. Thus, trend surface analysis (TSA) is used to model regional variation, regression is used to model medium-scale variation and kriging is used to model local variation. **John Collins** and **Curtis Woodcock** extend the application of geostatistics in their chapter to include a model of regularization that allows the changing of the spatial resolution (or scale of observation) of the analysis. Their particular focus is on estimating the percentage cover within individual pixels of snow cover.

In the next chapter, **James Carr** discusses texture classification as applied to the topographically varied regions of (i) Lake Tahoe, Nevada, and (ii) the fjords of Norway. Here, the variogram is used to derive texture images which may be used to increase the accuracy of classification where texture is an important discriminating variable. **Freek van der Meer** also discusses classification, but in his chapter makes use of more novel geostatistical approaches, including indicator kriging and cokriging, and conditional simulation (also referred to as stochastic imaging) to classify images into geological types. In recent years, conditional simulation has become an increasingly important tool in remote sensing and GIS (Journel, 1996).

Several chapters of this book focus on land cover, which is a natural property to estimate from remotely sensed images of the landscape. The classification of land use is much more problematic. **Stuart Barr** and **Mike Barnsley** describe interesting and novel graph-theoretic approaches to the derivation of second-order (land use) information from first-order (land cover) data. The chapter by **Victor Mesev** and **Paul Longley** continues the land cover and land use theme, and is concerned with their work on fractals and, specifically, how settlements can be considered scale-invariant and characterized using the calculation of fractal dimension (Mesev, 1997).

In the next chapter, **Jackie Hinton** argues for the adoption of integrated GIS in remote sensing and GIS research. Here, the advantages of an *object-based* approach to simple procedures such as classification in remote sensing are considered. However, the author argues further that a truly integrated GIS in which conversion between the raster and vector data models is no longer necessary will fundamentally change the nature of remote sensing and GIS research. **Paul Aplin, Peter Atkinson** and **Paul Curran** describe the advantages of an object-based approach to classification in remote sensing, but this time the focus is on very fine spatial resolution (approximately 4 m by 4 m in multispectral mode) satellite sensor imagery (which should be available by the time this book is published). This new source of imagery provides exciting new opportunities for remote sensing because of the huge increase in the information content of the data. However, a potential problem with such imagery is that classification accuracies are likely to decrease (not increase) per-pixel. Thus, the imagery is emulated using airborne multispectral scanner imagery to provide some clues about the magnitude of the classification accuracies which might be expected, and these accuracies are compared to those achievable by object (land cover parcel)-based classification.

Nick Drake, Xiaoyang Zhang, Eva Berkhout, Rogario Bonifacio, David Grimes, John Wainwright and **Mark Mulligan** consider some of the problems associated with producing a global soil erosion map from data obtained at local scales. In particular, soil erosion tends to be underestimated because of problems with (i) the model, (ii) the model implementation and (iii) errors in the calculation of the model parameters. They describe several different approaches to scaling the data in a soil erosion model, and propose several solutions to the problems identified. In the final chapter, **Peter Atkinson** and **Nick Tate** bring the many themes of this book together in an overview and a look towards the future.

References

ASPRS, 1994, *Remote Sensing and Geographic Information Systems: An Integration of Technologies for Resource Management* (Bethesda, MD: ASPRS).

Atkinson, P.M. and Tatnall, A.R., 1997, Introduction: neural networks in remote sensing, *International Journal of Remote Sensing*, **18**, 699–709.

Curran, P.J. and Atkinson, P.M., 1998, Remote sensing and geostatistics, *Progress in Physical Geography*, **22**, 61–78.

Journel, A.G., 1996, Modelling uncertainty and spatial dependence: stochastic imaging, *International Journal of Geographic Information Systems*, **10**, 517–522.

Legg, C.A., 1994, *Remote Sensing and Geographic Information Systems: Geological Mapping, Mineral Exploration and Mining* (Chichester: Wiley).

Mesev, V. (ed.), 1997, *Remote Sensing of Urban Systems*, Special Issue of *Computers, Environment and Urban Systems*, **21**(3/4).

Sample, V.A. (ed.), 1994, *Remote Sensing and GIS in Ecosystem Management* (Washington, DC: Island Press).

Star, J.L. (ed.), 1991, *The Integration of Remote Sensing and Geographic Information Systems*, Special Issue of *Photogrammetric Engineering and Remote Sensing*, **57**(b).

Star, J.L., Estes, J.E. and McGwire, K.C., 1997, *Integration of Geographic Information Systems and Remote Sensing* (Cambridge: Cambridge University Press).

2
Land Cover Classification Revisited

Paul M. Mather

2.1 Introduction

The problem of describing the nature of the Earth's surface is one which has attracted the attention of geographers, ecologists and other scientists interested in the spatial distribution of phenomena for many decades. In the period 1900–1950 the prevailing paradigm was the regional one; the land surface of the Earth was viewed as a set of discrete regions, each with specific characteristics and, possibly, a personality. Geographers became less convinced of the usefulness of this methodology during the 1950s spurred, no doubt, by the comments of Kimble (1951, p. 159), who described their endeavours as 'putting boundaries that do not exist around areas that do not matter'.

Following much the same approach as that used by geographers during the first half of the twentieth century, users of remotely sensed data have often seemed to embrace an attitude that envisages the world as a set of distinct and mutually exclusive regions, each composed of contiguous and mainly rectangular pixels. As Lillesand and Keifer (1994, p. 585) remark: 'The overall objective of image classification procedures is to automatically categorize all pixels in an image into land cover classes or themes.' This is the 'hard classification' approach, in which the aim is to place each pixel into a single land cover class. More recently there has been a move away from the attitude that land-surface cover necessarily consists of a patchwork of separate and homogeneous regions towards the view that a better way of describing the spatially varying character of land cover is to consider this phenomenon in terms of probability surfaces, much as geographers did several decades ago. Each pixel is allowed to have a 'class membership' probability rather than a single label representing one discrete and identifiable category of land cover, and the result of this operation is described as a 'soft' classification.

Advances in Remote Sensing and GIS Analysis. Edited by Peter M. Atkinson and Nicholas J. Tate.
© 1999 John Wiley & Sons Ltd.

In terms of aspiring towards a less rigid method of categorizing the Earth's surface, the 'soft' approach is similar to mixture modelling which has been used widely, particularly by geologists, to describe the spatial variability of the properties of land surfaces, usually in semi-arid areas. However, the mixture modelling approach uses the concept of continuous surfaces of variation. The 'soft' classification methodology appears to be based on a relaxation of the view that land cover regions exist. In practice, the techniques used to derive mixture proportions and soft classification membership probabilities may be similar, though the philosophical approach may differ.

The aims of this chapter are to consider the 'hard' and 'soft' approaches to classification, to compare the latter to mixture modelling, and to ask whether the term 'classification' itself is being misused by proponents of 'soft' techniques. The attributes of the competing methods are summarized briefly prior to a discussion of their applications. It is suggested that textbook descriptions of these methods are often idealized and that the relationship between such theoretical descriptions and practical experience is sometimes more tenuous than a survey of journal papers might suggest. An apocryphal remark, attributed to an economist, perhaps describes the situation in reverse: 'It works in practice; now let's see if it works in theory.'

2.2 Traditional Methods of Image Classification

The term *classification* is defined by *Chambers Twentieth Century Dictionary of the English Language* as the 'act of forming into classes'; a class is 'a rank or order of persons or things'. The *Pocket Oxford Dictionary* gives the definition of 'class' as 'one of the parts into which a kingdom can be divided', and gives examples: genus, species, variety. In its regular scientific use the word is generally taken to mean the placing of a discrete object into one of a predefined number of categories. For example, an animal may be classified as a dog, cat, cow, or goat. Sometimes the term classification is used to mean the definition and description of a schema consisting of particular categories (dog, cat, cow, goat). The placing of an individual animal into one of these categories is perhaps better described as 'identification'. In their classic text, Sokal and Sneath (1975, p. 3) state that: 'Classification is the ordering of organisms into groups (or sets) on the basis of their relationships.' These authors go on to state that 'organisms' can be replaced in their definition by 'concepts' or 'entities', and they offer the view that the term 'classification' has a restricted sense of putting entities into distinct classes. The alternative is to array them into a continuous spectrum showing no distinct divisions. Numerical taxonomists refer to the latter operation as *ordination* as distinct from *classification* and *identification*.

When the concept of classification is transferred to spatially varying and continuous phenomena, such as soils, it becomes more difficult to think of classification in the restricted sense. When we say 'this is a sandy soil' we are not determining that a particular patch or region of soil is uniformly sandy (and the same as other 'sandy' soils elsewhere in the world), but that the attributes of the soil at a particular point in geographical space resemble those of one of an unbroken sequence of idealized soils ranging in composition from clayey and silty to pebbly. In a sense, we are not classifying the soil but describing, or generalizing, its characteristics at a given spatial scale. Sharp boundaries do occur in nature, and the problem of description and classification becomes rather easier where they do; but are practical constructs such as 'sandy soil', 'dense population' or 'Mediterranean climate' merely simplifications that enable us to conceptualize the world at a particular scale? And can the description of an area in such terms be said to constitute its classification? It should be noted that the foregoing discussion assumes that a strict classification in dictionary definition terms is dealing with sets of similar objects – dogs, cats, cows and goats have been mentioned. In practice, it is the case that each such animal is unique (that is why we give names to our pets). But this is not the issue: I am suggesting that we can classify an animal as a cow because it is a discrete object, with the characteristics of a cow. The operation of identifying the class to which a particular object belongs is not a difficult one and is, in fact, an everyday experience (e.g., Toulmin (1953, p. 51) cites Wittgenstein's remark: 'What is or is not a cow is for the public to decide').

It is harder to translate the concept of classification to targets that are not discrete, which are part of a continuum. In other words, we should not confuse the real world with our perception of it as being made up of sets of objects that we call pixels. The uniqueness question ('everything is different and generalization is therefore impossible') is a separate one. It is answered by considering the scale of observation and the list of attributes that are used to determine class membership. Those attributes that make individual animals unique are perhaps irrelevant to the question 'is this a dog?'.

Perhaps the economist's pragmatic suggestion, noted earlier, holds a clue. We speak colloquially of 'classifying an image' rather than classifying the land cover of a certain area. The image does indeed consist of discrete objects, the pixels, and these can be grouped into sets, which may have some spatial and attributional correspondence with what we see on the ground. Most image processing and GIS packages offer several modules, which provide methods of testing this hypothesis. The best-known methods are described as supervised or unsupervised, depending on their need for ground reference data. For example, the maximum likelihood classifier (ML) requires samples of pixels which, via field observation or air photograph interpretation, are held to be representative of specific land cover types. ML relies upon the assumption that the populations from which these training samples are drawn are multivariate–normal in their distribution. Clearly this is not the case in remote sensing, for the image pixel values are non-negative, discrete and have an upper bound of 255, 1023 or some other value depending on the characteristics of the instrument which acquired the data, whereas the normal distribution relates to continuously measured and unbounded data. The possible detrimental effects of non-normality of class frequency distributions are either largely ignored by practitioners or seized upon by proponents of alternative methods as the Achilles' heel of the technique. A number of studies have addressed the normality problem in relation to the estimation of the sample correlation coefficient, upon which the ML technique is based (Carroll, 1961; Kowalski, 1972). Although ML gives results which can lead to claims of 'accuracy' of 80% or more, its reliance on the assumption of normality, which incidentally precludes the use of categorical data, is widely suggested as a reason for its abandonment.

The currently favoured alternative to the statistical classifier is the artificial neural network (ANN), particularly the feed-forward multi-layer perceptron using back-propagation (e.g., Kanellopoulos *et al.*, 1997). Such networks are able to handle any kind of numerical data, and are free of distributional assumptions. However, experience shows that while the use of ANN may improve land cover classification accuracy by a few per cent, they do not represent a breakthrough in terms of raising classification accuracies much beyond 80%. Furthermore, networks must be designed in terms of the number of hidden layers, the number of neurons in each hidden layer, and the nature of the transfer function associated with each neuron in the hidden layers. The choice of network architecture is largely a matter of trial and error, though this is costly in terms of the investigator's time. Little attention appears to have been paid by remote sensing users of ANN to the question of optimal networks; indeed, suggestions are made that 'network architecture doesn't really matter', presumably so long as the results conform to the user's preconceptions. It is undoubtedly the case that the design and training of an artificial neural network is time-consuming, though it is equally true to say that – once trained – the network is computationally fast compared to ML.

Whatever technique is used, classifier output is compared to field observations, existing maps, or data derived from air photographs to determine classification accuracy. This operation is normally done on a pixel-by-pixel basis to produce a contingency or error matrix. It has been suggested (C.G. Wilkinson, verbal presentation, COMPARES Workshop, York University July 1996) that classification accuracies of greater than 80% are rare, and that this value might represent an upper limit to the accuracies achievable from remotely sensed data. Two issues are raised by this comment. The first is that accuracy depends on the scale of observation relative to the spatial characteristics of the target, while the second is the question – discussed above – of the nature of the image pixel. To take an extreme case, it would be reasonable to expect that, with 20 m spatial resolution data, it would be possible to discriminate between land and water in a cloud-free SPOT HRV image. Accuracy in this case would approach 100%. However, if an attempt were made to distinguish between different kinds of fruit tree using the same imagery then classification accuracy would inevitably be much lower. Hence there must be a correspondence between the spatial resolution of the image and the geographical scale of the coverage of the target of interest. Here one is distinguishing between the assessment of the discriminability of the targets rather than their identification in terms of predefined categories. Discrimination and identification are the two basic steps in a classification. Spectral coverage and bandwidth also have an effect on classification accuracy; thus, the instrument should measure distinguishing spectral characteristics of targets, that is, it should measure in regions of the spectrum in which targets are best distinguished, and with an appropriate bandwidth. Equally significant is the use of second-order observations such as image texture. The second issue, the nature of the image pixel, is more fundamental and returns us to the questions raised earlier – is a pixel a classifiable object?

Critics of the traditional or 'hard' approach use this question as their starting point for their justification of a 'soft' or 'fuzzy' method of classification. In the following section, attention will be focused on the use of the two classifiers mentioned above (maximum likelihood and artificial neural network) as the basis for fuzzy classification.

2.3 Fuzzy/Soft Classification

Standard methods such as maximum likelihood and artificial neural networks have one thing in common. They use what is effectively a measure of similarity to determine the class allocation of a particular pixel. In the case of ML the probability that pixel i belongs to class j is measured for all k classes. The class for which the probability is highest is the one that is selected. The artificial neural network uses the activations of the output layer neurons as a measure of the resemblance between a pixel and each of the candidate classes, and a pixel is allocated to the class for which the associated output neuron activation is highest.

If a pixel is not a classifiable object then 'hard' methods will not work except in specific cases. For example, at a certain scale, appropriate to the spatial and spectral measurement characteristics of the instrument, a landscape may be composed of a sequence of compact and discriminable spatial objects, such as forest, water, and grassland or crops enclosed in agricultural fields. In such cases, each image pixel will be allocated to a specific class and the spatial pattern of the class labels will in general correspond closely to the characteristics of the real landscape, except for the small number of pixels that straddle real-world boundaries. If the number of mixed pixels is large then either the spatial resolution of the instrument is too great to measure adequately the spatial variability of the land-surface cover, or the spatial distribution of that cover is varying continuously.

Soft or fuzzy methods of classification provide for each pixel a measure of the degree of similarity (e.g, a membership probability) for every class (e.g., Bezdek et al., 1984; Fisher and Pathirana, 1990; Foody et al., 1992; Foody, 1996; Maselli et al., 1996). Pixels covering areas of uniform type should show a high degree of similarity to one class, with low similarity measures for all other classes. Pixels composed of two or more cover types should have membership probabilities related to the proportions of the corresponding cover classes. These methods have their uses in specific situations, for example in urban areas where a pixel which has mixture components of 'building' and 'vegetation' might well represent a dwelling house in a higher-status residential area. Even so, this example demonstrates a fundamental problem: if such mixed pixels are of interest, and if they are sufficiently numerous, then the implication is that the spatial resolution of the imagery is too coarse to satisfy the user's requirements. On the other hand, if mixed pixels are not numerous and if they occur along the boundaries of agricultural fields then they are probably not of great interest. One might be tempted to conclude that if the investigation is focused on the question of mixed pixels then the scale is wrong. It is also tempting to suggest a variable-scale approach in which the pixel size is proportional to the degree of homogeneity of the land surface. This approach is used implicitly when a quadtree is derived from raster data.

If the land surface exhibits continuous spatial variability in terms of the characteristics of interest to a specific investigation (such as vegetation cover or per cent bare soil) then the methods described above will not work. In other words, if the make-up of the landscape is varying continuously, then no boundary lines can be drawn and no 'classifiable' objects can exist except as a figment in the mind of the investigator. There is little point in trying to describe such areas in terms either of a set of non-overlapping land cover classes (as a 'hard' classifier would) or in terms of an overlapping set of such classes (as a 'soft' classifier would). The major argument in favour of the 'soft' classifier would, therefore, appear to be that it does not misallocate boundary (or mixed) pixels arbitrarily to one or other of the candidate land cover classes but, rather, provides a description of these pixels in terms of their membership probabilities for each of the possible classes to which they might belong.

The term 'mixed pixel' may imply that the scale of observation is inappropriate and does not match the spatial scale of variation in the landscape. Where this is clearly not the case (e.g., the use of 10 or 20 m pixels to measure the surface reflectance properties of an area such as eastern England or the mid-west of the United States, both of which are characterized by large agricultural fields) then a small proportion of pixels will be located in the boundary region between adjacent fields, and may, therefore, be described as 'mixed'. However, the proportion of such pixels should be relatively small, and should not detract from overall classification accuracy.

2.4 Landscape as a Continuum

In semi-natural landscapes the proportions of the gross land cover types (soil, vegetation) and the effects of illumination and shade conditions resulting from topographic variations change in a systematic fashion over the area in response to climatic, microclimatic and soil nutrient conditions. Hard classification is clearly inappropriate, as there are no 'objects' to be classified. The use of a 'soft' or fuzzy classifier implies that land cover classes do exist but that they are confused at the given scale of observation. The only logical conclusion is that the landscape is a continuum, formed from continuously varying proportions of idealized types which probably do not exist in the real world. Just as soils may be described in terms of their proportions of sand, silt and clay using the traditional triangular diagram so land cover can be described in terms of proportions of 'end-members' (Smith et al., 1990; Settle and Drake, 1993; Adams et al., 1995). The technique of linear mixture modelling is frequently used to unmix the signal measured at a given pixel into its component parts. It has been widely used by remote sensing geologists to estimate the proportions of different rock types in semi-arid areas. In some cases, laboratory spectra were available to provide descriptions of the 'pure types' which are rarely, if ever, found on an image. If the effects of atmosphere can be removed then it is reasonable to accept that the remotely sensed spectra of the ground surface should match laboratory spectra. The effects of differential illumination are removed by defining 'shade' as one of the end-members.

When applied to remote sensing of semi-vegetated areas the linear mixture model approach assumes that end-members can be defined. Frequently they must be recognized from the image itself ('image end-members'). Disregarding theoretical considerations, such as the fact that the model assumes a single-scattering approach, it is the difficulty in locating end-member spectra that presents the main difficulty to the user. Logic indicates that an end-member proportion cannot be negative and, if the model is properly specified, that the sum of the proportions of end-members at a given point must be less than or equal to unity. It is possible to build these constraints into the linear mixture model so that the results derived for every individual pixel satisfy these logical requirements. It is, however, more practical to consider the unconstrained model which simply computes, from a library of end-member spectra, the end-member proportions at a given point. If the model fits perfectly then there should be no end-member proportions less than zero or greater than unity, and the sum of the proportions should not exceed 1.0. Furthermore, a map of root mean squared error should not show any systematic pattern. Only by using an unconstrained model is it possible to check that these conditions are met.

One constraint imposed by linear unmixing is that the number of end-members cannot exceed the number of spectral bands available. Thus, if the Landsat TM reflective bands are used, the number of end-members cannot exceed six (or, practically, five because shade is normally included to take into account the fact that not all pixels lie on a horizontal and directly illuminated surface). The investigator may have a mental picture of the landscape which includes more than five components, and so generalization may be needed; for example, forest, crops, scrub and herbs may be collectively described as vegetation. How is such a melange to be described? Can a single vector of measurements on the TM bands adequately characterize 'vegetation'? Which pixel should be chosen to represent the 'vegetation' end-member? An average cannot be used as, by definition, some pixel spectra must necessarily be more 'extreme' than the average. Some practitioners have mapped image pixels to the space defined by the first two principal components of the image set, then surrounded the scatter of points with a polygonal boundary, and used the pixels closest to the polygon nodes to represent the end-members. The use of the first two principal components excludes information (which may have some significance) that is contained in the higher-order components. Consequently, the use of procedures such as non-linear mapping or multidimensional scaling (Mather, 1976) to map the measurements on all principal components to a two-dimensional space might be considered. Even so, the selection of end-members – which is crucial to the successful application of the linear mixing model – is fraught with difficulties. In addition, the use of the technique imposes a relatively simple view of the components of the landscape as a result of the low dimensionality of medium-resolution remotely sensed data.

Alternatives to the linear mixing model are empirical. The Spectral Angle Mapper technique has been introduced into remote sensing in recent years. This technique measures the similarity between a reference pixel and every other image pixel, so that for each reference pixel a 'similarity' image can be generated, with the similarity measure being an analogue of, or a surrogate for, the mixture proportion. The use of a coefficient of proportional similarity (cosine theta) was used more than 30 years ago (Imbrie, 1963) and so the method is not new. The multidimensional space in which the cosine is measured is defined by the spectral bands. The origin of the space is thus a point which has zero intensity and no colour (i.e., it is black). The reference pixel is atmospherically corrected and plotted as a vector in this space (e.g., in a three-dimensional space, a reference pixel is represented by a vector joining the points [0,0,0] and $[b_1,b_2,b_3]$). The length of this vector is related to the overall reflectance of the pixel and the location of the point $[b_1,b_2,b_3]$ provides a description of the shape of the spectrum of the reference pixel. The measurements on any other pixel can similarly be described in terms of a vector, and its similarity to the reference pixel is measured by the cosine of the angle between the two vectors. The smaller the angle the greater the similarity. Since all vectors must lie in the first quadrant the range of cosine theta is [0,1] and a simple multiplication by 255 is sufficient, after truncating to the nearest integer, to provide an 8-bit representation of the spatial distribution of cosine theta. Thus, the method is simple in concept and straightforward in its application.

This approach still leaves unresolved the question of how the reference vectors are to be identified. The work of Imbrie (1963) and Imbrie and van Andel (1964) can be utilized to attempt to find an answer. These authors were interested in the description of rock samples in terms of their composition. They took the 'Q-mode' factor analysis model, which is based on the analysis of similarities between objects (rock samples, pixels) rather than the analysis of correlations between the variables (elements, minerals, spectral bands) that is the basis of the 'R-mode' principal components model. The complexities of factor analysis are beyond the scope of this chapter and the reader is referred to Mather (1976, 1981) for a discussion. In Q-mode analysis the factor loadings represent the relationships between individual pixels and a set of artificial constructs called factors, just as in principal components analysis the loadings measure the similarities between the spectral bands and the principal components. Factor analysis, apart from utilizing a mathematical model that is conceptually different from that used by principal components analysis, involves an additional step: rotation of the factors. The initial rotation has the aim of producing a factor loadings matrix that has a simple structure, with a few high loadings on each factor and many near-zero loadings. The varimax rotation is commonly used to achieve this aim. Varimax retains the orthogonality property, so that factors are statistically uncorrelated. Imbrie (1963) describes a means of applying an oblique rotation which allows the factor axes to become correlated (thus opening up the prospect of an endless hierarchy of factor analyses). The pixel with the highest absolute loading on each selected varimax factor is chosen as a target, and the vector analysis method outlined by Imbrie (1963) is then applied so that each of the target pixels receives a loading of 1.0 and all other pixels are expressed as proportions of the target pixels. Thus, pixel number 100 may have the highest absolute loading on varimax factor number 2, so in the oblique vector rotation the loading of pixel number 100 becomes 1.0 and all other pixels are expressed in terms of their proportion of this value.

Apart from the question of selecting an appropriate number of Q-mode factors to rotate (usually this number can be estimated from the eigenvalue spectrum of the inter-pixel similarity matrix) the main operational problem is that the dimension of this similarity matrix is the number of pixels in the image, so that an image made up of 1024^2 pixels would generate a similarity matrix of 1 099 511 627 776 entries requiring 4×10^6 Mb of storage. This is clearly outside the capabilities of even the largest desktop machine. Some kind of sampling procedure, for example, using targeted samples drawn from 'typical' areas of the image, or systematic sampling at every nth line and mth pixel should be used to reduce the size of the problem to manageable proportions. If the sample adequately represents the range of variation present in the full image then the target pixels identified at the oblique vector rotation stage will be usable as reference pixels in the spectral angle mapping procedure. However, the potential of this technique remains to be tested and some significant computational problems relating to the efficient solution for eigenvalues and eigenvectors are under investigation.

2.5 Conclusions

The Earth's surface can be viewed and analysed at a range of scales. Remote sensing observations are obtained at several fixed scales. One of the problems alluded to in this chapter concerns the matching of the observer's scale of interest and the spatial resolution of remotely sensed images. At larger map scales the land surface becomes more generalized, and the identity of individual objects is lost while larger objects become more significant. At a 1:50 000 scale the agricultural field or individual woodland are identifiable objects in the UK East Midlands landscape, and so I use this map scale when taking walks near my home. At 1:1M scale these objects disappear and regions like the Pennine Uplands become the focus of interest. At some scales, therefore, and in certain landscape types, it is possible to use hard classification methods to identify the field, the woodland and the lake or river. At other scales, this identification is not feasible and attention is focused on the gradients of generalized surfaces representing change. Yet other landscapes do not admit of a discrete classification-based approach because, at human scales, they are formed of continua of overlaid surfaces.

Classification, whether hard or soft, is not an appropriate tool for the analysis of such landscapes, and alternatives (in the form of linear mixture modelling and spectral angle mapping) are suggested. These techniques have their own limitations, specifically relating to the determination of the number of end-members and their spectral characteristics. Q-mode factor analysis is presented as a possible solution to these problems.

Acknowledgements

I am most grateful for the comments of two anonymous referees whose suggestions have led to the clarification of many of the ideas expressed in this chapter.

References

Adams, J.B., Sabol, D.E., Kapos, V., Filho, R.A., Roberts, D.A., Smith, M.O. and Gillespie, A.R., 1995, Classification of multispectral images based on fractions of end-members: application to land cover changes in the Brazilian Amazon, *Remote Sensing of Environment*, **52**, 137–145.

Bezdek, J.C., Erlich, R. and Full, W., 1984, FCM: The fuzzy c-means clustering algorithm, *Computers and Geosciences*, **10**, 191–203.

Carroll, J.B., 1961, The nature of the data, or how to choose a correlation coefficient, *Psychometrika*, **26**, 347–372.

Fisher, P.F. and Pathirana, S., 1990, The evaluation of fuzzy membership of land cover classes in the suburban zone, *Remote Sensing of Environment*, **34**, 121–132.

Foody, G., 1996, Relating the land-cover composition of mixed pixels to artificial neural network classification outputs, *Photogrammetric Engineering and Remote Sensing*, **62**, 491–499.

Foody, G., Campbell, N.A., Trodd, N.M. and Wood, T.F., 1992, Derivation and applications of probabilistic measures of class membership from maximum likelihood classification, *Photogrammetric Engineering and Remote Sensing*, **58**, 1335–1341.

Imbrie, J., 1963, *Factor and Vector Analysis Programs for Analysing Geologic Data*. Technical Report No. 6 of ONR Task No. 389-135, Contract Nonr 1228(26), US Office of Naval Research, Geography Branch. Published by Northwestern University, Evanston, Illinois.

Imbrie, J. and van Andel, T.H., 1964, Vector analysis of heavy mineral data, *Geological Society of America Bulletin*, **75**, 1131–1156.

Kanellopoulos, I., Wilkinson, G.G., Roli, F. and Austin, J. (eds), 1997, *Neuro-Computation in Remote Sensing Data Analysis* (Berlin: Springer-Verlag).

Kimble, G.H.T., 1951, The inadequacy of the regional concept, in L.D. Stamp and S.W. Wooldridge (eds), *London Essays in Geography: Rodwell Jones Memorial Volume* (London: Longmans, Green & Co.), 151–174.

Kowalski, C.J., 1972, On the effects of non-normality on the distribution of the sample product–moment correlation coefficient, *Applied Statistics*, **21**, 1–12.

Lillesand, T.M. and Keifer, R.W., 1994, *Remote Sensing and Image Interpretation*, 3rd edition (New York: Wiley).

Maselli, F., Rodolf, A. and Conese, C., 1996, Fuzzy classification of spatially degraded Thematic Mapper data for the estimation of sub-pixel components, *International Journal of Remote Sensing*, **17**, 537–551.

Mather, P.M., 1976, *Computational Methods of Multivariate Analysis in Physical Geography* (Chichester: Wiley).

Mather, P.M., 1981, Factor analysis, in N. Wrigley and R.J. Bennett (eds), *Quantitative Geography: A British View* (London: Routledge & Kegan Paul), 144–150.

Settle, J.J. and Drake, N.A., 1993, Linear mixture modelling and the estimation of ground cover proportions, *International Journal of Remote Sensing*, **14**, 1159–1177.

Smith, M.O., Ustin, S.L., Adams, J.B. and Gillespie, A.R., 1990, Vegetation in deserts. I: A regional measure of abundance from multispectral images, *Remote Sensing of Environment*, **31**, 1–16.

Sokal, R.R. and Sneath, P.H.A., 1975, *Numerical Taxonomy* (San Francisco: Freeman).

Toulmin, S., 1953, *The Philosophy of Science* (London: Arrow Books).

3
Image Classification with a Neural Network: From Completely-Crisp to Fully-Fuzzy Situations

Giles M. Foody

3.1 Introduction

Remote sensing may be used to map, monitor and estimate the properties of environmental phenomena. Of these three areas of application, mapping, particularly thematic mapping, is perhaps the most common and may be a prerequisite for the others. Traditionally, classification techniques have been used as the tool for thematic mapping. This applies to mapping through visual as well as digital analysis of remotely sensed data. Thus, for example, classification is the basis of mapping through aerial photograph interpretation as well as from computer-based analyses of digital sensor data (Lo, 1986; Campbell, 1996). Numerous classification approaches have been used with varying degrees of success. Despite the considerable developments made recently, the accuracy with which thematic maps may be derived from remotely sensed data is, however, often still judged to be too low for operational use (Townshend, 1992; Wilkinson, 1996). Therefore, the enormous potential of remote sensing as a source of land cover data is not being realized. This also limits the value of land cover maps derived from remotely sensed data for GIS users as error will propagate into later analyses based upon them. Thus, despite mapping being one of the most common applications of remote sensing, with a history extending back over several decades, we have not yet reached the stage at which accurate land cover maps can be derived from remotely sensed data on an operational basis (Townshend, 1992). A range of reasons may be cited for this. These include issues such as the nature of the classes, the spectral

---

Advances in Remote Sensing and GIS Analysis. Edited by Peter M. Atkinson and Nicholas J. Tate.
© 1999 John Wiley & Sons Ltd.

and radiometric resolutions of the remotely sensed data and the methods used in mapping. Here, attention is focused only on the latter issue. Particular attention is given to the rôle of artificial neural networks in the resolution of some of the problems currently limiting the accuracy with which land cover may be mapped from remotely sensed data. Although having a wide range of applications in remote sensing, neural networks have been used most commonly for supervised image classification (Atkinson and Tatnall, 1997; Wilkinson, 1997). An aim of this chapter is to review briefly the rôle of neural networks in image classification and provide some references that may prove useful starting points for those new to the topic. A further key aim, however, is to show how the conventional neural-network-based approach to image classification may be modified to allow fuller use of the land cover information content of remotely sensed data to be made and derive a classification at any point along what will be defined as the continuum of classification fuzziness.

3.2 Overview of Neural Network Applications in Remote Sensing

This section aims to provide a brief, albeit somewhat superficial, introduction to neural networks. This should, however, provide a context for their application in remote sensing and may help act as a starting point for those new to the subject area.

A neural network is a form of artificial intelligence that imitates some functions of the human brain. Neural networks are general-purpose computing tools that can solve complex non-linear problems (Simpson and Li, 1993; Fischer, 1996). The network comprises a large number of simple processing units linked by weighted connections according to a specified architecture. All the long-term knowledge of the network is effectively stored in the strength of the weighted connections between units. Such networks may learn, generalize and typically are massively parallel in nature (Aleksander and Morton, 1990; Hammerstrom, 1993a, 1993b; Bishop, 1995). They have been used in a wide range of applications in remote sensing and image analysis including supervised classification (Benediktsson et al., 1990; Kanellopoulos et al., 1992), unsupervised classification (Baraldi and Parmiggiani, 1995; Hara et al., 1995), image segmentation (Austin, 1997; Clastres et al., 1997; Visa and Peura, 1997), geometric correction (Smith et al., 1995), image compression (Walker et al., 1994; Gaganis et al., 1997), model inversion or variable estimation (Smith, 1993; Pierce et al., 1994; Baret, 1995; Schweiger and Key, 1997; Wang and Dong, 1997) and multisource data analysis (Peddle et al., 1994), with many other applications possible.

There are a range of neural networks and hybrids that may be used (Aleksander and Morton, 1990; Simpson and Li, 1993; Bishop, 1995). In remote sensing, three types of network are most commonly encountered. These are, the multi-layered feedforward (sometimes referred to as multi-layer perceptron networks), Hopfield and Kohonen networks. Each type of network is very different from the others and consequently they vary in their appropriateness for different applications. Feedforward networks have been used widely for supervised image classification (e.g., Kanellopoulos et al., 1992) and model inversion (e.g., Pierce et al., 1994; Baret, 1995). Hopfield networks have been used in studies involving stereo-matching and feature tracking (Nasrabadi and Choo, 1992; Lee et al., 1994; Côté and Tatnall, 1995; Lewis et al., 1995). Kohonen networks are self-organizing and so are particularly attractive for unsupervised and semi-supervised classification (Lewis *et al.*, 1992; Pham and Bayro-Corrochano, 1994). Here attention will focus only on the feedforward neural networks as these are by far the most commonly used in remote sensing. Emphasis will also be placed on their utility for supervised image classification which has been their major area of application to date.

3.3 Supervised Classification with a Feedforward Neural Network

Before discussing the classification of remotely sensed data by a feedforward neural network it may be beneficial to put this approach in the context of generally adopted classification approaches. The most commonly used classification approaches in remote sensing are statistical classification algorithms such as the maximum likelihood classification, although recently a range of non-parametric alternatives have been advocated (e.g., Dymond, 1993; Peddle, 1993). Although widely used, conventional statistical classification techniques may not always be appropriate for mapping from remotely sensed data. For example, the requirements and assumptions of the maximum likelihood classification, one of the most widely used techniques, are often unsatisfiable. Four, somewhat interrelated, key problems may be identified. First, as a conventional parametric classifier, the data are assumed to be normally distributed. This may often not be the case and there may be significant inter-class differences in the distributions. Furthermore, even if the distribution of the data for each class could be defined it would not be to possible apply class-specific corrections to the remotely sensed data as class membership is, of course, unknown; it is the objective of the classification to determine class membership. Second, to define a representative sample on which to derive descriptive statistics (e.g., mean and variance) upon which the analysis is based a large training sample is required. Typically, it is recommended that the minimum training set size is some 10–30 times the number of discriminating variables (e.g., wavebands) per-class (Mather, 1987; Piper, 1992). Clearly a very large training set is required for mapping from multispectral data sets and this runs contrary to a major goal of remote sensing, which involves extrapolation over large areas from limited ground data. With high-dimensional data sets, such as those acquired by imaging spectrometers, the training set requirement for correct application of such a classification may be exorbitantly high. Related to this issue, the Hughes phenomenon, whereby classification accuracy may decline with an increase in the number of discriminating variables, may be observed with the maximum likelihood classification. Third, the classification can only make direct use of data acquired at a high level of measurement (e.g., ratio level) and cannot accommodate directly directional data. Unfortunately, there may be useful discriminatory information that is available, particularly to GIS users, at a low level of measurement precision (e.g., a nominal level soil map) or has a directional component (e.g., slope aspect) and to use this information the analyst typically has to stratify the data set by the low-level ancillary data. This will, however, also compound the training data requirement as each stratum requires its own training sets. Fourth, the maximum likelihood classification is computationally demanding and, therefore, relatively slow. The significance of this problem may become increasingly evident in the near future given the large data volumes anticipated from proposed sensing systems (Gershon and Miller, 1993) and increasing use of hyperspectral data sets. Although the analyst may proceed with a maximum likelihood classification when its assumptions are not satisfied, for example without correcting for non-normal distributions or with disregard to the training set size requirements for the data set, it is likely that the full information content of the remotely sensed data will not be utilized.

A neural network is less sensitive to some of the problems associated with conventional classifications. The network makes no assumptions about the nature and distribution of the data. As a consequence a large sample may not be required to estimate the properties of the distribution, although a representative training set is, of course, still required to provide an adequate description of the classes, and the training set properties, such as size and composition, require careful selection in relation to the properties of the classes and network (Baum and Haussler, 1989; Foody *et al.*, 1995; Staufer and Fischer, 1997). A neural network also learns the underlying relationships in the data and effectively weights the importance of the discriminating variables. It therefore has no limitations to data dimensionality which reduces the need for feature selection to identify, for instance, an optimal band combination for a classification. Pre-processing operations including feature selection may, however, still be beneficial, particularly in the reduction of network complexity and thereby training requirements in addition to the provision of potential increases in classification accuracy (Chang and Lippmann, 1991; Benediktsson and Sveinsson, 1997; Yu and Weigl, 1997). The neural network is also able to use directly data acquired at any level of measurement precision and accommodate directional data when appropriately scaled. These factors combined also enable the neural network to be used as a black box which may be attractive when there is little or no prior knowledge about the particular problem. Lastly, classification by a trained neural network is extremely rapid.

3.3.1 Data Processing

Feedforward neural networks are particularly attractive for supervised classification as a consequence of their ability to learn by example and generalize (Schalkoff, 1992). This type of network, together with a learning algorithm such as backpropagation (Benediktsson *et al.*, 1990), back impedance (Zhang and Scofield, 1994) or quickprop (Fahlman, 1988; Foody *et al.*, 1995), is the most commonly used form of neural computing in remote sensing. The networks may be envisaged as comprising a set of simple processing units arranged in layers, with each unit in a layer connected by a weighted channel to every unit in the adjacent layer(s). Combined, these elements transform the remotely sensed input data into a class allocation (Figure 3.1). The architecture of the artificial neural network is determined by a range of factors which relate, in part, to the nature of the remotely sensed data and desired classification. There is, for example, usually an input unit for every discriminating variable and an output unit associated with each class in the classification. The number of hidden units and layers is often defined on the basis of a series of trial runs. Alternatively, these architectural parameters could be defined with aid of 'rules of thumb' or optimized

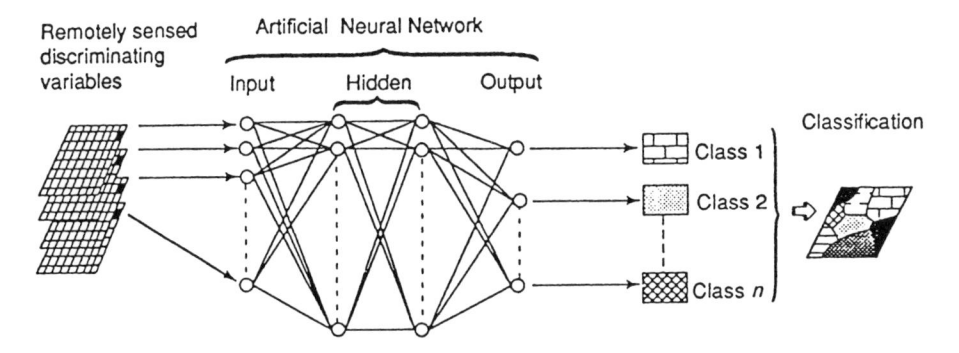

Figure 3.1 An overview of the classification of remotely sensed data with an artificial neural network. Typically a bias unit (not shown for clarity) is connected to the processing units by a weighted connection and included in the training process. For a remote sensing example see Foody et al. (1995) and for further details on its role in training see Bishop (1995)

with the use of methods that allow the network to add (or delete) units until a satisfactory structure is produced (Chauvin, 1989; Bishop, 1995; Bischof and Leonardis, 1998). In general, the larger the number of hidden units and layers used the more able the network will be to learn the training data but this may be achieved at the expense of a reduced capacity for generalization and an increase in processing time.

Although a neural network may be used to solve complex problems the processing of data within the network is based on a large number of simple calculations performed in parallel. Each unit in the artificial neural network consists of a number of input channels, an activation function and an output channel which may be connected to other units in the network (Figure 3.1). Signals impinging on a unit's inputs are multiplied by the weight associated with the interconnecting channel and are summed to derive the net input (net_j) to the unit

$$net_j = \sum_i w_{ji} o_i \qquad (1)$$

where w_{ji} is the weight of the interconnection channel to unit j from unit (or input) i and o_i is the output of unit i (or external input i). This net input is then transformed by the activation function to produce an output (o_j) for the unit (Schalkoff, 1992). Typically a sigmoid activation function such as

$$o_j = \frac{1}{1 + e^{-\lambda net_j}} \qquad (2)$$

where λ is a gain parameter, which is often set to 1, is often used (Schalkoff, 1992; Bishop, 1995).

The values for the weighted channels between units are not set by the analyst for the task at hand but rather determined by the network itself during training. The latter involves the network attempting to learn the correct output for the training data. Generally, a learning algorithm such as backpropagation (Rumelhart et al., 1986; Aleksander and Morton, 1990) is used to minimize iteratively an error function over the network outputs and a set of target outputs, taken from a training data set. This process begins with the entry of the training data to the network, in which the weights connecting network units were set randomly. These data flow forward through the network to the output units. Here the network error, the difference between the desired and actual network output, is computed. This error is then effectively fed backward through the network towards the input layer with the weights connecting units changed in relation to the error. The whole process is then repeated many times until the error rate is minimized or reaches an acceptable level. Conventionally the overall output error is defined as half the overall sum of the squares of the output errors, which for the pth training pattern is

$$E_p = 0.5 \sum_j (t_{pj} - o_{pj})^2 \qquad (3)$$

where t_{pj} is the desired output and o_{pj} the actual network output of unit j, and the total epoch (iteration) error is

$$E = \sum_p E_p \qquad (4)$$

On each iteration backpropagation recursively computes the gradient or change in error with respect to each weight, $\delta E/\delta w$, in the network and these values are used to modify the weights between network units. The weights are changed by a relation such as

$$\Delta_p w_{ji} = \eta e_{pj} o_{pi} \qquad (5)$$

where $\Delta_p w_{ji}$ is the change for the weight which connects the jth unit with its ith incoming connection, η is a constant that defines the learning rate, e_{pj} is the computed error and o_{pi} the value of the ith incoming connection. To this relation a momentum term is often added to help the network avoid local minima in the error surface and weight oscillations around the minimum (Schalkoff, 1992). If included, the momentum term adds information from the weight change determined in the preceding training iteration to the calculation. The addition of a momentum to the nth iteration modifies Equation 5 to

$$\Delta_p w_{ji}(n) = \eta e_{pj} o_{pi} + \alpha \Delta_p w_{ji}(n-1) \qquad (6)$$

where α is the momentum term. For training by epoch an overall correction to a weight is made after the presentation of all the training data and is

$$\sum_p \Delta_p w_{ji} \tag{7}$$

The calculation of the error, e_{pj}, varies for output and hidden units. Since the desired output is known for the training data, the error for the output units may, assuming the use of a sigmoid activation function with $\lambda = 1$, be calculated from

$$e_{pj} = (t_{pj} - o_{pj})o_{pj}(1 - o_{pj}) \tag{8}$$

whereas for a hidden unit, whose outputs are connected to k other units, the error is defined in proportion to the sum of the errors of all k units as modified by the weights connecting these units by

$$e_{pj} = \left(\sum_k e_{pk} w_{kj}\right) o_{pj}(1 - o_{pj}) \tag{9}$$

Once the overall output error has declined to an acceptable level, which is typically determined subjectively, training ceases and the network is ready for the classification of cases of unknown class membership. For this, the data for each case of unknown class membership are input to the trained neural network and the case allocated to the class associated with the most highly activated unit in the output layer (Kanellopoulos et al., 1992). Further details on artificial neural network learning may be found in Rumelhart et al. (1986), Aleksander and Morton (1990), Schalkoff (1992) and Bishop (1995).

3.3.2 Evaluation Relative to Other Classification Approaches

Numerous comparative studies have been undertaken to assess the accuracy of neural network based classifications relative to those derived from more conventional classification approaches such as maximum likelihood classification, linear discriminant analysis and evidential reasoning (e.g., Benediktsson et al., 1990; Peddle et al., 1994; Paola and Schowengerdt, 1995). As a generalization, these studies have revealed that a neural network may be used to classify data at least as accurately, but often more accurately than conventional classification approaches. For instance, Peddle et al. (1994) show that classifications derived from a neural network were generally, but not always, more accurate than those derived using maximum likelihood and evidential reasoning approaches. Moreover, the use of additional discriminatory information (e.g., texture) in the analyses increased the accuracy of neural network classifications while that of the conventional maximum likelihood classification declined. Neural networks, therefore, have considerable potential for accurate land cover mapping and in the realization of the potential of remote sensing as a source of land cover and other thematic data.

3.3.3 Limitations of the Conventional Neural Network Approach

Despite a somewhat up-beat presentation in some of the literature, neural networks are not a panacea. Although emulating some aspects of the human eye–brain system, which is very effective for pattern recognition, the neural networks generally used have the computing power of lower lifeforms such as earthworms (Simpson and Li, 1993). The relative simplicity of artificial neural networks typically used can be illustrated with reference to one major factor determining the capacity of a network, the number of weighted connections (which is a function of the number of units) and their arrangement. The human brain, for instance, contains some 10^{11} units or neurons (Aleksander and Morton, 1990), while the artificial neural networks used in remote sensing are much smaller, with typically only some 10^1 to 10^3 units. There are many other problems associated with neural networks for image classification. Here, some are briefly discussed and possible solutions indicated.

Although neural networks may generally be used to classify data at least as accurately as other classification approaches there are a range of factors that limit their use (Wilkinson, 1997). A major limitation associated with artificial neural networks is that they are semantically poor. Thus, while an artificial neural network may be able to perform a certain task it is difficult to explain the results or gain any understanding about how the result was achieved. In terms of image classification, it is, for example, difficult to identify the relative contribution of different wavebands for inter-class discrimination. Some information may be gleaned from an analysis of the weighted connections, but if the analyst wished to understand how, for example, a particular class allocation was achieved it may be preferable to adopt an alternative technique and those based on genetic programming and fuzzy logic (Corne et al., 1996) may be particularly attractive.

The accuracy of a classification is also not always the only concern of the analyst. Other criteria of classification performance may be important. In many of the comparative studies undertaken, performance criteria other than accuracy, notably training time, have been evaluated and revealed that neural networks may be less attractive than other classification approaches. Numerous studies have shown that while a trained network may be used to classify data sets very rapidly and accurately, the training of the network can be computationally demanding and slow. This is one of the main drawbacks of neural network classifications that is generally noted and consequently considerable research has been directed at speeding up the training of neural networks for classification applications (Dawson et al., 1993; Manry et al., 1994). This problem may, however, be resolved through developments in computing, particularly in parallel hardware. Alternatively, if training time is a major constraint, different network types, such as the 'one-shot' binary diamond (Salu and Tilton, 1993) and radial basis function (Bishop, 1995) networks, or faster learning algorithms (e.g., Fahlman, 1988), may be used instead of the more commonly used feedforward network employing a backpropagation, or similar, learning algorithm.

There are a number of other problems which are more fundamental and somewhat conceptual in nature. Two key issues will be considered here. First, the subjectivity of the approach must be recognized and second the appropriateness of classification itself questioned.

Perhaps one of the most important problems is that classification, by whatever method, is highly subjective (Johnston, 1968). Despite the apparent objectivity of the method, the analyst has, for instance, control over a range of network parameters that strongly influence network performance, especially in terms of speed and accuracy (Table 3.1), and even if these various parameters are selected judiciously there is no guarantee that the neural network will provide an acceptable let alone optimal solution. The analyst is, therefore, often left to select and build a neural network with

Table 3.1 A few of the issues that the analyst may have control over and that influence the performance of a neural network classification. Many of these issues are interrelated and, in the absence of definite rules, the analyst generally designs the approach to adopt on the basis of past experience and through analysis of empirically derived information from trial runs in which various network parameters are altered

Parameter/issue	Comment
Number of hidden units and layers	Determines the capacity of the network to learn and generalize. In general, large networks may learn more accurately but have a poorer generalization ability than a small network. Larger networks are also slower to train. How many hidden units and layers should be used?
Learning algorithm	There are a range of learning algorithms available. Backpropagation is the most widely used but can be slow and faster variants, which make assumptions about the error surface, are popular. Which should be used?
Learning parameters	Learning algorithms such as backpropagation have parameters (e.g. momentum and learning rate) that must be selected. These can significantly influence the performance of a network. What values should be selected and should they be varied in training?
Data input and scaling	There is usually one input unit associated with each discriminating variable but other approaches may be used. Also the data input to the neural network generally have to be rescaled for the analysis, typically to a 0 to 1 or −1 to 1 scale. What method should be used to achieve this and what allowance should be made for data to extend beyond the range observed in the training set?
Number of training iterations	The training error is generally negatively related to the number of training iterations. The accuracy of generalizations may be non-monotonically related to the intensity of training; typically the accuracy of generalization increases as the network gradually learns the underlying relationships with greater accuracy but will eventually decline as the network becomes overtrained. How many iterations of the learning algorithm should be used?
When/how to terminate training	There is a need to ensure that the network has learnt to correctly identify class membership from the training data but is not overtrained and so has acceptable generalization ability. How is this to be assessed? Should verification sets be used?
Initial weights	The initial weight settings of the pre-trained network can significantly influence network performance. Typically, these are generally set randomly, but within what range?

little theoretical assistance. Thus, for instance, the definition of the number of hidden units and layers that should be used is a commonly encountered problem. A number of rules of thumb may be used to define the network architecture but these are only guides to network definition as the problem is highly data-set specific. Alternatively the analyst may attempt to define an appropriate architecture empirically on the basis of a series of trial runs which either begin with a large network from which units are gradually removed or a small one to which units are added until a satisfactory performance is obtained. The key issue here is the capacity of the network to learn accurately in training and maintain a high generalization power. A large network will be able to learn the training data accurately but possibly to such an extent that it cannot generalize appropriately. It could, therefore, have difficulty identifying the correct class of membership of cases that were only slightly dissimilar to cases in the training set. Conversely, a smaller network will have greater difficulty in learning but may have a greater generalization ability. Sometimes a large network is constructed, but insignificant components later pruned (Jiang et al., 1994) and this has been shown to increase the accuracy and efficiency of neural network classifications (Dreyer, 1993). To complicate matters, the learning and generalization capacity of a network is also a function of the number of training iterations. In general, the more iterations used the more accurate training may be but the lesser the generalization ability. Thus, large networks trained over many thousands of iterations may be overtrained, producing low errors in training but having a poor generalization ability. Conversely, a smaller network trained for fewer iterations may be less accurate in training, but have a greater generalization ability. Since generalization ability is more important than the accuracy of learning in a supervised classification, the analyst, therefore, has to be cautious not to overtrain the network or to slavishly follow the accuracy of training as an index of network quality. The analyst may, therefore, elect to use a small network and limit the number of iterations of the learning algorithm in training. A small network also contains fewer weights and so is more rapid to train than a large network. Although the neural network may weight the information content of the input data, feature selection may, therefore, still be desirable to identify the most important discriminating variables and exclude the others as this would reduce the size of the network and facilitate training. This could be done in advance, as with conventional classifiers. Thus, for example, mutual information content may be assessed to identify the most appropriate discriminating variables (Battiti, 1994). None the less, issues such as the selection of the network architecture and properties together with the avoidance of overfitting to the training data while deriving a sufficient generalization capacity are important issues in the classification of remotely sensed data (Fischer et al., 1997), but selections are largely based on subjective decisions.

Although the adoption of a neural network avoids problems with the assumptions made by other classification techniques it does not free the analyst from a range of basic problems that are common to all supervised classifications. Thus, even if the network architecture and parameters are well defined there are many other factors that will influence classification performance. For instance, the accuracy of the classification may be constrained largely by the quality of the training data. Issues such as the size and composition of the training sets have a considerable effect on the accuracy of a neural network classification (Zhuang et al., 1994; Foody et al., 1995; Blamire, 1996;

Staufer and Fischer, 1997) as they do on other classification approaches. The analyst must also accept that the use of a neural network does not guarantee that the classification will be sufficiently accurate. Although generally more accurate than other classifications, the accuracies reported in the literature, like those of other classifiers, often fall short of an operationally acceptable level (Wilkinson, 1996). Moreover neural network classifications do not always provide the highest classification accuracy, on a per-class and/or overall basis. Indeed, there is an argument that neural networks are the second best way of performing a task. Thus, if, for example, the data set to be classified does satisfy the requirements of the maximum likelihood classification then that approach rather than a neural network should be used. In such circumstances, the fundamental model underlying the maximum likelihood classification is a major advantage over the distribution-free black-box approach of the neural network. By recognizing that different classification approaches vary in their ability to separate the classes in an image, however, it may be appropriate to adopt a multi-classifier approach to make the best use of each method of classification (Wilkinson et al., 1995; Roli et al., 1997).

Lastly, the appropriateness of classification as a tool for mapping is debatable. The classification techniques used are 'hard', with each image pixel allocated to a single class. Such approaches are only appropriate for the mapping of classes that are discrete and mutually exclusive. On many occasions this will not be the case. Many land cover classes are continuous and so intergrade. This is in part a consequence of class definition. For example, how much tree cover is required for a patch of land to be classed as forest? Furthermore, classification is only appropriate if the basic spatial unit used, typically the pixel, is pure. This is rarely the case, with many pixels of mixed land cover composition contained within a remotely sensed image (Crapper, 1984; Campbell, 1996). These mixed pixels may occur whatever the nature of the classes. For instance, with continuous classes, mixed pixels will occur in the inter-class transition zones where the classes coexist spatially, whereas for discrete classes the area represented by a pixel will often enclose or straddle class boundaries. The exact proportion of mixed pixels in an image will vary with a range of factors, notably the land cover mosaic on the ground and the sensor's spatial resolution, but is often very large (Campbell, 1996). For coarse spatial resolution data sets used in mapping land cover at regional to global scales, and where remote sensing has perhaps its greatest potential role, mixed pixels may vastly dominate imagery (Foody et al., 1997). The mixed pixel problem is not, however, confined to coarse spatial resolution data. At fine spatial resolutions mixing still occurs. Here, however, instead of the concern being the relative extent of different classes within the area represented by a pixel in a coarse spatial resolution image, it generally becomes the extent of components, such as soil, leaves and shadow, of an individual class in a fine spatial resolution image. Mixed pixels will, therefore, be evident in fine spatial resolution data sets, particularly for heterogeneous classes such as urban areas (Townshend, 1981). Thus, the range of sensors that will soon be providing fine spatial resolution data (Barnsley and Hobson, 1996) may be as inappropriate as a source of data for hard classification analyses as coarse spatial resolution systems. If the full potential of remote sensing as a source of land cover data is to be realized, alternative approaches to conventional 'hard' classifications may be required.

3.4 Need for Fuzzy Classification

Since a 'hard' classification output can fail to represent appropriately land cover, alternative approaches, which should allow for partial and multiple class membership, have been sought (Wang, 1990a). These have focused on fuzzy or soft classification techniques in which the full class membership of each pixel is partitioned between all classes. In such a classification a pixel can display any possible membership scenario, from full membership of one class through to having its membership divided, in any permutation, between all classes.

Fuzzy classifications may be derived in a number of ways. There are, for example, a range of fuzzy classifiers (e.g., Cannon et al., 1986). Alternatively a fuzzy classification may be achieved by 'softening' the output of a 'hard' classification. This is possible because the conventional hard classification approaches are wasteful of the class membership information generated to achieve a class allocation. For example, measures of the strength of class membership, rather than just the code of the most likely class of membership, may be output from a classification. Thus, for example, with a probabilistic technique such as the maximum likelihood classification a probability vector containing the probability of membership a pixel has to each defined class could be output. In this probability distribution the partitioning of the class membership probabilities between the classes would, ideally, indicate the land cover composition of a mixed pixel (Foody, 1996; Foody and Arora, 1996). This type of output makes fuller use of the information on class membership generated in the classification and may be considered to be fuzzy, as an imprecise allocation may be made and a pixel can display membership of all classes. The remotely sensed data must still, however, satisfy the assumptions and requirements of the classification technique used, which is often unlikely with the widely used probability based classifiers.

Although generally used to produce a hard classification (Kanellopoulos et al., 1992), the output of an artificial neural network may be softened to provide a more appropriate representation of fuzzy land cover than a 'hard' classification (Foody, 1996). This is because the backpropagation learning algorithm produces a network that has an ability to interpolate (Pham and Bayro-Corrochano, 1994) and the magnitude of the activation level of an output unit may be taken as a measure of the strength of membership of the class associated with that unit (Foody, 1996, 1997). Thus, rather than deriving only the code of the class associated with the most activated network output unit, the magnitude of the activation level of each output unit could be output. These have a range of potential applications. They may, for instance, be used like other fuzzy membership values for the definition of fuzzy boundaries (Wang and Hall, 1996) which may be more appropriate than the crisp ones that are a common problem in GIS (Burrough, 1986), be integrated with other data that may be available in a GIS for the production of a refined land cover classification (Foody, 1995) or other post-classification processing (e.g., Barnsley and Barr, 1996), and, of particular relevance here, be used to indicate the sub-pixel land cover composition. The output unit activation levels may be used to reflect the sub-pixel land cover composition in the same way that probabilities are mapped from the maximum likelihood classification or fuzzy membership values from the fuzzy c-means algorithm (Foody, 1996). This type of approach also makes fuller use of the information content of the remotely sensed data. Furthermore, it may enable a more accurate and appropriate representation of land cover. This applies to both relatively discrete (Foody et al., 1997) and continuous classes (Foody and Boyd, 1999).

Although fuzzy classifications have been used to provide a more appropriate representation of land cover that may be considered to be fuzzy, the fuzziness of the land cover being represented has often been overlooked. It is important to note that the use of fuzzy classification techniques, therefore, does not fully resolve the mixed pixel problem, it only provides a means of appropriately representing land cover that may be considered fuzzy at the scale of the pixel. Thus, while the class allocation made by a fuzzy classification may appropriately accommodate mixed pixels it must be stressed that this is only one component of the classification process. Relatively little attention has, however, focused on the accommodation of fuzziness caused by mixed pixels in the training and testing stages of a supervised classification. In both of these stages of the classification the ground data on class membership are generally related to the remotely sensed data at the scale of the pixel and so may be fuzzy. Moreover, there may be other sources of uncertainty and error in the ground data. Consequently, ground data are rarely 'ground truth'. The adoption of a fuzzy classification algorithm in recognition of the need for a class assignment that allows for multiple and partial class membership does not in itself take any account of mixed pixels in the training and testing stages of the classification. Since a large proportion of image pixels may be mixed it is important that they be accommodated throughout the classification.

There are essentially three ways of accommodating mixed pixels in the classification. First, their effect can be ignored and the analysis proceed as if they were not present, making an implicit assumption that the pixels are in fact pure. It must be recognized that this will degrade all three stages of the classification: mixed pixels will degrade the class responses derived in training, cannot be appropriately allocated by a hard classification and may vary in the accuracy of their allocation. Second, the analyst could harden the data. With this approach each pixel is given the code of the dominant class. Although recognizing the existence of mixed pixels, no real accommodation for their presence is made and the classification will be degraded in the same way as if the mixed pixels had been ignored. Third, the analyst could seek to exclude mixed pixels, although their identification may be difficult, and this could result in the loss of most of the data set. All these approaches are wasteful of the information content of the remotely sensed data set. A more appropriate approach would be to recognize the existence of mixed pixels and use techniques which accommodate their presence throughout the analysis (Foody, 1997).

Frequently, the only way the fuzziness of the land cover on the ground is accommodated in the training and testing of a classification is by deliberately avoiding mixed pixels (Metzler and Cicone, 1983). Thus, although an image may be dominated by mixed pixels only pure pixels are selected for training. This typically involves selecting training sites from only very large homogeneous regions of each class, to avoid contamination of training sites by other classes (Campbell, 1996). Moreover, research on refining training sets has often focused on removing potentially mixed pixels from the training set (Arai, 1992). With these approaches it may be difficult to acquire a training set of an appropriate size. Moreover, the training statistics defined may not be fully representative of the classes. In testing the classification, pure pixels only are again generally used and the conventional measures of accuracy assessment were designed for application with 'hard' classifications. Even if uncertainty and multiple class membership is recognized, many studies may be constrained to 'harden' the data to the dominant class to enable a conventional assessment of accuracy. Such approaches may not provide an appropriate index of classification accuracy and are wasteful of information. As the majority of pixels may be mixed, failure to include them and accommodate for their effects in the accuracy assessment may, therefore, result in an inappropriate and inaccurate estimation of classification accuracy. Since fuzziness may be a characteristic feature of *both* the classification output and ground data, the use of a fuzzy classification algorithm alone may be insufficient for the resolution of the mixed pixel problem (Foody and Arora, 1996; Foody, 1997). The lack of methods that accommodate fuzziness has been a hindrance to aspects of the exploitation of remotely sensed data (Goodchild, 1994) and consequently much research has been aimed at accommodating fuzziness in the various stages of the classification (Wang, 1990b; Gopal and Woodcock, 1994; Maselli *et al.*, 1994; Foody, 1996).

3.5 The Continuum of Classification Fuzziness

As in the class allocation stage, mixed pixels are undesirable in the training and testing stages of a supervised classification, but if land cover is to be mapped accurately and the map evaluated appropriately they may be unavoidable. The inclusion of mixed pixels, deliberately or not, into any of the three stages of the conventional 'hard' supervised classification is inappropriate and likely to be a major source of error. Similarly the inclusion of mixed pixels in the training and/or testing stages of a fuzzy classification algorithm is inappropriate if conventional approaches which assume pure pixels are used in training and testing. Mixed pixels may, therefore, need to be accommodated in the whole classification process. Through a consideration of fuzziness applying to all three stages of the supervised classification it is possible to define a continuum of classification fuzziness (Foody, 1999). There is, therefore, not a simple distinction between hard and fuzzy classifications, but rather a range of classification approaches of variable fuzziness. At the hard end of the continuum are what may be termed completely-crisp classifications. These are based on the conventional approach to classification in which a pixel is associated with a single class at each stage of the classification and so the data may be considered to belong to crisp rather than fuzzy sets. Most supervised classifications of remotely sensed data adopt this approach but its application may be inappropriate due to the presence of mixed pixels. At the other extremity of the continuum are fully-fuzzy classifications. In these, fuzziness is accommodated in all three stages of the classification. This type of approach may provide a more realistic and accurate representation of the land cover of a site and use more fully the information content of the remotely sensed data. Between these extremes lie classifications of varying fuzziness, including those generally referred to in the literature as fuzzy classifications, in which only the class allocation stage actually accommodates fuzziness.

Although more difficult to derive than a completely-crisp classification, a number of approaches exist for the production of classifications of variable fuzziness. For instance, the conventional approach to maximum likelihood classification can be modified to accommodate fuzziness in any or all three stages of the analysis (Foody and Arora, 1996). However, neural networks are particularly attractive for direct accommodation of fuzziness inherent to many classifications. A major attraction of neural networks is their ability to accommodate the fuzziness of the training data directly into the classification (Foody, 1997; Foody et al., 1997); the accommodation of fuzziness into the testing stage of the classification is effectively independent of the approach used to generate the fuzzy classification. Neural networks are also capable of providing a classification at any point along the continuum of classification fuzziness. They can, therefore, be used in the derivation of classifications varying from the completely-crisp (Kanellopoulos et al., 1992), through classifications of intermediate fuzziness (Foody, 1996) to fully-fuzzy (Foody, 1997, 1999).

It is important to note that classifications may be arranged along a continuum of fuzziness and should not be thought of as a small number of classifications that can be listed in order of fuzziness defined by the number of stages based on crisp and fuzzy data. This is because the fuzziness of each stage of the classification is itself variable. This can be briefly illustrated with reference to each stage in turn. In training, the data may be crisp if the pixels are pure but for mixed pixels the training stage may vary in fuzziness. For example, the training data used may vary from the code of the dominant class to individual class proportions. In the class allocation stage, a pixel could be allocated to a single class or the output might also contain the second most likely class through to the provision of membership grades to all classes. Lastly, the data used in testing the classification may, like the training data, vary from the dominant class to individual class proportions. In each stage there are, therefore, different ways of accommodating the fuzziness which may utilize the information to differing extents. For example, in evaluating the accuracy of the output from a fuzzy classification the analyst could degrade the data to enable the calculation of a conventional measure of classification accuracy, use, for instance, the two most likely classes of membership (Woodcock and Gopal, 1995; Zhang and Foody, 1998), or use all the data on class membership derived in comparison against the ground data set (Kent and Mardia, 1988; Foody, 1996). Therefore, two classifications which may accommodate for fuzziness in the same number of stages of the classification may still differ in their respective fuzziness.

It may at first seem that fuzzy classifications, particularly, the fully-fuzzy ones, require more precise ground data for their application than a conventional hard classification. This is not, however, the case. In a hard classification the actual class of membership for each training and testing pixel must be determined to the same degree as in a fuzzy classification. The only difference is that in a hard classification each pixel should represent an area characterized by homogeneous cover of a single class whereas in the fuzzy classification each pixel may represent an area comprising any number of classes. In both, the analyst has to know the land cover composition of the entire area represented by each pixel; in any classification the ground data for a pixel are supposed to represent the actual land cover observed on the ground. Thus, the fundamental requirement of ground data detail is *exactly* the same for classifications at any point along the continuum of classification fuzziness. The information may, however, be presented and used differently in classifications along the continuum. With a completely-crisp classification the data may, for example, be hardened to the dominant class label but comprise the individual class proportions for a fully-fuzzy classification. None the less, while fully-fuzzy classifications do require precise ground data for their application, the same level of ground data detail is actually required for the correct application of any classification including the completely-crisp classifications; the ground data should describe the class membership properties of the area represented by the pixel. The analyst may, of course, decide or be constrained to use less precise ground data in training or testing a classification, but must accept that this could degrade the real and apparent accuracy of the classification. Using less precise ground data in training a hard classification is, for example, possible, but as this will cause the class training statistics to be contaminated by other classes it may be a source of misclassification. It may also indicate that the analyst is prepared, or constrained, to ignore the presence of mixed pixels and so accept error in ground data when using a hard classification.

3.6 Conclusions

Neural networks have many potential applications in remote sensing, but are particularly attractive and used for supervised image classification. Although they have many advantages over conventional classification approaches, and have often been noted to provide more accurate classifications, they are not without their problems. A key problem in mapping from remotely sensed data is that conventional 'hard' classification techniques may be inappropriate due especially to the need to accommodate for fuzziness caused particularly by mixed pixels. The recognition of the effect and significance of mixed pixels has, therefore, led to the derivation and application of a range of fuzzy classifications. However, the term 'fuzzy classification' has often been used in the literature to describe an analysis in which a fuzzy or soft classification output is derived but, usually, no explicit reference is made to the training and testing stages. Indeed, the training data are generally selected from large homogeneous regions to maintain training sample purity. The accuracy of a classification is also generally assessed with conventional accuracy assessment procedures which require pure pixels. In such an analysis, therefore, only one of the three stages of the classification is fuzzy; the training and testing stages are hard. The use of such a fuzzy classification, therefore, does not fully resolve the mixed pixel problem as their effects on the training and testing stages have not been accounted for. Recognizing that fuzziness may require accommodation in the training and testing stages of the classification as well as the allocation stage enables a continuum of classification fuzziness, from completely-crisp to fully-fuzzy, to be defined. Although less easy to perform than a conventional 'hard' image classification, the fuzzy classifications may provide a more accurate and appropriate representation of land cover from remotely sensed data. These approaches are not, however, without their problems. One criticism that may be raised against the fuzzy classifications, particularly the fully-fuzzy classifications, is the requirement for a high level of precision in the ground data. A fully-fuzzy classification, for example, requires the class composition of all training and testing pixels. However, this is *exactly* the same requirement for the proper application of any classification and so the apparent demand for a greater precision in the ground data is illusory. Since neural networks may be used to derive a classification at any point along the continuum of classification fuzziness they, therefore, have considerable potential for thematic mapping applications and play a significant role in the realization of the potential of remote sensing as a source of land cover information.

Acknowledgements

This chapter has drawn upon material presented at a number of meetings held during 1993–96 and I am grateful to all those associated with these meetings for their comments and advice.

References

Aleksander, I. and Morton, H., 1990, *An Introduction to Neural Computing* (London: Chapman & Hall).
Arai, K., 1992, A supervised Thematic Mapper classification with a purification of training samples, *International Journal of Remote Sensing*, 13, 2039–2049.
Atkinson, P.M. and Tatnall, A.R.L., 1997, Neural networks in remote sensing, *International Journal of Remote Sensing*, 18, 711–725.
Austin, J., 1997, High speed image segmentation using a binary neural network, in I. Kanellopoulos, G.G. Wilkinson, F. Roli and J. Austin (eds), *Neuro-computation in Remote Sensing Data Analysis* (Berlin: Springer-Verlag), 202–213.
Baraldi, A. and Parmiggiani, F., 1995, A neural network for unsupervised categorisation of multivalued input patterns: an application to satellite image clustering, *IEEE Transactions on Geoscience and Remote Sensing*, 33, 305–316.
Baret, F., 1995, Use of spectral reflectance variation to retrieve canopy biophysical characteristics, in F.M. Danson and S.P. Plummer (eds), *Advances in Environmental Remote Sensing* (Chichester: Wiley), 33–51.
Barnsley, M.J. and Barr, S.L., 1996, Inferring urban land use from satellite sensor images using kernel-based spatial reclassification, *Photogrammetric Engineering and Remote Sensing*, 62, 949–958.
Barnsley, M. and Hobson, P., 1996, Making sense of sensors, *GIS Europe*, 5(5), 34–36.
Battiti, R., 1994, Using mutual information for selecting features in supervised neural net learning, *IEEE Transactions on Neural Networks*, 5, 537–550.
Baum, E.B. and Haussler, D., 1989, What size net gives valid generalizations, in D.S. Touretzky (ed.), *Advances in Neural Information Processing Systems I* (San Mateo, CA: Morgan Kaufmann), 81–90.
Benediktsson, J.A. and Sveinsson, J.R., 1997, Feature extraction for multisource data classification with artificial neural networks, *International Journal of Remote Sensing*, 18, 727–740.
Benediktsson, J.A., Swain, P.H. and Ersoy, O.K., 1990, Neural network approaches versus statistical methods in classification of multisource remote sensing data, *IEEE Transactions on Geoscience and Remote Sensing*, 28, 540–551.
Bischof, H. and Leonardis, A., 1998, Finding optimal neural networks for land use classification, *IEEE Transactions on Ecoscience and Remote Sensing*, 36, 337–341.
Bishop, C.M., 1995, *Neural Networks for Pattern Reconition* (Oxford: Clarendon Press).
Blamire, P.A., 1996, The influence of relative sample size in training artificial neural networks, *International Journal of Remote Sensing*, 17, 223–230.

Burrough, P.A., 1986, *Principles of Geographical Information Systems for Land Resources Assessment* (Oxford: Oxford University Press).

Campbell, J.B., 1996, *Introduction to Remote Sensing*, 2nd edition (London: Taylor and Francis).

Cannon, R.L., Dave, J.V., Bezdek, J.C. and Trivedi, M.M., 1986, Segmentation of a thematic mapper image using the fuzzy c-means clustering algorithm, *IEEE Transactions on Geoscience and Remote Sensing*, **24**, 400–408.

Chang, E.I. and Lippmann, R.P., 1991, Using genetic algorithms to improve pattern classification performance, in R.P. Lippmann, J.E. Moody and D.S. Touretzky, (eds), *Advances in Neural Information Processing Systems 3* (San Mateo, CA: Morgan Kaufmann), 797–803.

Chauvin, Y., 1989, A back-propagation algorithm with optimal use of hidden units, in D.S. Touretzky, (ed.) *Advances in Neural Information Processing Systems 1* (San Mateo, CA: Morgan Kaufmann), 519–526.

Clastres, X., Samuelides, M. and Tarr, G.L., 1997, Dynamic segmentation of satellite images using pulsed coupled neural networks, in I. Kanellopoulos, G.G. Wilkinson, F. Roli and J. Austin (eds), *Neuro-computation in Remote Sensing Data Analysis*, (Berlin: Springer-Verlag), 160–167.

Corne, S., Murray, T., Openshaw, S., See, L. and Turton, I., 1996, Using artificial intelligence techniques to model subglacial water systems, *Proceedings 1st International Conference on Geocomputation*, **1**, 135–155.

Côté, S. and Tatnall, A.R.L., 1995, A neural network-based method for tracking features from satellite sensor images, *International Journal of Remote Sensing*, **16**, 3695–3701.

Crapper, P.F., 1984, An estimate of the number of boundary cells in a mapped landscape coded to grid cells, *Photogrammetric Engineering and Remote Sensing*, **50**, 1497–1503.

Dawson, M.S., Fung, A.K. and Manry, M.T., 1993, Surface parameter retrieval using fast learning neural networks, *Remote Sensing Reviews*, **7**, 1–18.

Dreyer, P., 1993, Classification of land cover using optimized neural nets on SPOT data, *Photogrammetric Engineering and Remote Sensing*, **59**, 617–621.

Dymond, J.R., 1993, An improved Skidmore/Turner classifier, *Photogrammetric Engineering and Remote Sensing*, **59**, 623–626.

Fahlman, S.E., 1988, Faster-learning variations of backpropagation: an empirical study, *Proceedings of the 1988 Connectionist Models Summer School*, 17–26 June (San Mateo, CA: Morgan Kaufmann), 38–51.

Fischer, M.M., 1996, Computational neural networks: a new paradigm for spatial analysis, *Proceedings 1st International Conference on Geocomputation*, **1**, 297–314.

Fischer, M.M., Gopal, S., Staufer, P. and Steinnocher, K., 1997, Evaluation of neural pattern classifiers for a remote sensing application, *Geographical Systems*, **4**, 195–225.

Foody, G.M., 1995, Land cover classification by an artificial neural network with ancillary information, *International Journal of Geographical Information Systems*, **9**, 527–542.

Foody, G.M., 1996, Approaches for the production and evaluation of fuzzy land cover classifications from remotely-sensed data, *International Journal of Remote Sensing*, **17**, 1317–1340.

Foody, G.M., 1997, Fully fuzzy supervised classification of land cover from remotely sensed imagery with an artificial neural network, *Neural Computing and Applications*, **5**, 238–247.

Foody, G.M., 1997, The continuum of classification fuzziness in thematic mapping, *Photogrammetric Engineering and Remote Sensing* (in press).

Foody, G.M. and Arora, M.K., 1996, Incorporating mixed pixels in the training, allocation and testing stages of supervised image classifications, *Pattern Recognition Letters*, **17**, 1389–1398.

Foody, G.M. and Boyd, D.S., 1999, Fuzzy mapping of tropical land cover along an environmental gradient from remotely sensed data with an artificial neural network, *Geographical Systems*, **1**, 23–25.

Foody, G.M., McCulloch, M.B. and Yates, W.B., 1995, Classification of remotely sensed data by an artificial neural network: issues related to training data characteristics, *Photogrammetric Engineering and Remote Sensing*, **61**, 391–401.

Foody, G.M., Lucas, R.M., Curran, P.J. and Honzak, M., 1997, Non-linear mixture modelling without end-members using an artificial neural network, *International Journal of Remote Sensing*, **18**, 937–953.

Gaganis, V., Zervakis, M. and Christodoulou, M., 1997, Neural nets and multichannel image processing applications, in I. Kanellopoulos, G.G. Wilkinson, F. Roli and J. Austin (eds), *Neuro-computation in Remote Sensing Data Analysis* (Berlin: Springer-Verlag), 57–70.

Gershon, N.D. and Miller, C.G., 1993, Dealing with the data deluge, *IEEE Spectrum*, **30**(7), 28–32.

Goodchild, M.F., 1994, Integrating GIS and remote sensing for vegetation analysis and modelling: methodological issues, *Journal of Vegetation Science*, **5**, 615–626.

Gopal, S. and Woodcock, C., 1994, Theory and methods for accuracy assessment of thematic maps using fuzzy sets, *Photogrammetric Engineering and Remote Sensing*, **60**, 181–188.

Hammerstrom, D., 1993a, Neural networks at work, *IEEE Spectrum*, **30**(6), 26–32.

Hammerstrom, D., 1993b, Working with neural networks, *IEEE Spectrum*, **30**(7), 46–53.

Hara, Y., Atkins, R.G., Shin, R.T., Kong, J.A., Yueh, S.H. and Kwok, R., 1995, Application of neural networks for sea ice classification in polarimetric SAR images, *IEEE Transactions on Geoscience and Remote Sensing*, **33**, 740–748.

Jiang, X., Chen, M-S., Manry, M.T., Dawson, M.S. and Fung, A.K., 1994, Analysis and optimization of neural networks for remote sensing, *Remote Sensing Reviews*, **9**, 97–114.

Johnston, R.J., 1968, Choice in classification: the subjectivity of objective methods, *Annals of the Association of American Geographers*, **58**, 575–589.

Kanellopoulos, I., Varfis, A., Wilkinson, G.G. and Megier, J., 1992, Land-cover discrimination in SPOT HRV imagery using an artificial neural network – a 20–class experiment, *International Journal of Remote Sensing*, **13**, 917–924.

Kent, J.T. and Mardia, K.V., 1988, Spatial classification using fuzzy membership models, *IEEE Transactions on Pattern Analysis and Machine Intelligence*, **10**, 659–671.

Lee, J.J., Shim, J.C. and Ha, Y.H., 1994, Stereo correspondence using the Hopfield neural network of a new energy function, *Pattern Recognition*, **27**, 1513–1522.

Lewis, D.J., Corr, D.G., Gent, C.R. and Sheppard, C.P., 1992, Semi-supervised artificial neural networks for classification of remotely sensed images, *Remote Sensing from Research to Operation* (Nottingham: Remote Sensing Society), 489–497.

Lewis, H.G., Cote, S. and Tatnall, A.R.L., 1995, Shape, motion and contextual descriptors for a neural network cloud classifier, *RSS'95 Remote Sensing in Action* (Nottingham: Remote Sensing Society), 3–10.

Lo, C.P., 1986, *Applied Remote Sensing* (Harlow: Longman).

Manry, M.T., Dawson, M.S., Fung, A.K., Apollo, S.J., Allen, L.S. and Lyle, W.D., 1994, Fast training of neural networks for remote sensing, *Remote Sensing Reviews*, **9**, 77–96.

Maselli, F., Conese, C. and Petkov, L., 1994, Use of probability entropy for the estimation and graphical representation of the accuracy of maximum likelihood classifications, *ISPRS Journal of Photogrammetry and Remote Sensing*, **49**, 13–20.

Mather, P.M. 1987, *Computer Processing of Remotely-Sensed Images* (Chichester: Wiley).

Metzler, M.D. and Cicone, R.C., 1983. Assessment of technologies for classification of mixed pixels, *Proceedings of the Seventeenth International Symposium on Remote Sensing of Environment*, Ann Arbor, Michigan: ERIM, **3**, 1015–1021.

Nasrabadi, N.M. and Choo, C.Y., 1992, Hopfield network for stereo vision correspondence, *IEEE Transactions on Neural Networks*, **3**, 5–13.

Paola, J.D. and Schowengerdt, R.A., 1995, A detailed comparison of backpropagation neural network and maximum likelihood classification for urban land use classification, *IEEE Transactions on Geoscience and Remote Sensing*, **33**, 981–996.

Peddle, D.R., 1993, An empirical comparison of evidential reasoning, linear discriminant analysis, and maximum likelihood algorithms for alpine land cover classification, *Canadian Journal of Remote Sensing*, **19**(1), 31–44.

Peddle, D.R., Foody, G.M., Zhang, A., Franklin, S.E. and LeDrew, E.F., 1994, Multisource image classification II: An empirical comparison of evidential reasoning and neural network approaches, *Canadian Journal of Remote Sensing*, **20**(4), 397–408.

Pham, D.T. and Bayro-Corrochano, E.J., 1994, Self-organising neural-network-based pattern clustering method with fuzzy outputs, *Pattern Recognition*, **27**, 1103–1110.

Pierce, L.E., Sarabandi, K. and Ulaby, F.T., 1994, Application of an artificial neural network in canopy scattering inversion, *International Journal of Remote Sensing*, **15**, 3263–3270.

Piper, J., 1992, Variability and bias in experimentally measured classifier error rates, *Pattern Recognition Letters*, **13**, 685–692.

Roli, F., Giacinto, G. and Vernazza, G., 1997, Comparison and combination of statistical and neural network algorithms for remote-sensing image classification, in I. Kanellopoulos, G.G. Wilkinson, F. Roli and J. Austin (eds), *Neuro-computation in Remote Sensing Data Analysis* (Berlin: Springer-Verlag), 117–124.

Rumelhart, D.E., Hinton, G.E. and Williams, R.J., 1986, Learning internal representation by error propagation, in D.E. Rumelhart and J.L. McClelland (eds), *Parallel Distributed Processing: Explorations in the Microstructure of Cognition* (Cambridge, MA: MIT Press), 318–362.

Salu, Y. and Tilton, J., 1993, Classification of multispectral image data by the binary diamond neural network and by non-parametric, pixel-by-pixel methods, *IEEE Transactions on Geoscience and Remote Sensing*, **31**, 606–617.

Schalkoff, R.J., 1992, *Pattern Recognition: Statistical, Structural and Neural Approaches* (New York: Wiley).

Schweiger, A.J. and Key, J.R., 1997, Estimating surface radiation fluxes in the Arctic from TOVS brightness temperatures, *International Journal of Remote Sensing*, **18**, 955–970.

Simpson, G. and Li, K., 1993, *Artificial Neural Networks: Solutions to Problems in Remote Sensing*, Report to DRA Farnborough on contract FRN1b/150, Earth Observation Sciences, Farnham.

Smith, J.A., 1993, LAI inversion using a back-propagation neural network trained with a multiple scattering model, *IEEE Transactions on Geoscience and Remote Sensing*, **31**, 1102–1106.

Smith, M.J., Moore, T. and Dumville, M., 1995, The performance of neural nets in image rectification, *RSS'95 Remote Sensing in Action* (Nottingham: Remote Sensing Society), 11–18.

Staufer, P. and Fischer, M.M., 1997, Spectral pattern recognition by a two-layer perceptron: effects of training set size, in I. Kanellopoulos, G.G. Wilkinson, F. Roli and J. Austin (eds), *Neuro-computation in Remote Sensing Data Analysis* (Berlin: Springer-Verlag), 105–116.

Townshend, J.R.G., 1981, The spatial resolving power of Earth resources satellites, *Progress in Physical Geography*, **5**, 32–55.

Townshend, J.R.G., 1992, Land cover, *International Journal of Remote Sensing*, **13**, 1319–1328.

Visa, A. and Peura, M., 1997, Generalisation of neural network based segmentation results for classification purposes, in I. Kanellopoulos, G.G. Wilkinson, F. Roli and J. Austin (eds), *Neuro-computation in Remote Sensing Data Analysis* (Berlin: Springer-Verlag), 255–261.

Walker, N.P., Eglen, S.J. and Lawrence, B.A., 1994, Image compression using neural networks, *GEC Journal of Research*, **11**, 66–75.

Wang, F., 1990a, Improving remote sensing image analysis through fuzzy information representation, *Photogrammetric Engineering and Remote Sensing*, **56**, 1163–1169.

Wang, F., 1990b, Fuzzy supervised classification of remote sensing images, *IEEE Transactions on Geoscience and Remote Sensing*, **28**, 194–201.

Wang, F. and Hall, G.B., 1996, Fuzzy representation of geographical boundaries in GIS, *International Journal of Geographical Information Systems*, **10**, 573–590.

Wang, Y. and Dong, D., 1997, Retrieving forest stand parameters from SAR backscatter data using a neural network trained by a canopy backscatter model, *International Journal of Remote Sensing*, **18**, 981–989.

Wilkinson, G.G., 1996, Classification algorithms – where next?, in E. Binaghi, P.A. Brivio and A. Rampini (eds), *Soft Computing in Remote Sensing Data Analysis* (Singapore: World Scientific), 93–99.

Wilkinson, G.G., 1997, Open questions in neurocomputing for Earth observation, in I. Kanellopoulos, G.G. Wilkinson, F. Roli and J. Austin (eds), *Neuro-computation in Remote Sensing Data Analysis* (Berlin: Springer-Verlag), 3–13.

Wilkinson, G.G., Fierens, F. and Kanellopoulos, I., 1995, Integration of neural and statistical approaches in spatial data classification, *Geographical Systems*, **2**, 1–20.

Woodcock, C.E. and Gopal, S., 1995, Remote sensing of forests: new data layers for GIS,

Proceedings 1995 ACSM/ASPRS Annual Convention and Exposition (Bethesda, MD: ACSM/ASPRS), **2**, 420–428.

Yu, S. and Weigl, K., 1997, A hybrid method for preprocessing and classification of SPOT images, in I. Kanellopoulos, G.G. Wilkinson, F. Roli and J. Austin (eds), *Neuro-computation in Remote Sensing Data Analysis* (Berlin: Springer-Verlag), 134–141.

Zhang, J. and Foody, G.M., 1998, A fuzzy classification of sub-urban land cover from remotely sensed imagery, *International Journal of Remote Sensing*, **19**, 2721–2738.

Zhang, M. and Scofield, R.A., 1994, Artificial neural network techniques for estimating heavy convective rainfall and recognising cloud mergers, *International Journal of Remote Sensing*, **15**, 3241–3261.

Zhuang, X., Engel, B.A., Lozano-Garcia, D.F., Fernandez, R.N. and Johannsen, C.J., 1994, Optimisation of training data required for neuro-classification, *International Journal of Remote Sensing*, **15**, 3271–3277.

4
Cloud Motion Analysis

Hugh G. Lewis, Franz T. Newland, Stéphane Côté and Adrian R.L. Tatnall

4.1 Introduction

Knowledge of the motion and development of clouds over a sequence of satellite images provides a rich source of meteorological information, from weather and rainfall prediction to wind information. Automatic determination of cloud motion is, however, a non-trivial task. Satellite-derived wind data from cloud motion have been used operationally in forecasting and meteorological modelling since the 1970s. In fact, cloud motion is the sole source of wind data for large parts of the Southern Hemisphere. With advances in numerical weather models, which are currently able to estimate cloud motion in many instances to greater accuracy than the motion vectors are able to describe, satellite-derived cloud motion is offering diminishing benefits (Hayden *et al.*, 1994).

Global and sectional images from satellites in geostationary orbit are available in visible and infrared (IR) wavebands, and either water vapour or carbon dioxide absorption wavebands, depending on the instrument. In conventional and operational approaches to cloud motion wind generation these images are initially segmented into square search windows (typically 32 × 32 pixels), then smaller template windows from each segment are cross-correlated with the corresponding segment in the next image. The maximum cross-correlation (MCC) method then consists of identifying the pair of template windows presenting the highest value of cross-correlation, which is interpreted as the best match. A cloud motion vector is then generated between these two template windows. Vector pairs across three successive images are compared for automatic quality checking and a final manual quality check is performed before the data are disseminated to numerical weather prediction (NWP) modellers and other users as cloud motion winds. Variants of the MCC method developed by different satellite operators are described in EUMETSAT (1993) and EUMETSAT (1996).

The cloud motion vectors produced by the MCC method are used to indicate the wind flow. However, interpreting the vectors in this way assumes that clouds are passive tracers of wind, and that the maximum cross-correlation identifies the cloud motion related to the wind rather than the cloud dynamics. In practice, the MCC method produces accurate vectors where the cloud is a good tracer, but their quality is less where this is not the case. For example, upslope fog, stratus, and mountain waves may show no cloud movement, yet have strong associated winds, while trade cumulus and cirrus in the jetstream on the other hand are good cloud tracers (Purdom and Dills, 1993). Further, motion at scales smaller or larger than the template correlation window, and motion related to rotation, are poorly represented by the cloud motion vector produced by the MCC method (Côté and Tatnall, 1995a; Mahrt and Sun, 1995). Therefore, the MCC method does not offer a suitable framework with which to analyse the life cycle of clouds or the motion of synoptic-scale cloud systems.

The nature of the MCC method is basic; pixels identified in windows are matched over sequences of images, with limited consideration to the context in which these windows appear. For instance, some areas might require small template windows to isolate the fine local variations that are present while in some other areas, where the spatial and temporal variations are very smooth, a larger template window might be preferable to minimize noise. Considering the various spatial and temporal scales observed in cloud dynamics, it would be much more appropriate to use a more flexible method that adapts to these situations and that provides a better solution for each case.

The mechanisms that influence the life cycle, dynamics and motion of cloud operate at many different spatial and temporal scales and a single method for analysing cloud motion is unlikely to be applicable to all of them. Alternatively, applying many methods, each directed at different aspects of cloud motion but without contextual knowledge, will be likely to produce disparate results that are hard to integrate. An approach that uses contextual knowledge to analyse and interpret the motion generated from many methods is presented here. In this approach, the context of the cloud motion for cloud *pixels*, cloud *objects* and cloud *elements* found within a cloud *region* (see Figure 4.1) is included.

Cloud regions are represented by synoptic-scale cloud systems at sizes of thousands of kilometres. These regions define much of what we perceive as weather; fronts, airmasses, depressions and anticyclones often appear as synoptic-scale features in remotely sensed imagery. The motion within these regions is complex and dependent upon the region type and can indicate the presence of unstable air and processes such as subsidence, cyclogenesis, frontogenesis, or frontolysis (UK Met. Office, 1978). Once regions are identified, they can be used for directing suitable motion analyses to any segments within them. For example, within frontal regions, curvilinear features defined on the body accurately describe the region dynamics while texture analysis may

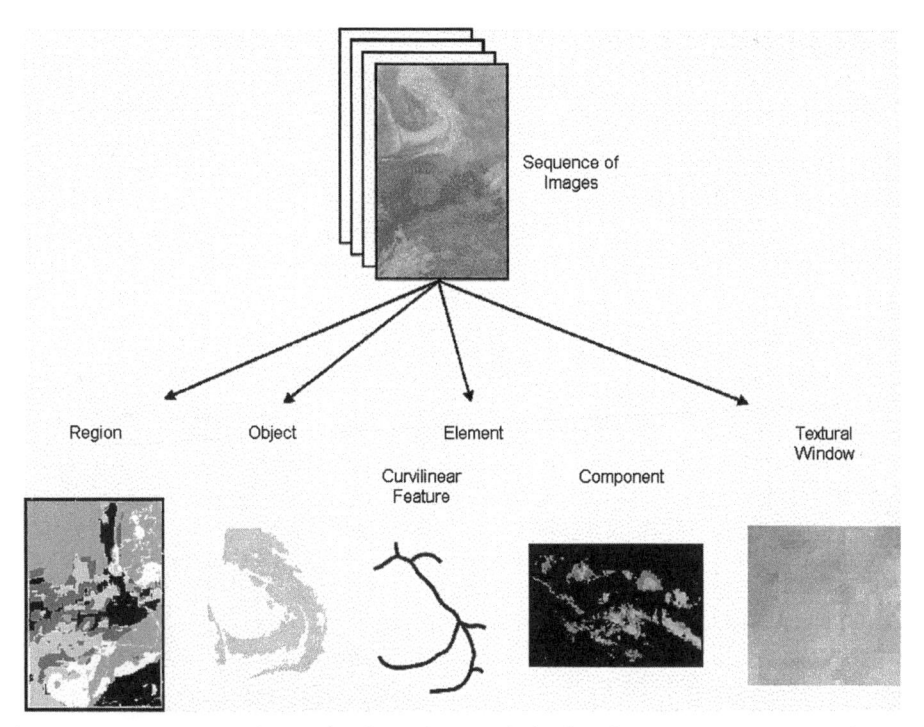

Figure 4.1 An approach to cloud motion analysis that incorporates segmentation and matching methods for the identification and analysis of cloud regions, objects, elements and pixels

indicate secondary processes acting in a different direction to the frontal motion. The regional knowledge also provides a meteorological context within which to interpret such conflicting motion if present.

Cloud objects represent mesoscale features and whole clouds. The motion of these cloud objects can be described in terms of shape changes, rotation and skewing and can be interpreted in terms of their life cycle and cloud motion direction. Microscale phenomena and larger features within these cloud objects are represented by *components* and include convective updraughts (UK Met. Office, 1978). *Curvilinear features* represent more abstract information extracted from cloud objects, such as edges and skeletons. The final aspect of motion analysis considered provides motion at the scale of a few cloud pixels, organized within correlation windows, and pixel-level vectors using optical flow techniques.

Several segmentation methods suitable for generating these cloud regions, objects and elements are discussed. Techniques for matching the variety of cloud features generated by the segmentation methods are summarized and a powerful Hopfield neural network matching method is described. Finally, the combination of the resulting motion vectors from different types of segment is briefly explored and examples are presented.

4.2 Segmentation

The purpose of segmentation is to define features to be matched. Typically, segmentation is achieved by grouping together pixels sharing common properties defined in the spectral, spatial or temporal domains. Conventional segmentation algorithms for digitized imagery have included grey-level thresholding, edge detection, region growing, active contours (snakes) and differencing (Sonka et al., 1993). The segmentation methods described below are appropriate for generating feature types that are relevant to cloud motion analysis.

4.2.1 Cloud Regions

The generation and analysis of cloud motion vectors are simplified if images are initially segmented into regions showing the same type of motion. Secondary segmentation and subsequent matching techniques can be aimed at the specific types of motion which have been identified within the region. Two kinds of methods for achieving this initial segmentation are presented. The first method uses *a priori* meteorological knowledge to identify regions of static stability and instability (UK Met. Office, 1978) in which different types of motion can be expected. The second method uses temporal and spatial properties of image sequences to identify regions having common motion types.

4.2.1.1 *Convective Cell and Complete Cover Regions*

Methods for the identification and classification of synoptic-scale cloud regions in remotely sensed imagery have been proposed in two bottom-up classification schemes, by the UK Meteorological Office (Pankiewicz, 1995a) and the US Naval Research Laboratory (Peak and Tag, 1992; Bankert, 1994). Both of these proposed methods employ supervised classification techniques. Many methods of supervised classification exist, including techniques traditional to remote sensing, such as maximum likelihood, and more recent techniques involving neural networks. Commonly, the approach taken is to partition or cluster the feature space using feature vectors from labelled samples and then to assign new feature vectors to a class based upon their location in the partitioned feature space. Non-parametric methods, which include neural networks, make no assumptions about the distribution of the labelled data, while parametric methods assume a known distribution. Neural networks are superseding traditional statistical pattern recognition methods for cloud classification because many of the prior assumptions made for traditional algorithms have been rarely met in practice. Neural networks offer more representative algorithms because they learn from interactions with their environment rather than by explicit programming and few restrictions are placed on the type of functional relationships that can be learnt. In addition, they have the ability to define non-linear decision boundaries and the ability to generalize (interpolate and extrapolate) the training information to similar situations. The feed-forward neural network trained using backpropagation of error (Rumelhart et al., 1986) has been a popular recent choice for the classification of clouds (Lee et al., 1990; Pankiewicz, 1997).

Commonly, the backpropagation training strategy is used iteratively to find the minimum of the energy function

$$E = \frac{1}{2} \sum_{p=1}^{N} \sum_{i=1}^{m} \left(t_{pi} - y_{pi} \right)^2 \tag{1}$$

over all labelled samples N in a training set and all m output nodes of the network, where t_{pi} is the target output activation and y_{pi} the actual output activation of node i. Training proceeds with the presentation of the training data to the network, which has weights randomly set. Data are propagated in a feed-forward direction to the output layer, the network error $t_{pi} - y_{pi}$ is calculated and propagated backwards toward the input layer, and the weights are modified in response to the error. This process is repeated until the error reaches a predefined tolerance level. Classification accuracies superior to those obtained from traditional methods have been achieved and consequently this method was used to identify cloud regions that were statically stable and cloud regions that were statically unstable.

Four categories of cloud region were identified in Meteosat images: cells at low altitude, cells at high altitude, complete cover at low altitude and complete cover at high altitude. *Cumuliform* clouds, which include open and closed convective cells, are associated with an unstable atmosphere, while *stratiform* clouds, which are generally in a widespread sheet, are associated with a stable atmosphere (UK Met. Office, 1978). Therefore, regions containing convective cells were assumed to be unstable and regions containing complete cloud cover were assumed to be stable. A fifth category was introduced for areas of clear sky. Samples from these categories, selected randomly from four images and comprising nine statistical features (Pankiewicz, 1995b) as inputs and a 1-out-of-5 binary encoded output (Bishop, 1995), were used to train a feed-forward neural network. Each cloud region category was represented by a single output neuron in this encoding scheme. The backpropagation learning strategy with an added momentum term (Haykin, 1994) was used to modify the weights of the network after each presentation of the training samples (epoch) and the generalizing power of the classifier was tested every 50 epochs using previously unseen samples. The best classification accuracy calculated for the testing set of previously unseen data was 97.2% after 550 epochs.

The synoptic-scale features identified in Meteosat images using the neural network classifier show good correspondence with those seen in surface frontal analyses (see Figure 4.2). Regions labelled as complete cloud cover represented layered cloud typically associated with cold, warm and occluded fronts, while regions labelled as cells represented areas of convective development often associated with showery convection or fair-weather cumulus. These regions were also similar in nature to those defined manually by Lewis et al. (1997) to interpret imagery using cloud shape-change descriptors.

4.2.1.2 Fuzzy Motion Regions

Using motion itself to segment image sequences improves the subsequent motion estimation because the regions that are identified have homogeneous motion vectors.

44   Advances in Remote Sensing and GIS Analysis

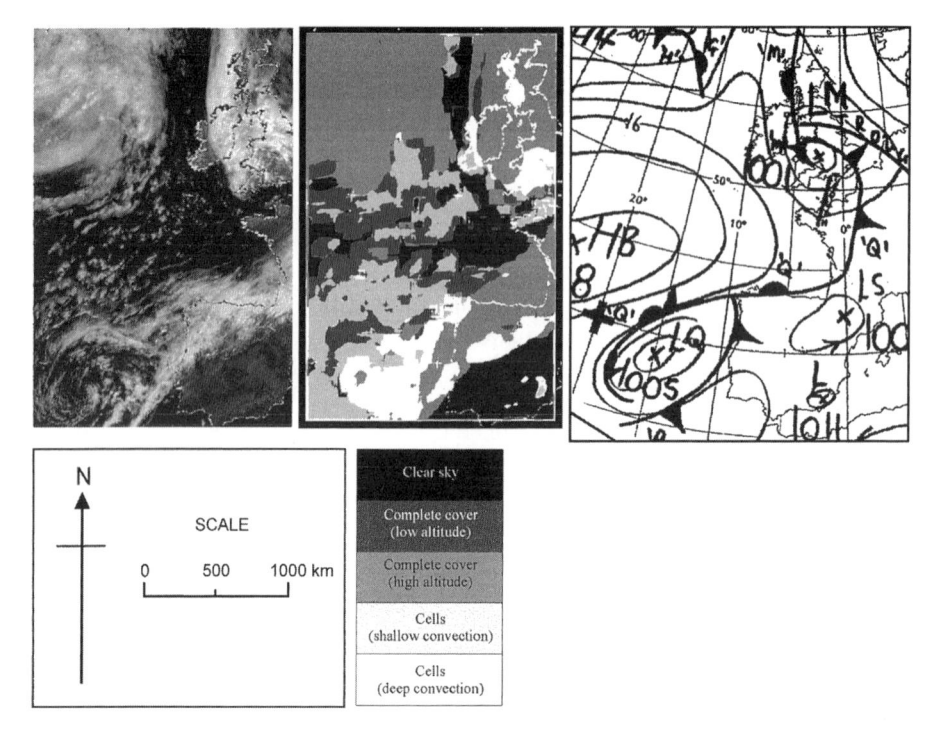

Figure 4.2 *Classification of a corrected Meteosat image (visible waveband generated on 6 August 1996 at 12:00 GMT) into five synoptic-scale cloud region categories by a neural network classifier. The left panel shows the corrected Meteosat image, the centre panel shows the synoptic-scale cloud regions, and the right panel shows the surface analysis for 00:00 GMT (data provided by the UK Meteorological Office)*

Typically a bottom-up approach is used and local motion is first estimated. Segmentation at ambiguous motion edges or clustering of similar local vectors then produces crisp motion edges (Thompson, 1980; Murray and Buxton, 1987; Smith and Brady, 1995b). This bottom-up approach requires a reasonable estimate of the local motion to ensure sufficient discrimination between neighbouring regions. Segmenting image sequences by motion *type* represents a top-down approach, and has the advantage that appropriate techniques can be applied to features within the resulting regions (Faugeras *et al.*, 1998).

Different motion types often coexist in cloud motion, resulting in a significant overlap between regions when defined by motion *type*. A fuzzy system can determine the *degree* to which a motion type is present in a region, in the form of a *membership*. Three motion types that can be considered for cloud are:

- *frontal*, where the cloud body is typically an extruded feature that has quite strong dynamics,
- *textural*, where the cloud body does not move significantly over time, but the internal texture of the cloud displays motion, and
- *small object*, consisting of small clouds and broken cloud.

The fuzzy system employed for the cloud motion segmentation used a rule base of fuzzy rules (Table 4.1) for interpreting image parameters in terms of motion regions (Newland *et al.*, 1996). The parameters generated from image sequences consisted of local grey level, spatial texture strength and grey-level time persistence. The fuzzy rule base interpreted the parameterized sequence data into the three classes above (see Figure 4.3). The fuzzy membership for each region defines the relevance of an appropriate motion analysis, and the output from the different analyses can, therefore, be weighted accordingly.

4.2.2 Cloud Objects

The first stage in the identification of cloud objects is the identification of clear and cloudy pixels. It is difficult to determine whether a pixel is cloudy because the variety of cloud types and illumination produce a large range of grey levels. Further, where the cloud is thin (transparent), pixel grey levels represent the radiative properties of the cloud *and* the underlying surface. Hence, a cloudy pixel can be partly filled or fully filled. Two methods are presented for the segmentation of cloud objects; the first performs a binary, or *crisp*, classification that labels pixels as either cloud or clear sky, and the second uses fuzzy theory to assign pixels a degree of cloudiness.

4.2.2.1 Crisp Cloud Objects

Numerous detection methods have been developed to identify clear and cloudy pixels in satellite images. A series of tests is usually employed, all of which must be passed successfully in order for a pixel to be labelled as cloud free. The techniques used for these tests include thresholding (Sakellariou *et al.*, 1993), a comparison of two channels (Kelly, 1985), spatial coherence (Saunders and Kriebel, 1988), and temporal coherence (Key and Barry, 1989).

Table 4.1 Examples of fuzzy rules for fuzzy motion region segmentation (Newland et al., 1996). The antecedent must be satisfied to some degree for the rule to become active. The precedent is then fired to some degree. The confidence in the rule is used to combine the precedents from all the active rules

IF ANTECEDENT 1 (Time GLD is A)	AND ANTECEDENT 2 (Spatial GLD is B)	AND ANTECEDENT 3 (Cloud μ is C)	THEN PRECEDENT (Output is D)	(Confidence)
High	Medium	High	Texture	1
Medium	Low	High	Front	1
Medium	Medium	High	Front	1
Medium	High	High	Front	1
Low	High	High	Small object	1
Medium	High	High	Small object	1
High	High	High	Small object	1
High	Medium	Medium	Texture	0.8
Medium	Medium	Medium	Front	0.8
Medium	High	Medium	Small object	0.8

46 *Advances in Remote Sensing and GIS Analysis*

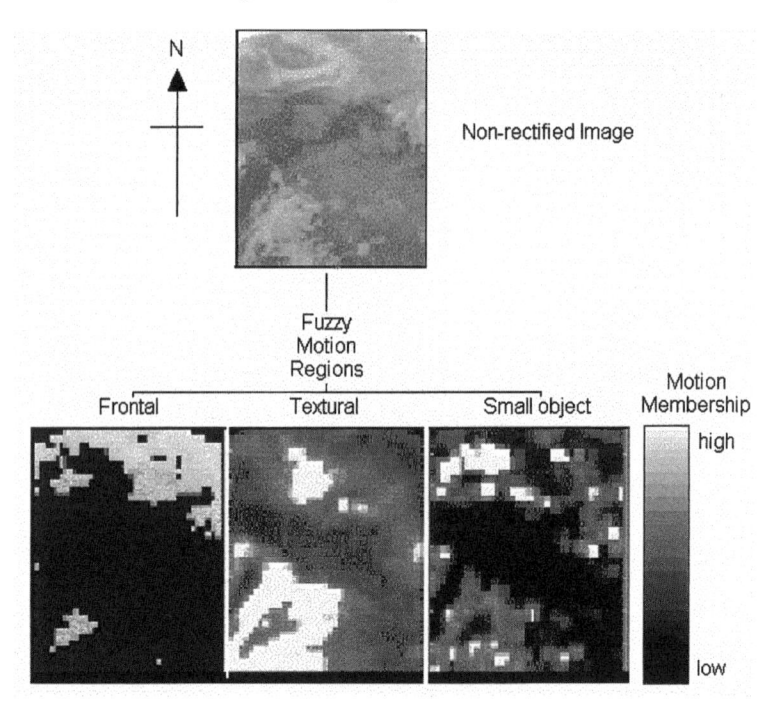

Figure 4.3 *Fuzzy motion region memberships for a sequence of Meteosat images (infrared waveband). From the left: frontal, textural and small objects. The grey levels shown are proportional to the membership of the motion type*

The threshold test offers potentially the simplest and computationally most inexpensive method for identifying clear and cloudy pixels. However, it has limitations that mean that thresholds can be inappropriate for, and sensitive to, changes in the background. The dynamic reflectance threshold test (Saunders and Kriebel, 1988), the hierarchical threshold segmentation (Peak and Tag, 1994) and the adaptive average brightness threshold (Leung and Jordan, 1995) attempt to address these problems. These methods are based upon the selection of a scene-adaptive, or a locally adaptive, threshold, as opposed to a static method which applies a single threshold to the entire image. Lewis *et al.* (1997) determined optimal subset threshold grey levels for the Meteosat visible waveband by identifying the grey-level intersection of the brightest background mode (usually land) and the darkest cloud mode of a locally defined histogram (see Figure 4.4) using a feed-forward neural network. The thresholding task was presented to the neural network as a regression problem, where the goal was to determine the value of a threshold grey level given a particular histogram distribution. Both regression and classification problems can be viewed as particular cases of function approximation and many aspects of one are common to the other (Bishop, 1995). The advantages offered by neural networks for classification tasks (see section 4.2.1.1) are shared, therefore, with neural networks for regression tasks.

Recent advances in the method described by Lewis *et al.* (1997) have shown that histograms generated from windows of 80 by 80 pixels produce the best estimates of threshold grey level. Validation of a revised neural network for windows of this size

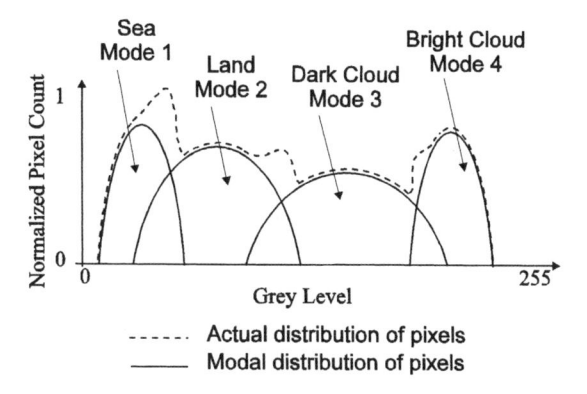

Figure 4.4 An ideal multimodal distribution of pixel grey levels within an image window

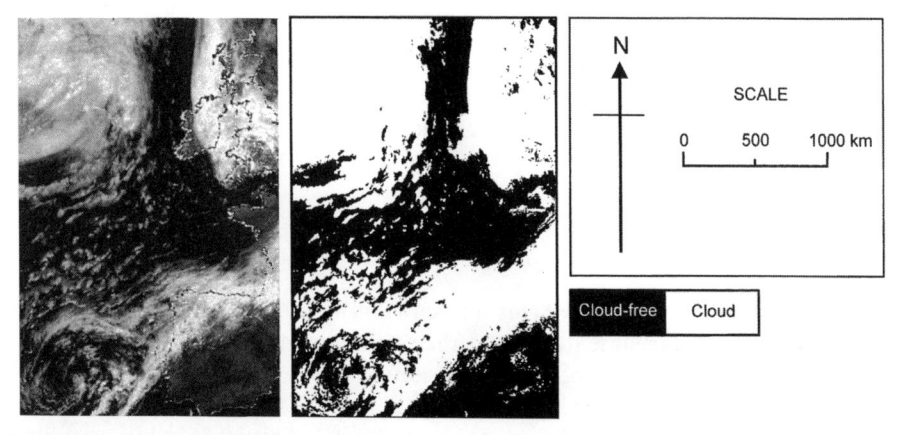

Figure 4.5 Detection of cloud and clear sky in a corrected Meteosat image (visible waveband generated on 6 August 1996 at 12:00 GMT) using an adaptive, local threshold method. The left panel shows the corrected Meteosat image and the centre panel shows the cloud objects (white)

indicated a root mean square error of ± 8 grey levels for the estimation of visible waveband thresholds from a set of 40 previously unseen samples (see Figure 4.5).

4.2.2.2 Fuzzy Cloud Objects

Crisp thresholding of images or image regions at a single grey level produce cloud objects with precise edges. Where the cloud is thin enough for the radiative properties of the underlying terrain to be partly displayed, or where the cloud does not fully fill the pixel, it becomes difficult to assign individual pixels to cloud or background. Analysis of pixels within the context of a cloud object or cloudy region makes this identification easier, because the properties of the region aid discrimination. Multispectral analyses offer further assistance (Schmetz et al., 1993).

A *fuzzy threshold* assigns cloudy edge pixels a degree of cloudiness. Fuzzy clouds give smoother object and component analyses, due to the gradual decay at their edges. Previous studies of these fuzzy *thresholded coherent structures* have been considered for motion analyses in fluid dynamics, and have shown greater motion stability than their corresponding crisp structures (Feher and Zabusky, 1996). *Fuzzy connectedness* of objects in medical imagery has similarly been considered to overcome the inherent inaccuracies in biomedical scanner image capture (Udupa and Samarasekera, 1996). A neural region-growing technique (Gaughan and Flachs, 1990) could be adapted to generate such fuzzy connected regions, but, to reduce computational expense, much simpler implementations still offer greater motion stability than non-fuzzy approaches.

To generate fuzzy clouds, cloud *cores* were initially extracted using one of two techniques; locally defined hard thresholds were applied at a suitable grey level (see above); alternatively, use of a background template also produced similarly good cloud cores. This template was generated by integrating non-cloudy pixels from a sequence of satellite images. The sequence was composed of images taken at the same time of day as the image under analysis, over the preceding week. Cloud cores were then identified in the image by extracting clusters of pixels that differed from its template. All the pixels in the resulting cores of both methods contain cloud. A fuzzy system operating on grey level and gradient was then applied to pixels in the neighbourhood of a core edge to determine the degree of cloud content. The resulting fuzzy cloud objects are coherent structures with smooth edges.

4.2.3 Cloud Elements

Cloud elements consist of *components* and *curvilinear features*. The segmentation of components and curvilinear features, and their subsequent matching, generates information about the motion associated with convection and fronts respectively. The method presented for segmenting components used *a priori* knowledge of cloud dynamics to isolate rising cloud tops associated with convection, while curvilinear features were segmented using techniques from image processing and machine vision.

4.2.3.1 *Components*

It is difficult to find a segmentation technique that will identify components directly. A multi-level thresholding algorithm defines features at various ranges of temperature, or albedo, but the segments defined in this manner do not necessarily correspond to actual cloud structures. However, some links can be made. For example, a correspondence between visible albedo, temperature and the presence of convective updraughts exists. Convective updraughts rise more quickly than the surrounding air, and the corresponding rise in the cloud top is evident as an increase in visible albedo and a decrease in temperature. A limitation is imposed upon the identification of convective updraughts by the spatial resolution of the imagery. The images used for this method had a spatial resolution of 4 km per pixel that was too coarse to identify individual updraughts. However, the rising tops of large cumulus and cumulonimbus clouds are apparent at this spatial resolution. These act as tracers for convective updraughts.

To segment images of cloud into components, the grey levels at which the visible and infrared contrast changes occur need to be identified. These grey levels, corresponding to intersections of cloud modes, are not well defined in one-dimensional brightness histograms (Lewis et al., 1997). Two-dimensional histograms based on the grey-level co-occurrence of pixels in the full image are typically used to generate texture features for classification (Haralick et al., 1973), but have also been used for segmentation (Haddon et al., 1996).

Co-occurrence histograms represent a different level of information compared to one-dimensional brightness histograms, because the spatial structure of grey levels in an image is used to compute them (Haralick et al., 1973). The co-occurrence histogram represents the frequency with which a pixel with intensity j occurs at a displacement Δ relative to a pixel of intensity i. Pixels making up remotely sensed imagery typically occur in small or large clusters of equal or similar grey level, and as a consequence most of the modal distributions in co-occurrence histograms are naturally centred along the leading diagonal $i = j$ (Haddon and Boyce, 1990). Variation in the form of noise, texture or a combination of both, within these pixel clusters will affect the spread of the modal distributions along, and perpendicular to, the leading diagonal. The disadvantages arising from the use of co-occurrence histograms include a high computational expense (compared to the one-dimensional brightness histogram), and a requirement for a large number of pixels. However, the advantage offered by this approach is that the histogram modes can be emphasized by changing the displacement and number of pixels contributing to the co-occurrence.

To identify cloud components produced by convective updraughts, modes on the leading diagonal of the co-occurrence histogram of an image were enhanced in two ways; first, by calculating co-occurrence values for pixel neighbourhoods spanning the components of interest and, second, by combining diagonal co-occurrence values and near-diagonal values (see Figure 4.6).

Modal intersections were assumed to exist at local minima on the diagonal and these points identified a series of grey level thresholds, which were then used for

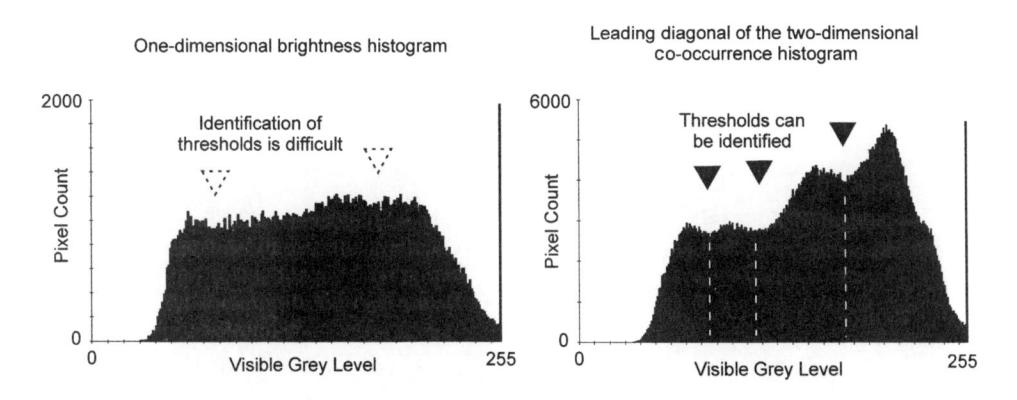

Figure 4.6 Contrasting modal distributions obtained from one and two-dimensional histograms of a corrected Meteosat image (visible waveband generated on 4 October 1996 at 12:00 GMT). The left panel shows the cloud modes of the one-dimensional brightness histogram, and the right panel shows the cloud modes on the enhanced leading diagonal of the two-dimensional co-occurrence histogram. Clear-sky modes were identified and screened out of both histograms using an adaptive threshold method

segmentation. Segmented images contain a reduced number of grey levels (see Figure 4.7), each level corresponding to a mode partitioned on the leading diagonal of the co-occurrence histogram. The rising tops of cumulus and cumulonimbus clouds were then identified as the brightest components of clouds.

4.2.3.2 Curvilinear Features

Canny (1986) defined three criteria for edge detection, namely that the edge enhancing filter should maximize the signal-to-noise ratio, that the correct position should be reported and that only one response should be returned from a single edge. Smith and Brady (1995a) added a speed criterion: the algorithm should be fast enough to be suitable for the overall image processing system. In Sonka et al. (1993), three categories of edge extraction operator are identified:

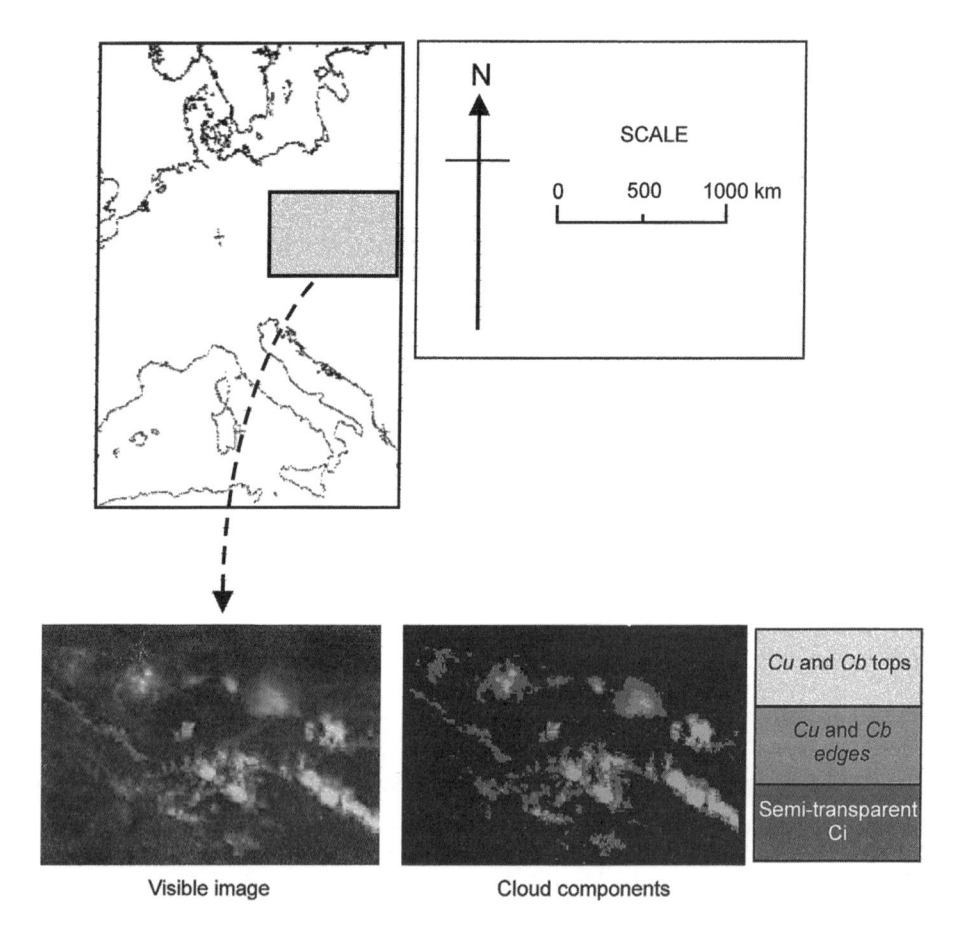

Figure 4.7 Components of cumulonimbus clouds identified in a corrected Meteosat image (visible waveband generated on 16 May 1997 at 12:30 GMT). The left panel shows part of the corrected image and the right panel shows the corresponding cloud components. The brightness of components in the right panel is proportional to the grey level of pixels in the left panel

- Operators approximating image gradients. Some are rotationally invariant (e.g., Laplacian).
- Operators based on zero crossings of the second derivative (including the Laplacian of Gaussian (LoG) and Canny edge detector).
- Operators that attempt to match an image function to a parametric model of edges. One interpretation of this is in the Hough Transform, where objects of known shape and size can be identified.

Thus for simple line extraction where noise tolerance is high (e.g., line extraction for fitting to an object model), simple gradient analyses may suffice. More complex extraction schemes offer the possibility of line association and *direction*, but at a higher computational cost. Straight-line extraction techniques are typically not flexible enough to handle cloud edge features, however.

Curve analysis is more appropriate for cloud objects. Perhaps one of the computationally cheapest implementations of curve extraction involves localized curve fitting. In Côté and Tatnall (1995a) a technique for fitting quadratics to multi-level thresholded cloud data is used, whereby the minima and maxima of curvature and the degree of curvature along the edge can be quickly and easily ascertained. A local approximation to a quadratic provides the necessary parameters for the problem domain without the need for great computational expense. Where a little more precision and completion is required, curves can be fitted to a set of points on the gradient identified. Goshtasby and Shyu (1995) used the Rational Gaussian (RaG) curve to achieve greater precision without the need to generate second-order image derivatives.

Another particularly powerful object-based curvilinear feature is the *skeleton*, otherwise known as symmetric or medial axis transformation, which marks the centreline of an object. There are two principal definitions of the skeleton, namely the locus of maximal circular disks (disks that fit inside the object but in no other disks within the object), and the points of first contact for a wavefront propagating from the object boundary (Dill et al., 1987). Knowing the skeleton of an object and the *corresponding radius* to the edge for every point along the skeleton allows the object to be reproduced while retaining the object's connectivity.

A cloud skeleton is particularly relevant to frontal analysis, where the change in skeletal shape and position is strongly related to the frontal motion. Rotation and elongation of a cloud can also be seen clearly from skeletal development. Once a skeleton is matched across an image sequence, the change in corresponding radius at each point on the skeleton can be used to determine where growth is occurring within the cloud. Similarly, other cloud components can be analysed relative to the skeleton.

Cloud skeletons can be generated from a crisp or fuzzy cloud object. A circle is drawn at every point on the object such that at least a part of it is touching its edge. If the circle touches two or more points on the edge of the object simultaneously, and the separation of two of the points is at least the radius of the circle, then the circle's centre is on the object's skeleton (Figure 4.8). This is geometrically equivalent to the wavefront propagation method described by Dill et al. (1987). In the case of a fuzzy object, the circle must touch two points in the fuzzy edge region that have a similar degree of cloud membership and that are again separated by at least the radius of the circle.

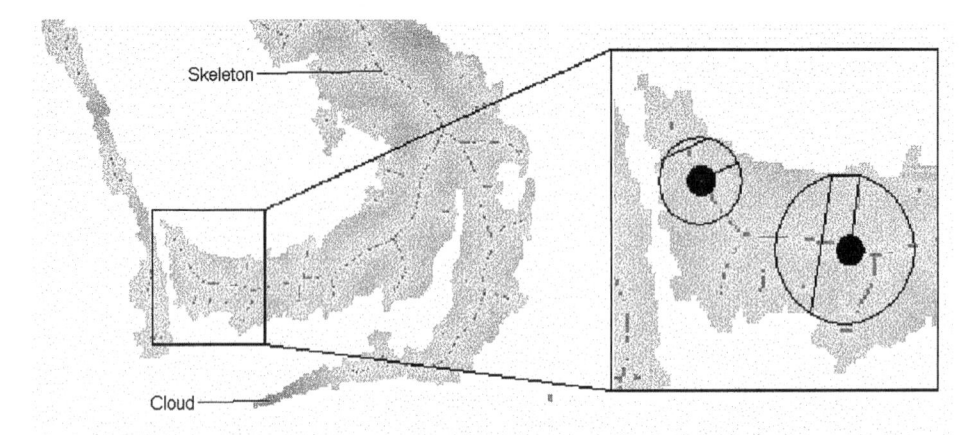

Figure 4.8 *Cloud skeleton. If the circle touches the edge at two points greater than one radius apart, the circle's centre is taken as on the cloud's skeleton*

By varying the required separation of touching points from the circle radius, a skeleton with greater or lesser complexity can be produced. A variable point separation criterion is required for non-elongated clouds to increase skeletal stability. Dill *et al.* (1987) discussed the analysis of skeletons at many different spatial resolutions to take account of local structure in non-elongated objects. For cloud analyses, small-scale skeletal structures are, however, of potentially less interest. They require separate interpretation from the larger-scale stable skeletal components that display cloud life-cycle motion. Other approaches to skeleton generation are given in Brandt and Algazi (1992) where a Voronoi diagram is used to determine the location of the skeleton, and Feher and Zabusky (1996), who provide a skeletal representation of a vorticity field for fluid flow analysis.

4.2.4 Pixels in Correlation Windows

Correlation windows isolate a square of texture within a cloud field. They are typically abstract, having little meteorological meaning when taken out of context, and are used in the MCC method, for example, to determine how squares of texture have moved. Ideally, the correlation window size would be representative of the homogeneity of the cloud region being sampled: large window sizes for homogeneous areas of cloud, and small window sizes for areas having a high variance, but conventional methods simply use a fixed window size. A variable size correlation window was proposed by Côté (1996) for the estimation of sea surface velocities from AVHRR infrared image sequences. This technique was found to provide improved motion estimates, in a combined MCC–neural network approach (see below), over the traditional MCC method.

4.3 Matching

The segmentation methods described above generate cloud features which require matching over image sequences to obtain cloud motion vectors. Many matching methods exist, but typically they are only applicable to specific problem domains and specific feature types. One of these methods, similarity assessment, usually takes the form

$$\text{Similarity value} = f(C_2 - C_1) \qquad (2)$$

where C_1 and C_2 quantify some characteristic of two features in image one and image two of a sequence of images. A set of matches is made between corresponding features when the function f is minimized over all possible matches. Many characteristics can be generated from segmented features, but most are only applicable to subsets of features (Sonka et al., 1993). Shape parameters have been used to characterize cloud features for matching across image sequences. In Lewis et al. (1997) the length of cloud edges (i.e., the perimeter) was used for similarity assessment, but a measurement of area has since been found to be more stable for deforming cloud features. In both cases, the similarity assessment for the shape characteristic was combined with a further assessment of the separation between corresponding cloud features. Extrema on curvilinear features were characterized by their curvature and distance to a central point in Côté and Tatnall (1995b) to generate cloud motion vectors.

Similarity assessment schemes, however, do not use contextual information to find a set of matches, and the best set of matches might not be the set for which the similarity function (2) is minimized. The context in which matchings are made is an important consideration, because this takes into account *a priori* knowledge that is dependent upon the matching problem. In general, a feature being matched out of its context has more chances of mismatch because of deformation and image noise (Côté, 1996). Côté and Tatnall (1995a) suggested that winds are locally coherent, that cloud features close to each other should move in a similar direction and that this contextual information should be included as part of the matching process. An ideal matching method would, therefore, consider both the similarity of features and the context in which they were moving. The compromise between a goal (to match similar cloud features) and a constraint (to maintain the local wind coherence) can be represented as a summation

$$\text{Energy} = \text{objective} + \text{constraints} \qquad (3)$$

and the minimization of this energy function will result in a solution that offers the best solution to the cloud feature matching problem.

Artificial neural networks are among several recent developments in stereo image correspondence and object matching (Nasrabadi and Choo, 1992). Côté and Tatnall (1995a) developed a method that utilized a Hopfield neural network (Hopfield, 1982) to minimize an energy function describing a cloud feature matching problem (Côté and Tatnall, 1995a), and this method has since been applied to sea surface feature tracking (Côté, 1996), and to image rectification using digitized map coastlines (Côté and Tatnall, 1997).

Like the more familiar feed-forward neural networks, the Hopfield neural network is composed of neurons linked to other neurons by weighted connections. However, its main distinction comes from the fact that it is recurrent; neuron outputs are fed back to neuron inputs. Consequently, while the dynamics of feed-forward networks are trivial, they play an important rôle in the Hopfield network. Therefore, the Hopfield network is characterized by an *equation of motion* describing its behaviour over time. This equation of motion can be simulated numerically on a digital computer (Côté and Tatnall, 1995a), allowing the simulation of the temporal behaviour of the network.

The main characteristic of Hopfield neural networks is their convergence to a stable state: from a set of initial neural states, the status of the network varies with time until convergence to a stable state. These states correspond to the minima of an energy function defined from neural weights and biases (Hopfield and Tank, 1985). The energy function is completely determined by these values. Therefore, new energy functions can be created easily by carefully setting those values. Since the Hopfield neural network always converges to a minimum of its energy function, the set of weights and biases can be arranged in such a configuration so that the energy function represents an actual problem to be solved, where the minimum values represent the solutions of the problem. The energy minimization property of the network can then be used to find an energy minimum of the function, or a solution to the problem.

A Hopfield network coded with an energy function describing the contribution of an objective and constraints as a summation (3) will, therefore, converge to a solution offering a compromise between the objective and the constraints when satisfied (i.e., contributing a low energy to the function). The advantage of this formulation, using the Hopfield neural network, is that the cost of matching features can be included with the constraints on the matching, and this removes the need to treat constraints separately (Côté, 1996).

The problem of feature matching can be coded easily as an energy function. The objective term quantifies the similarity between potentially corresponding features using values obtained from a similarity assessment of the form shown in (3). Constraints which have been applied to cloud object motion include terms that limit the number of possible matches to the number of features identified in the first image and ideally match every feature (Lewis *et al.*, 1997), and terms that ensure the coherence of neighbouring matches (Côté and Tatnall, 1995a).

Recent advances in the estimation of sea surface velocities have combined the best features of cross-correlation with those of the Hopfield neural network (Côté, 1996). The instability introduced to the *maximum* cross-correlation method by image noise and feature deformation generating spurious matching vectors has been removed by identifying the *optimum* correlation value within the search window for maintaining neighbourhood coherence. This identification is achieved using the Hopfield network, where the maximization of the correlation value is the objective, and the minimization of the differences between the matching vectors of local neighbours represents the constraints. The main advantage of this approach is that contextual information is used in the matching process. This is significantly superior to the MCC method alone, because, provided that its energy function is coded intelligently, the Hopfield network method can adapt appropriately to each cloud dynamics context. While this approach has yet to be applied to the matching of cloud correlation windows, it offers the potential for improving the accuracy of vectors for cloud motion analysis.

Alternative methods are usually applied when the matching solution does not require *a priori* knowledge. Optical flow techniques, otherwise known as differential methods, are one approach to the correspondence problem for pixels. These techniques look for local motion by analysing grey level variation and higher derivatives (Nagel, 1987).

For skeletons and well-separated edges, a simple match was achieved by finding the shortest distance between points on the skeleton or edge in the current image and points in the next image. In the case of one point being matched to many (the aperture problem), a mean vector was generated from all the matches as an approximate solution. This is not the most stable approach, but provided a fast matching mechanism that worked unambiguously in a majority of cases (see Figure 4.9).

4.4 Motion Analysis

Most analyses of multiple motion vectors consist of combining vectors of different spatial resolutions to produce some optimal vector field at the finest spatial resolution. Mathematical verification is available for such optimization (Nam *et al.*, 1995; Xie *et al.*, 1996; Lee and Zhang, 1996). While it may appear that the highest temporal or spatial precision should be used for the analysis of the cloud motion, this is not the case. Multi-resolution vector data represent the same type of motion at every spatial resolution, but vectors produced by matching the different types of cloud feature do not. For example, the motion of cloud objects defined in terms of rotation, growth and elongation, etc. describe life cycle, but *cloud life cycle* cannot be expressed at the *pixel*

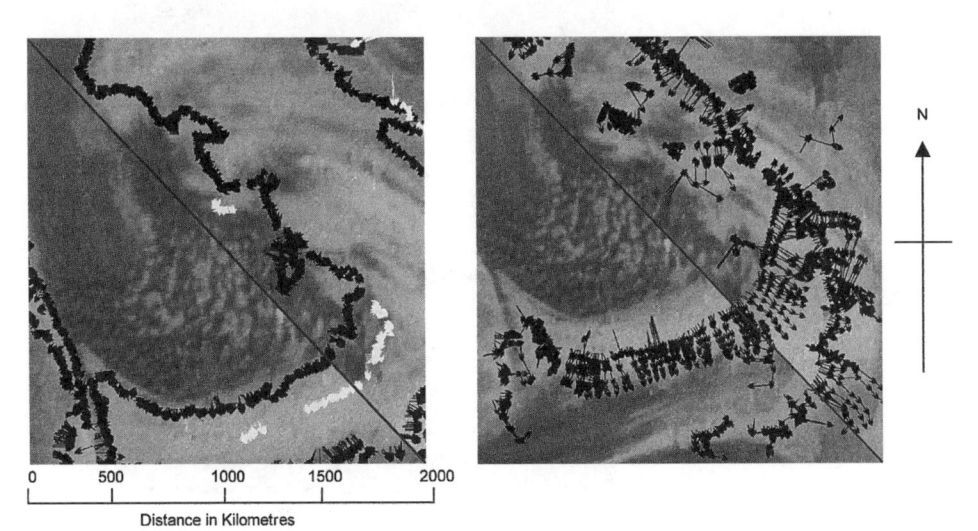

Figure 4.9 Edge and skeleton vectors for a sequence of seven Meteosat images (infrared waveband generated on 18 February 1996). The left panel shows edge vectors for the first to second image in black, and the corresponding vectors for the sixth to seventh image in white. The right panel shows the sequence of skeleton vectors over the whole image sequence. In both images, the lower-left half consists of the first image and the upper-right half the last image

56   Advances in Remote Sensing and GIS Analysis

level without interpretation. However, if such interpretation is made, the motion described by different types of cloud segment can be combined sensibly, as illustrated in Figure 4.10.

In Figure 4.10, the cloud skeleton, representing the main axis of the front, is moving to the south-east. The pixel motion, representative of the wind in this sequence, was south-westerly. A multi-resolution analysis might identify these two motion types, but would not combine them correctly. By recognizing that the skeletal vectors were due to a process different from the process producing the pixel vectors, their interpretation revealed the action of a conveyor-belt relative to the frontal weather system (Bader *et al.*, 1995).

The initial segmentation of images into regions containing similar motion types, such as those illustrated in Figures 4.2 and 4.3, helped to identify the subsequent segmentation methods that were appropriate. In the example described above, the application of the skeleton segmentation method to the image in Figure 4.10 was as a result of the initial identification of frontal motion. In a similar manner, the regions classified as convective cells in Figure 4.2 were suitable for segmentation into components because they were associated with an unstable atmosphere; in unstable atmospheres, convection, once initiated in saturated air, results in building cumulus and cumulonimbus and the rising tops of these clouds are identified as components. By matching components in image sequences, and by analysing their change in shape and temperature, information about the convection process is obtained.

The interpretation and analysis of the cloud motion vectors derived under this

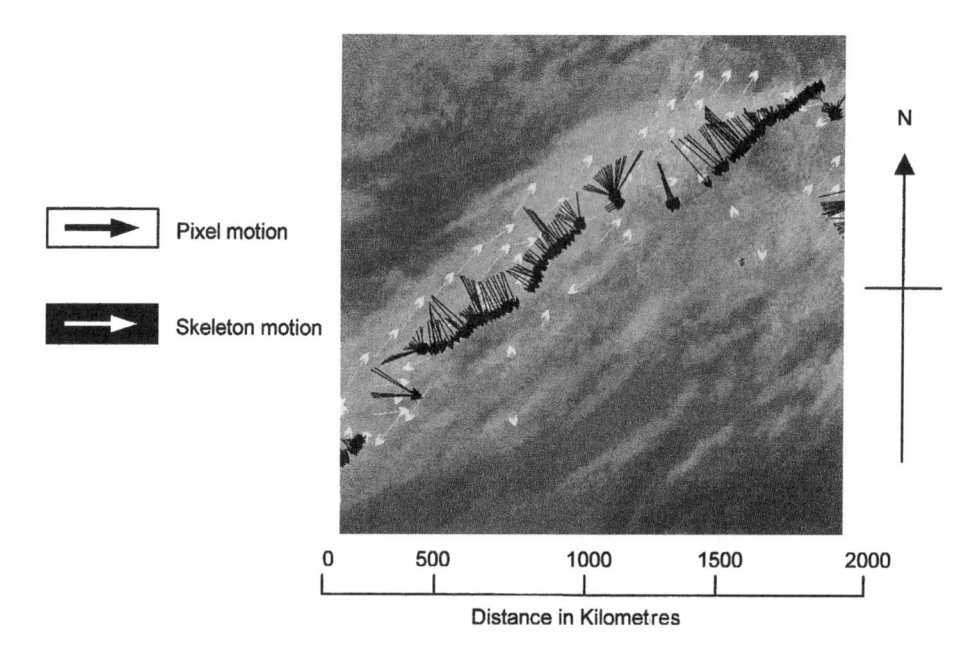

Figure 4.10 *Two motion types present in one cloud object. The black vectors represent skeletal motion and indicate the translation of the front in a south-easterly direction. The white vectors represent pixel motion and indicate a wind from the south-west*

scheme are currently performed manually and are consequently subjective. Automation of the analysis, as a future development, could be achieved using expert and fuzzy systems.

4.5 Conclusions

The maximum cross-correlation method is commonly used to analyse cloud motion, and some of the vectors produced by this method are used as cloud motion winds. However, the information that can be obtained using this method is limited because little contextual and *a priori* knowledge is included. The MCC method uses correlation windows that can only offer information about the motion of texture within a fixed window size.

An approach which addresses the limitations of the MCC method for motion analysis has been presented. This approach uses several segmentation methods to identify different cloud features: regions, objects, elements and pixels. Matching corresponding segments using a Hopfield neural network method produces vector fields that represent the motion associated with these different types of cloud feature. An analytical approach that combines the information from the multiple vector fields is then able to describe aspects of motion that MCC and texture motion analysis methods cannot. An application of this type of analysis to the motion in a frontal region was presented to show its benefits over such conventional analyses.

References

Bader, M.J., Forbes, G.S., Grant, J.R., Lilley, R.B.E. and Waters, A.J., 1995, *Images in Weather Forecasting* (Cambridge: Cambridge University Press).
Bankert, R.L., 1994, Cloud classification of AVHRR imagery in maritime regions using a probabilistic neural network, *Journal of Applied Meteorology*, 33, 909–918.
Bishop, C.M., 1995, *Neural Networks for Pattern Recognition* (Oxford: Clarendon Press).
Brandt, J.W. and Ralph Algazi, V., 1992, Continuous skeleton computation by voronoi diagram, *CVGIP: Image Understanding*, 55, 329–338.
Canny, J.F., 1986, A computational approach to edge detection, *IEEE Transactions on Pattern Analysis and Machine Intelligence*, 8, 679–698.
Côté, S., 1996, Measurement of sea-surface velocities from satellite sensor images using the Hopfield neural network, PhD thesis, University of Southampton, Southampton, UK.
Côté, S. and Tatnall, A.R.L., 1995a, A neural network-based method for tracking features from satellite sensor images, *International Journal of Remote Sensing*, 16, 3695–3701.
Côté, S. and Tatnall, A. R. L., 1995b, Estimation of ocean surface currents from satellite imagery using a Hopfield neural network, *Proceedings of the Third Thematic Conference on Remote Sensing for Marine and Coastal Environments* Vol. I (Ann Arbor, MI: Environmental Research Institute of Michigan), 538–549.
Côté, S. and Tatnall, A.R.L., 1997, The Hopfield neural network as a tool for feature tracking and recognition from satellite sensor images, *International Journal of Remote Sensing*, 18, 871–885.
Dill, A.R., Levine, M.D. and Noble, P. B., 1987, Multiple resolution skeletons, *IEEE Transactions on Pattern Analysis and Machine Intelligence*, 9(4), 495–504.

EUMETSAT, 1993, *Proceedings of the Second International Wind Workshop* (EUMETSAT: EUM P 14).
EUMETSAT, 1996, *Proceedings of the Third International Wind Workshop* (EUMETSAT: EUM P 18).
Faugeras, O., Torr, P.H.S., Kanade, T., Hollinghurst, N., Lasenby, J., Sabin, M. and Fitzgibbon, A., 1998, Geometric motion segmentation and model selection – Discussion. *Philosophical Transactions of The Royal Society of London Series A – Mathematical, Physical and Engineering Sciences*, **356**, 1388–1340.
Feher, A. and Zabusky, N.J., 1996, An interactive imaging environment for scientific visualization and quantification (Visiometric), *International Imaging Systems Journal, Special Edition*, **7**, 121–130.
Gaughan, P.T. and Flachs, G.M., 1990, Region growing and object classification using a neural network, *Proceedings of the SPIE Applications of Artificial Neural Networks*, **1294**, 187–198.
Goshtaby, A. and Shyu, H-L., 1995, Edge detection by curve fitting, *Image and Vision Computing*, **13**, 169–178.
Haddon, J.F. and Boyce, J.F., 1990, Image segmentation by unifying region and boundary information, *IEEE Transactions on Pattern Analysis and Machine Intelligence*, **12**, 929–948.
Haddon, J.F., Boyce, J.F. and Strens, M., 1996, Autonomous segmentation and neural network texture classification of IR image sequences, *IEE Colloquium on Image Processing for Remote Sensing* (London: IEE), 39–44.
Haralick, R.M., Shanmugam, K. and Dinstein, I., 1973, Textural features for image classification, *IEEE Transactions on Systems, Man, and Cybernetics*, **3**, 610–621.
Hayden, C.M., Menzel, W.P., Nieman, S.J., Schmit, T.J. and Velden, C.S., 1994, Recent progress in methods for deriving winds from satellite data at NESDIS/CIMSS, *Advanced Space Research*, **14**, 99–110.
Haykin, S., 1994, *Neural Networks: A Comprehensive Foundation* (New York: Macmillan College Publishing Company).
Hopfield, J.J., 1982, Neural networks and physical systems with emergent collective computational abilities, *Proceedings of the National Academy of Sciences*, **79**, 2554–2558.
Hopfield, J.J. and Tank, D.W., 1985, Neural computation of decisions in optimization problems, *Biological Cybernetics*, **52**, 141–152.
Kelly, K.A., 1985, Separating clouds from ocean in infrared images, *Remote Sensing of Environment*, **17**, 67–83.
Key, J. and Barry, R.G., 1989, Cloud cover analysis with Arctic AVHRR data: 1. Cloud detection, *Journal of Geophysical Research*, **94**, 18521–18535.
Lee, J., Weger, R.C., Sengupta, S.K. and Welch, R.M., 1990, A neural network approach to cloud classification, *IEEE Transactions on Geoscience and Remote Sensing*, **28**, 846–855.
Lee, X. and Zhang, Y-Q., 1996, A fast hierarchical motion compensation scheme for video coding using block feature matching, *IEEE Transactions on Circuits and Systems for Video Technology*, **6**, 627–635.
Leung, I.J.H. and Jordan, J.E., 1995, Image processing for weather satellite cloud segmentation, *Proceedings of the 1995 Canadian Conference on Electrical and Computer Engineering*, Vol. 2, (Canada: IEEE), 953–956.
Lewis, H.G., Côté, S. and Tatnall, A.R.L., 1997, Determination of spatial and temporal characteristics as an aid to neural network cloud classification, *International Journal of Remote Sensing*, **18**, 899–915.
Mahrt, L. and Sun, J., 1995, Dependence of surface exchange coefficients on averaging scale and grid size, *Quarterly Journal of the Royal Meteorological Society*, **121**, 1835–1852.
Murray, D.W. and Buxton, B.F., 1987, Scene segmentation from visual motion using global optimisation, *IEEE Transactions on Pattern Analysis and Machine Intelligence*, **9**, 220–228.
Nagel, H-H., 1987, On the estimation of optical flow: relations between different approaches and some new results, *Artificial Intelligence*, **33**, 299–324.
Nam, K.M., Kim, J-S., Park, R-H. and Shim, Y.S., 1995, A fast hierarchical motion vector estimation algorithm using mean pyramid, *IEEE Transactions on Circuits and Systems for Video Technology*, **5**, 344–351.

Nasrabadi, N.M. and Choo, C.Y., 1992, Hopfield network for stereo vision correspondence, *IEEE Transactions on Neural Networks*, **3**, 5–13.
Newland, F.T., Tatnall, A.R.L. and Brown, M., 1996, Neurofuzzy extraction of wind data from remotely sensed images, *Proceedings of the Third International Wind Workshop* (EUMETSAT: EUM P 18), 257–264.
Pankiewicz, G., 1995a, Pattern recognition techniques for the identification of cloud and cloud systems, *Meteorological Applications*, **2**, 257–271.
Pankiewicz, G., 1995b, A neural network cloud classifier trained with AVHRR data for use on Meteosat imagery. *Proceedings of the 1995 Meteorological Satellite Data Users' Conference* (EUMETSAT: EUM P 17), 393–400.
Pankiewicz, G., 1997, Neural network classification of convective airmasses for a flood forecasting system, *International Journal of Remote Sensing*, **18**, 887–898.
Peak, J.E. and Tag, P.M., 1992, Toward automated interpretation of satellite imagery for Navy shipboard applications, *Bulletin of the American Meteorological Society*, **73**, 995–1008.
Peak, J.E. and Tag, P.M., 1994, Segmentation of satellite imagery using hierarchical thresholding and neural networks, *Journal of Applied Meteorology*, **33**, 605–616.
Purdom, J.F.W. and Dills, P.N., 1993, Cloud motion and height measurements from multiple satellites including cloud heights and motions in polar regions, *Proceedings of the Second International Wind Workshop* (EUMETSAT: EUM P 14), 245–248.
Rumelhart, D.E., Hinton, G.E. and Williams, R.J., 1986, Learning internal representations by error propagation, in D.E. Rumelhart and J.L. McClelland (eds), *Parallel Distributed Processing*, Vol. 1 (Boston, MA: MIT Press), 318–362.
Sakellariou, N.K., Leighton, H.G. and Li, Z., 1993, Identification of clear and cloudy pixels at high latitudes from AVHRR radiances, *International Journal of Remote Sensing*, **14**, 2005–2024.
Saunders, R.W. and Kriebel, K.T., 1988, An improved method for detecting clear sky and cloudy radiances from AVHRR data, *International Journal of Remote Sensing*, **9**, 123–150.
Schmetz, J., Holmlund, K., Hoffman, J., Strauss, B., Mason, B., Gärtner, V., Koch, A. and Van de Berg, L., 1993, Operational cloud-motion winds from METEOSAT infrared images, *Journal of Applied Meteorology*, **32**, 1206–1225.
Smith, S.M. and Brady, J.M., 1995a, *SUSAN – A New Approach to Low Level Image Processing*, DRA Technical Report TR95SMS1c, Defence Research Agency, UK.
Smith, S.M. and Brady, J.M., 1995b, Real-time motion segmentation and shape tracking, *IEEE Transactions on Pattern Analysis and Machine Intelligence*, **17**, 814–820.
Sonka, M., Hlavac, V. and Boyle, R., 1993, *Image Processing, Analysis and Machine Vision* (London: Chapman & Hall).
Thompson, W.B., 1980, Combining motion and contrast for segmentation, *IEEE Transactions on Pattern Analysis and Machine Intelligence*, **2**, 543–549.
Udupa, J.K. and Samarasekera, S., 1996, Fuzzy connectedness and object definition: theory, algorithms, and applications in image segmentation, *Graphical Models and Image Processing*, **58**, 246–261.
UK Met. Office., 1978, *A Course in Elementary Meteorology*, 2nd edition (London: Her Majesty's Stationery Office).
Xie, K., Van Eycken, L. and Oosterlinck, A., 1996, Hierarchical motion estimation with smoothness constraints and postprocessing, *Optical Engineering*, **35**, 145–155.

5
Methods for Estimating Image Signal-to-Noise Ratio (SNR)

Geoffrey M. Smith and Paul J. Curran

5.1 Introduction

All remote sensing instruments have the signal they record contaminated by noise, but the estimation of noise within remotely sensed images has received little attention. Many researchers have identified noise (e.g., Anuta *et al.*, 1984; Bierwirth, 1990; Adams *et al.*, 1995; Simpson and Keller, 1995), some have tried to remove it (e.g., Quarmby, 1987; Green *et al.*, 1988; Lee *et al.*, 1990; Centeno and Haertel, 1995; Harris and Saunders, 1996), while others have suggested desired noise levels (e.g., Goetz and Calvin, 1987; Peterson and Hubbard, 1992; Dekker, 1993; Smith and Curran, 1996), but few have tried to estimate the noise within images or develop procedures for noise estimation. This is probably because noise will not normally be large enough to impair instrument function or produce unintelligible images. Also, it is usually assumed by the users of remotely sensed images that the conditions for recording data are optimized (Duggin and Robinove, 1990). When remotely sensed images are to be used to estimate environmental variables or to perform classifications between areas with only subtly different spectral properties, then the amount of noise in the data can be very important. If there is too much noise then the results of image processing (i.e., the estimates or classifications) will be of an unacceptable quality.

The output of remote sensing instruments is, therefore, made up of the signal (information that is related to the input) and noise (variation unrelated to the input). At optical wavelengths, noise in the surface signal is variation in the measurement of radiance that is not related to radiance leaving the target (Landgrebe and Malaret, 1986). This variation in the measured radiance may be caused by the atmosphere, the

---

Advances in Remote Sensing and GIS Analysis. Edited by Peter M. Atkinson and Nicholas J. Tate.
© 1999 John Wiley & Sons Ltd.

instrument, or the platform on which it is flown. Before analysing remotely sensed data an appreciation of the amount of noise present in the instrument output should be made with respect to the analysis that is to be performed. There are three approaches to estimating the amount of noise in an instrument: in the laboratory during instrument calibration, as part of an inflight calibration procedure, or by analysing the end products. Images are the end products of remote sensing data acquisition and are the data from which useful information is extracted. Therefore, from a user's perspective, the most useful estimates of noise within the data will be derived from images.

The aim of this chapter is to assess five methods which use the spatial nature of images to estimate the inherent noise and therefore 'quality' of remotely sensed images.

5.2 What is Noise?

Noise can be divided into two types: coherent and random. Coherent, sometimes called system or periodic noise, is describable with a known source within the system (Wrigley *et al.*, 1984) and can be removed in most cases. Random noise cannot be predicted in advance (Carbon, 1968) and therefore cannot be removed from the signal easily. Its source is generally unknown and can be caused by random variations in any part of the system. Even so, random noise can be measured as it exhibits a given level of variability over time and also space in the case of remotely sensed data. If repeated measurements of the same surface are made then the average of these measurements will be an estimate of the true measurement, or signal (Shanmugan and Breipohl, 1988). The variation of the individual measurements from the true measurement is an estimate of the random noise present (Duggin *et al.*, 1985).

The impact of noise in the output of an instrument will depend on the relative amount of signal that is to be used as information. Therefore, any estimate of noise should be made in the context of the strength of the signal (Lillesand and Kiefer, 1994). Noise will be represented here as the signal-to-noise ratio (SNR) (Lo, 1986), which is given by the equation

$$\text{SNR} = \frac{\text{SIGNAL}}{\text{NOISE}}. \qquad (1)$$

The SNR is proportional to data quality, as the larger the SNR the easier it will be to distinguish the useful information, the signal, from the noise. The SNR of an instrument is a function of the spatial, spectral and radiometric resolution of the instrument (Magner and Huegal, 1990; Lillesand and Kiefer, 1994) and can be described by

$$\text{SNR}(\lambda) \propto D(\lambda) \beta^2 \left(\frac{H'}{V}\right)^{0.5} \Delta\lambda \, L(\lambda) \qquad (2)$$

where $\text{SNR}(\lambda)$ is the signal-to-noise ratio at wavelength λ, D is the detector quality, β is the instantaneous field-of-view of the instrument, H' is the height of the instrument platform, V is the platform velocity, $\Delta\lambda$ is the spectral resolution of the instrument and L is the radiance of the target area. The SNR of the instrument output is also related to the quality of the optics and electronics within the instrument and the attenuation of the atmosphere at the time of image acquisition. Current electro-optical remote sensing instruments have been found to have a range of SNRs (Table 5.1) determined by the factors listed above and the signal levels encountered during normal operations.

5.3 Methods

The imagery produced by remote sensing instruments can be used to assess the amount of noise present. Measurements of signals from adjacent spatial locations can be assessed to estimate signal and noise, but the difference between such measurements will also include spatial variation in the target area, unless the target area is spatially homogeneous. This spatial variation in the target area cannot be deconvolved and will be included in and therefore inflate the estimate of random noise.

The five methods described here represent different approaches to the estimation of image noise, which aim to remove or minimize the effects of spatial variation. They vary in the amount of data they use, their processing times and degree of user input. The SNR is calculated by dividing an estimate of the signal by the estimate of the noise. This process is then repeated for each waveband in the image. The methods are called homogeneous area, nearly homogeneous area, geostatistical, homogeneous block and multiple waveband.

5.3.1 Homogeneous Area (HA)

This is the simplest method and can be applied to any portion of the image. It uses a small window of pixels within which the land cover is assumed to be homogeneous. The size of window is determined by the variability of the surface in the area of the image under consideration. For areas where the homogeneous surface is spatially extensive the window size can be large (tens of pixels by tens of pixels) but for areas where the homogeneous surface is restricted to small parcels of land the window must be small (i.e., 2 pixels by 2 pixels) and carefully positioned to minimize the effect of surface variability on the estimate of noise. The signal (R_a) at the point xy is estimated by averaging the pixel responses in the window using

$$R_a = \frac{\sum_{i=x-\frac{w}{2}}^{x+\frac{w}{2}} \sum_{j=y+\frac{w}{2}}^{y-\frac{w}{2}} R_{ij}}{w^2} \qquad (3)$$

Table 5.1 Some signal-to-noise ratios (SNR) for electro-optical remote sensing instruments on orbital platforms (Curran and Hay, 1986; Pease, 1991)

Platform	Instrument	SNR
NOAA	Advanced Very High Resolution Radiometer	33:1
Landsat	Multispectral Scanner System	98:1
SPOT	High Resolution Visible (Panchromatic)	190:1
Landsat	Thematic Mapper	341:1
SPOT	High Resolution Visible (Multispectral)	410:1

where R_{ij} is the pixel response at point ij within the window of size w. The noise component can be estimated by calculating the standard deviation (R_{sd}) of the pixel responses within the window using

$$R_{sd} = \sqrt{\frac{\sum_{i=x-\frac{w}{2}}^{x+\frac{w}{2}} \sum_{j=y-\frac{w}{2}}^{y+\frac{w}{2}} (R_{ij} - R_a)^2}{w^2}}. \qquad (4)$$

The SNR is calculated with

$$\text{SNR} = \frac{R_a}{R_{sd}}. \qquad (5)$$

This technique was used by Fujimoto *et al.* (1989) to estimate the SNR of SPOT HRV imagery of the area around Tokyo using a window size of 32 pixels by 32 pixels over various land cover types. This window size was appropriate when used to estimate the SNR in an area of estuary but in urban areas the variability within the land cover increased the estimate of noise and, therefore, decreased the SNR. Duggin *et al.* (1985) calculated the mean and standard deviation of the radiance from small uniform areas within Landsat TM imagery. These values were then used to determine that the error due to random variation in radiance was larger than error due to systematic variation in the radiance. It would have been a simple step to then calculate SNR. Swanberg (1988) used this method to estimate the SNR of National Aeronautics and Space Administration (NASA) Airborne Visible/Infrared Imaging Spectrometer (AVIRIS) (Vane *et al.*, 1993) data for forest stands. To minimize the effects of forest heterogeneity a 3 pixel by 3 pixel window was selected and located in dense stands where the leaf area index was known to be large.

5.3.2 Nearly Homogeneous Area (*NH*)

Boardman and Goetz (1991) appreciated the difficulty of finding large homogeneous areas within remotely sensed images for the estimation of noise. The method they proposed allowed for some natural variation in the area selected for analysis. A large nearly homogeneous area is located within the image. The standard deviation of a small group of pixels within the nearly homogeneous area is calculated. The pixels in the nearly homogeneous area are then averaged in groups of increasing size and the standard deviation calculated again for each group size. The standard deviations (R_{sd}) are then plotted against the reciprocal of the square root of the number of pixels (n) used in the groups (Figure 5.1). It is assumed that as the pixels are averaged in increasingly larger groups the component of the standard deviation due to instrument noise should decrease. The points on Figure 5.1 are used to estimate the intercept on the vertical axis as this represents the standard deviation for an infinite number of samples. This value is assumed to be made up of spatial variation only. Therefore ΔR_{sd} is related to instrument noise. The signal is estimated from the average pixel value in the nearly homogeneous area and the SNR is estimated by dividing it by ΔR_{sd}.

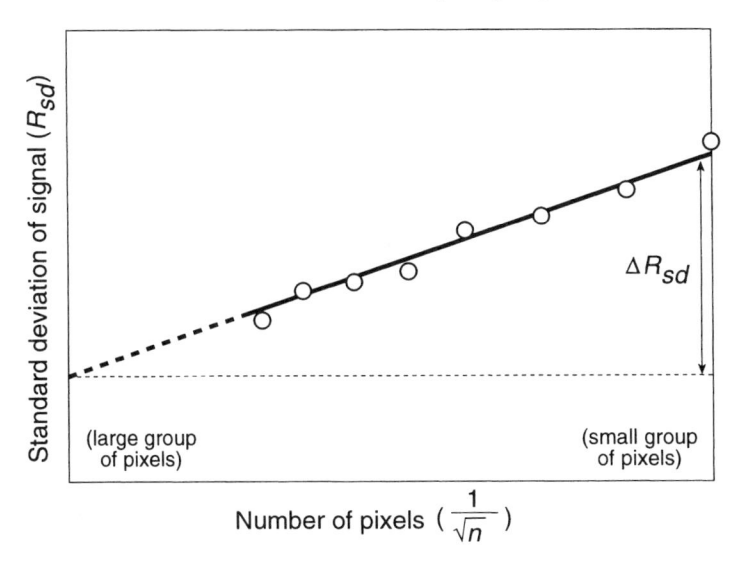

Figure 5.1 The relationship between the number of pixels averaged per group (n) and the standard deviation in the signal in each group (R_{sd}) showing the extrapolation of the relationship to estimate (ΔR_{sd}) which is assumed to represent noise

Boardman and Goetz (1991) used this method to estimate the SNR of AVIRIS imagery of the Rattlesnake Hills, Wyoming, prior to mapping sedimentary lithofacies.

5.3.3 Geostatistical (*GS*)

In an attempt to estimate the within-pixel variation attributable to noise, Curran and Dungan (1989) developed a method for the estimation of image noise which used pixels along a traverse. The technique employed the variogram (Isaaks and Srivastava, 1989; Curran and Atkinson, 1998) (Figure 5.2), a plot of the semivariances ($\overline{S_h^2}$) of the responses of the pixel pairs from the traverse against their separation or lag (*h*). The semivariance is estimated using

$$\overline{S_h^2} = \left(\frac{1}{2m_h}\right) \sum_{i=1}^{m_h} [R(x_i) - R(x_i + h)]^2 \qquad (6)$$

where *m* is the number of pixel pairs with lag *h*, *R* is the response of the pixel and x_i is the location of the first pixel of the *i*th pair. The variogram can only be plotted down to a lag of one from the data available as at least two pixels are required in the estimation of semivariance. To determine the variance within a pixel that we can attribute primarily to noise it is necessary to estimate values of semivariance at lags of less than one, which can be done by extrapolating the variogram model below one lag until it reaches a lag of zero. The semivariance of the model when the lag is equal to zero is known as the nugget variance (c_0), which is an unbiased estimate of the aspatial variation within each pixel. It was shown by Curran and Dungan (1989) that the square root of the nugget variance can be used as an estimate of the standard deviation

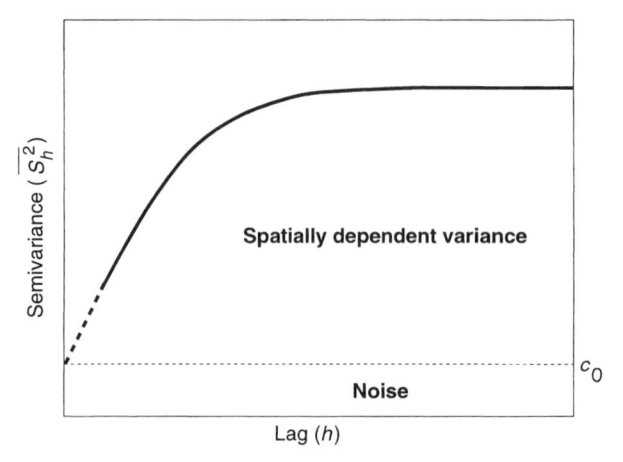

Figure 5.2 A variogram calculated from a transect of pixels within a single land cover unit showing the estimation of the nugget variance (c_0) which can be used to estimate noise

and, therefore, the noise component of the pixel responses within the traverse. The signal is estimated from the average pixel response along the traverse ($\bar{R}$) and the SNR is estimated using

$$\text{SNR} = \frac{\bar{R}}{\sqrt{c_0}}. \qquad (7)$$

This method assumes that the spatial variation in the pixel response is related to the lags between pixels and not their locations, which is the case when all the pixels in the traverse come from a similar land cover type. Also, it assumes that the orientation of the traverse will not affect the variogram but this will be dependent on the instrument operation. To estimate accurately the variogram a traverse must be selected that has sufficient pixels to allow an adequate number of pixel pairs with the same lag. Webster (1985) suggests that the maximum lag used in the variogram should be less than a fifth of the traverse length. The method of extrapolating the variogram should conform to one of the authorized models. Curran and Dungan (1989) estimated the SNR of AVIRIS data for a range of land cover types by applying the method to a traverse within each land cover type. The estimates of noise were found to be very similar, but as expected the estimates of SNR were related to land cover type. A similar geostatistical approach was applied by Legg (1990) to estimate the SNR of imagery from Soviet multispectral scanners. The nugget variance was calculated in the same way, but the signal was taken erroneously as twice the standard deviation of the pixels within the traverse.

5.3.4 Homogeneous Block (*HB*)

Gao (1993) suggested a development of the homogeneous area method for images where large homogeneous areas are difficult to find. The homogeneous block method attempts to use many small homogeneous areas to estimate the noise by dividing the image into blocks for which local means and standard deviations can be calculated, as in the homogeneous area method. As a number of these local means and standard deviations will be from homogeneous areas they can be used to estimate the noise in the image. The local standard deviations are combined to form a histogram of frequency against local standard deviation (Figure 5.3). From experiments performed by Gao (1993), using simulated images with known amounts of noise added, the peak of the histogram was found to be within a few per cent of the added noise. The local standard deviation corresponding to the peak of the histogram is then used as an estimate of noise. The signal is estimated from the average response of all pixels in the waveband. The homogeneous block method was applied by Gao (1993) to AVIRIS and Geophysical and Environmental Research Imaging Spectrometer (GERIS) imagery of Rogers Dry Lake, California, and Cuprite, Nevada, respectively.

5.3.5 Multiple Waveband (MW)

Roger and Arnold (1996) proposed a method for estimating image noise using not only the responses of pixels in adjacent spatial locations, but also the responses of the same pixel in adjacent spectral wavebands. Consequently, this method is only applicable to data recorded by imaging spectrometers, where the adjacent wavebands will be sufficiently correlated and so provide a realistic prediction of the waveband of interest. As with the homogeneous block method, the image is divided into several blocks. For each block, regression equations are generated to estimate the response at each pixel (R) from the responses in the two adjacent spectral wavebands and an adjacent spatial location (R) using

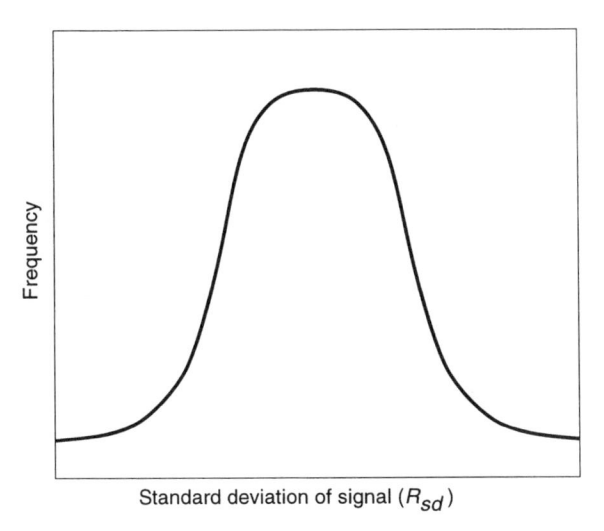

$$\hat{R}_{ijk} = aR_{ijk-1} + bR_{ijk-1} + cR_{pk} + d \qquad (8)$$

Figure 5.3 A histogram of block standard deviation of the signal. The peak of the distribution is assumed to represent noise

where

$$R_{pk} = \begin{cases} R_{i-1jk} & i > 1 \\ R_{ij-1k} & i = 1, j > 1 \end{cases} \qquad (9)$$

and where i and j are the location of the pixel in waveband k (Figure 5.4). One pixel in the block is not calculated so that an adjacent spatial location is always available for the prediction equation. The coefficients a, b, c and d are computed to minimize the sum of the squared residuals (S^2) which is given by

$$S^2 = \sum_{i=1}^{w} \sum_{j=1}^{h} (R_{ijk} - \hat{R}_{ijk})^2 \qquad (10)$$

where w and h define the block size. S^2 is then used to give an estimate of the noise variance of each block using

$$\sigma^2 = \frac{S^2}{(M-4)} \qquad (11)$$

where

$$M = wh - 1. \qquad (12)$$

A homogeneous set of the estimates of block noise variance is then determined and averaged to give an estimate of noise in the waveband. The signal is the average response of all the pixels in the waveband. Roger and Arnold (1996) tested their method on AVIRIS data for several study sites and dates. It appeared to give more

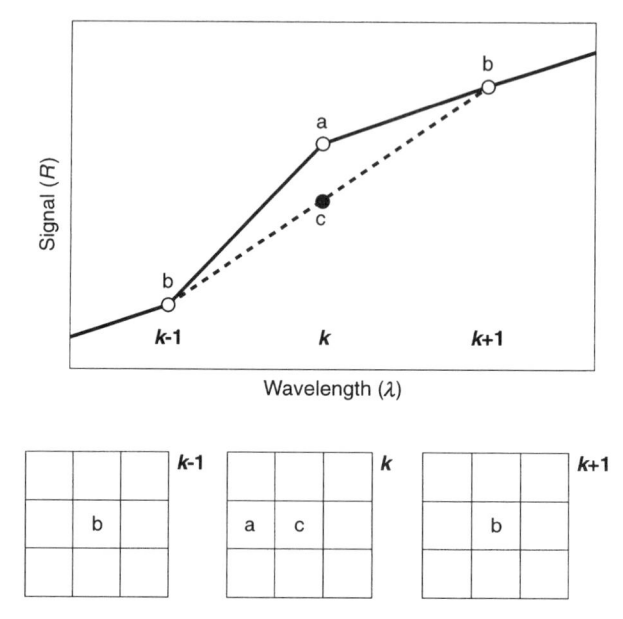

Figure 5.4 A diagram representing the spatial (a) and spectral (b) data which are used to estimate the true signal (c), where k is a given wavelength

consistent results between images from the same flying season than the homogeneous block method.

5.4 A Comparison of the Methods for Estimating Image SNR

To compare the performance of the methods described above they were applied to an image acquired by the AVIRIS. Imaging spectrometers have relatively small SNRs as the narrow wavebands over which data are recorded limit the amount of radiation falling on the detector and, therefore, the amount of signal (Curran, 1994). This small signal is reduced further by the small instantaneous field-of-view normally employed on imaging spectrometers, the small time available to record each response and the small radiances encountered when viewing dark targets (e.g., vegetation). Also, the use of AVIRIS data allowed the multiple waveband method to be used, although the other methods could have been applied to imagery from any optical instrument. The image used was recorded in July 1992 for an area in north-central Florida, north-east of Gainesville. The area is dominated by plantation forests with lakes, wetlands and agricultural areas. A 250 pixel by 250 pixel subset image, centred on a forest study site which had been used in attempts to map foliar biochemical contents from imaging spectrometer data (Curran and Kupiec, 1995; Smith and Curran, 1995; Curran *et al.*, 1997), was used. The localized methods used pixels extracted from the plantation forest, while the others used the whole subset. The result for each method was an SNR spectrum.

For the homogeneous area (HA) method a window size of 2 pixels by 2 pixels was used. A visual examination of the image indicated considerable spatial variations within the plantation forest and, therefore, the windows were located using a pseudo true colour image to identify the least variable areas. Three window locations were used and their SNR spectra were averaged.

For the nearly homogeneous area (NH) method a 25 pixel by 25 pixel window was positioned within an area of plantation forest. The pixels within the window were averaged in groups of 2, 4, 6, 9, 12, 16, 20 and 25.

For the geostatistical (GS) method a traverse approximately 60 pixels long was located within the plantation forest which allowed the creation of a variogram with 12 usable lags. The nugget variance was estimated using a linear model based on the semivariances of the eight shortest lags.

For the homogeneous block (HB) method, a block size of 4 pixels by 4 pixels was used to divide the image, providing approximately 4500 local means and standard deviations. This allowed the creation of a large number of homogeneous blocks, even though there were a large number of roads, railroads and boundaries and a large amount of spatial variation in the land covers within the image.

For the multiple waveband (MW) method the suggested block size of 16 pixels by 16 pixels was reduced to 10 pixels by 10 pixels due to the heterogeneity of the image identified above and to allow easier division of the subset image. As suggested by Roger and Arnold (1996) the upper and lower 15% of the estimates of block noise variance were excluded when forming the homogeneous set of estimates of block noise variance.

70   Advances in Remote Sensing and GIS Analysis

The results of the five methods are shown in Figure 5.5 and are summarized in Table 5.2. The estimates of average SNR in Table 5.2 are for four sections of the spectrum recorded by the AVIRIS. This division was based on the average vegetation spectrum, the location of major atmospheric absorption features and the regions in the spectrum covered by the AVIRIS spectrometers.

The resulting SNR spectra contain a large amount of variation. The minor variations are generally unrelated between methods, but the major variations caused by viewing a mostly vegetated area through a water-vapour-rich atmosphere affect the SNR plot produced for each of the methods. Below 700 nm and above 1900 nm each of the methods produces a similar result, possibly due to the low signal levels in these regions. Between 700 nm and 1900 nm the high signal levels help to enhance the differences between the SNR spectra caused by the different approaches to the estimation of noise. The nearly homogeneous area method produces the smallest estimates of SNR, suggesting that it may be overestimating the amount of noise, while the multispectral method produces the largest estimates of SNR and may be underes-

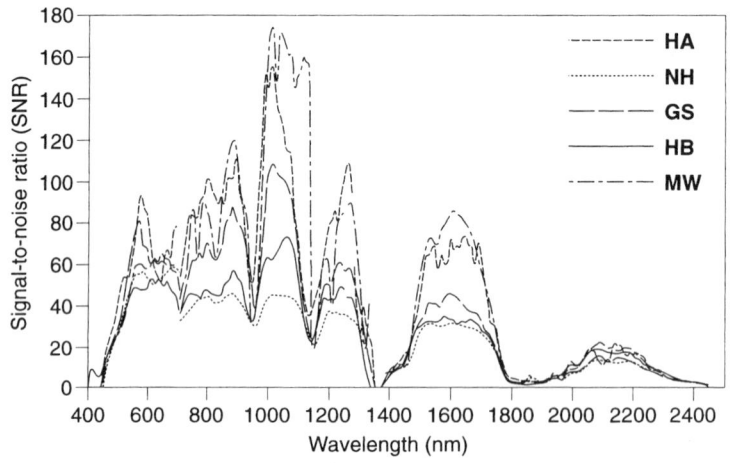

Figure 5.5 *The resulting SNR spectra from the application of the five methods for the estimation of SNR from image data (HA: homogeneous area; NH: nearly homogeneous area; GS: geostatistical; HB: homogeneous block; MW: multiple waveband)*

Table 5.2 *A summary of the SNR estimates produced by the five methods (HA: homogeneous area; NH: nearly homogeneous area; GS: geostatistical; HB: homogeneous block; MW: multiple waveband) used to estimate the SNR of an AVIRIS image and the estimate of AVIRIS SNR derived from the on-board calibrator (OBC) (Smith and Curran, 1996)*

Waveband (nm)	HA	NH	GS	HB	MW	OBC
<700	61:1	47:1	54:1	42:1	49:1	50:1
700–1400	86:1	38:1	64:1	47:1	93:1	65:1
1400–1900	54:1	27:1	35:1	29:1	61:1	30:1
>1900	10:1	8:1	9:1	11:1	11:1	10:1

timating the amount of noise. The remaining methods generally occupy the ranges set out by two methods mentioned above. There appears to be no grouping of the spectra based on criteria such as amount of data used or complexity of approach.

5.5 Conclusions

Five different image-based methods were used to estimate the SNR of a single AVIRIS image. As each image is unique there is no independent data set against which we can evaluate the accuracy of these SNR estimates. What is clear, however, is that different methods result in different SNRs.

The methods provided similar estimates of SNR when the AVIRIS viewed vegetated or mostly vegetated targets. This was surprising because they represented a range of sample sizes and distributions from 12 pixels up to a 250 pixel by 250 pixel image and also the addition of data from adjacent wavebands in the case of the multiple waveband method. The homogeneous area, nearly homogeneous area and geostatistical methods used data that were selected from within a plantation forest and were, therefore, expected to have smaller signal levels. Also, the homogeneous block method used data from a complete waveband and the multiple waveband method used almost all the data from three wavebands. The latter two methods were expected to have a larger average signal. However, for the image used here, most of the land was covered by agricultural crops or forest, therefore the average signal used by each method was very similar. It was clear that when considering the SNR of a whole image, the methods which use data from a single cover type may not be representative.

Each method uses a different approach to the estimation of noise, from an informed positioning of a window for the measurement of variance to the use of multispectral and multispatial information to estimate the true signal against which the noise can be estimated. All the methods go some way to achieving the aim of extracting estimates of noise without the inclusion of spatial variation, but it is not possible to say which ones have the optimal performance. Therefore, when considering the use of one of these methods their advantages and disadvantages must be borne in mind (Table 5.3). The homogeneous area method will have its noise estimates inflated by inter-pixel variability which decreases the estimated SNR. Also, the small number of pixels used in heterogeneous images reduces the accuracy of the estimates of signal and noise. The

Table 5.3 Comparisons of advantages and disadvantages of five methods for estimating image SNR (see Table 5.2 for definition of methods)

Method	Amount of data used	Number of cover types used	Processing time	User input	Effect of spatial variation
HA	Small	One	Short	Great	High
NH	Medium	One	Medium	Medium	Medium
GS	Medium	One	Short	Medium	Medium
HB	Large	Many	Long	Little	Medium
MW	Large	Many	Long	Little	Low

nearly homogeneous area method tries to minimize the effect of inter-pixel variation, but requires the use of linear extrapolation to estimate the scene variance. The geostatistical method attempts to estimate aspatial within-pixel variation, but does not always generate a variogram that fits a selected authorized model. In this study the use of the linear model usually inflated the nugget variance slightly, which inflated the noise and decreased the estimated SNR. The homogeneous block method, as it is based on a development of the homogeneous area method, will also introduce inter-pixel variation into the estimated noise and decrease the SNR. Also, in heterogeneous images the peak of the histogram cannot always be assumed to represent the standard deviation of homogeneous blocks. In many cases the true peak can be difficult to estimate and may require curve fitting. The multiple waveband method uses only a third of the data from adjacent pixels and, therefore, should be the least affected by spatial variation. In images where the spatial and spectral differences in signal are large over small areas and between spectral wavebands the accuracy of the block noise variance may be low.

The homogeneous area and nearly homogeneous area methods were fast but required user input and produced noisy results. The geostatistical method was quite fast but needed a large amount of user input to select the traverse and monitor the observed and modelled variograms. The whole image and multiple waveband methods were the slowest due to the large number of calculations performed, but required no user action and produced the most consistent results in multiple tests.

The message is clear; the SNR is dependent on the method employed to calculate it and so great care is needed in the choice of a method and the interpretation of SNRs published by others.

Acknowledgements

The authors wish to acknowledge the Natural Environment Research Council for a studentship to GMS (GT4/90/TLS/59) and a research grant to PJC (GR3/7647, 1990–93), the National Aeronautics and Space Administration for AVIRIS flight time and the University of New Hampshire for a research grant to PJC (Accelerated Canopy Chemistry Program, 1992–95). The authors are indebted to many individuals who made this work possible, notably Henry Gholz (University of Florida), John Kupiec (Environment Agency), John Aber and Mary Martin (University of New Hampshire) and Jennifer Dungan and David Peterson (NASA/Ames Research Center).

References

Adams, J.B., Sabol, D.E., Kapos, V., Almeida, R., Roberts, D.A., Smith, M.O. and Gillespie, A.R., 1995, Classification of multispectral images based on fractions of endmembers – application to land-cover change in the Brazilian Amazon, *Remote Sensing of Environment*, **52**, 137–154.

Anuta, P.E., Bartolucci, L.A., Dean, M.E., Lozano, D.F., Malaret, E., McGillem, C.D., Valdes, J.A. and Valenzuela, C.R., 1984, LANDSAT-4 MSS and Thematic Mapper data quality and information content analysis, *IEEE Transactions in Geoscience and Remote Sensing*, **22**, 222–235.

Bierwith, P.N., 1990, Mineral mapping and vegetation removal via data-calibrated pixel unmixing, using multispectral images, *International Journal of Remote Sensing*, **11**, 1999–2017.

Boardman, J.W. and Goetz, A.F.H., 1991, Sedimentary facies analysis using AVIRIS data: A geophysical inverse problem, *Proceedings of the Third AVIRIS Workshop*, JPL Publication 91–28 (Pasadena, CA; Jet Propulsion Laboratory), 4–13.

Carbon, A.B., 1968, *Communication Systems: An Introduction to Signals and Noise in Electrical Communication* (New York: McGraw-Hill).

Centeno, J.A.S. and Haertel, V., 1995, Adaptive low-pass fuzzy filter for noise removal, *Photogrammetric Engineering and Remote Sensing*, **61**, 1267–1272.

Curran, P.J. 1994, Imaging spectrometry, *Progress in Physical Geography*, **18**, 247–266.

Curran, P.J. and Atkinson, P.M., 1998, Geostatistics and remote sensing, *Progress in Physical Geography*, **21**, 61–78.

Curran, P.J. and Dungan, J.L., 1989, Estimation of signal-to-noise: A new procedure applied to AVIRIS data, *IEEE Transactions on Geoscience and Remote Sensing*, **27**, 620–628.

Curran, P.J. and Hay, A.M., 1986, The importance of measurement error for certain procedures in remote sensing at optical wavelengths, *Photogrammetric Engineering and Remote Sensing*, **52**, 229–241.

Curran, P.J. and Kupiec, J.A., 1995, Imaging spectroscopy: a new tool for ecology, in F.M. Danson and S.E. Plummer (eds), *Advances in Environmental Remote Sensing* (Chichester: Wiley), 71–88.

Curran, P.J., Kupiec, J.A. and Smith, G.M., 1997, Remote sensing the biochemical composition of a slash pine canopy, *IEEE Transactions on Geoscience and Remote Sensing*, **35**, 415–420.

Dekker, A.G., 1993, Detection of optical water quality parameters for eutrophic water by high resolution remote sensing, PhD thesis, Proefschrift Vrije Universiteit, Amsterdam.

Duggin, M.J. and Robinove, C.J., 1990, Assumptions implicit in remote sensing data acquisition and analysis, *International Journal of Remote Sensing*, **11**, 1669–1694.

Duggin, M.J., Sakhavat, H. and Lindsay, J., 1985, Systematic and random variations in Thematic Mapper digital radiance data, *Photogrammetric Engineering and Remote Sensing*, **51**, 1427–1434.

Fujimoto, N., Takahashi, Y., Moriyama, T., Shimada, M., Wakabayashi, H., Nakatani, Y. and Obayashi, S., 1989, Evaluation of SPOT HRV image data received in Japan, *Quantitative Remote Sensing: An Economic Tool for the Nineties* (Vancouver: IEEE), 463–466.

Gao, B.-C., 1993, An operational method for estimating signal to noise ratios from data acquired with imaging spectrometers, *Remote Sensing of Environment*, **43**, 23–33.

Goetz, A.F.H. and Calvin, W.M., 1987, Imaging spectrometry: Spectral resolution and analytical identification of spectral features, *Proceedings, Society of Photo-optical Instrumentation Engineers: Imaging Spectroscopy II, SPIE* 834 (Bellingham, WA), 158–165.

Green, R.O., Vane, G. and Conel, J.E., 1988, Determination of inflight AVIRIS spectral, radiometric, spatial and signal-to-noise characteristics using atmospheric and surface measurements from the vicinity of the rare-earth-bearing carbonatite at Mountain Pass, California, *Proceedings, Airborne Visible/Infrared Imaging Spectrometer (AVIRIS) Workshop*, JPL Publication 88–38 (Pasadena, CA: Jet Propulsion Laboratory), 162–184.

Harris, A.R. and Saunders, M.A., 1996, Global validation of the Along-Track Scanning Radiometer against drifting buoys, *Journal of Geophysical Research – Oceans*, **101**, 12127–12140.

Isaaks, E.H. and Srivastava, R.M., 1989, *Applied Geostatistics* (New York: Oxford University Press).

Landgrebe, D.A. and Malaret, E., 1986, Noise in remote-sensing systems: The effect on classification error, *IEEE Transactions on Geoscience and Remote Sensing*, **24**, 294–299.

Lee, J.B., Woodyatt, A.S. and Berman, M., 1990, Enhancement of high spectral resolution remote sensing data by a noise-adjusted principal components transform, *IEEE Transactions on Geoscience and Remote Sensing*, **28**, 295–304.

Legg, C.A., 1990, Digital satellite imagery from the Soviet Union; An important new tool for environmental monitoring, *Proceedings of the 16th Annual Conference of the Remote Sensing Society* (Nottingham: Remote Sensing Society), 65–74.

Lillesand, T.M. and Kiefer, R.W., 1994, *Remote Sensing and Image Interpretation*, 3rd edition (New York: Wiley).

Lo, C.P., 1986, *Applied Remote Sensing* (Harlow: Longman).

Magner, T.J. and Huegal, F.G., 1990, The moderate resolution imaging spectrometer-tilt (MODIS-T), *Proceedings, Society of Photo-optical Instrumentation Engineers: Remote Sensing of the Biosphere, SPIE 1300*, (Bellingham, WA), 145–155.

Pease, C.B., 1991, *Satellite Imaging Instruments* (New York: Ellis Horwood).

Peterson, D.L. and Hubbard, G.S., 1992, Scientific issues and potential remote-sensing requirements for plant biochemical content, *Journal of Imaging Science and Technology*, **36**, 446–456.

Quarmby, N.A., 1987, Noise removal for SPOT HRV images, *International Journal of Remote Sensing*, **8**, 1229–1234.

Roger, R.E. and Arnold, J.F., 1996, Reliably estimating the noise in AVIRIS hyperspectral images, *International Journal of Remote Sensing*, **17**, 1951–1962.

Shanmugan, K.S. and Breipohl, A.M., 1988, *Random Signals: Detection, Estimation and Data Analysis* (Chichester: Wiley).

Simpson, J.J. and Keller, R.H., 1995, An improved fuzzy-logic segmentation of sea-ice, clouds and ocean in remotely-sensed arctic imagery, *Remote Sensing of Environment*, **54**, 290–312.

Smith, G.M. and Curran, P.J., 1995, The estimation of foliar biochemical content of a slash pine canopy from AVIRIS imagery, *Canadian Journal of Remote Sensing*, **21**, 234–244.

Smith, G.M. and Curran, P.J., 1996, The signal-to-noise ratio (SNR) required for the estimation of foliar biochemical concentrations, *International Journal of Remote Sensing*, **17**, 1031–1058.

Swanberg, N.A., 1988, AVIRIS data quality for coniferous canopy chemistry, *Proceedings of the Airborne Visible/Infrared Imaging Spectrometer (AVIRIS) Performance Evaluation Workshop*, JPL publication 88–38 (Pasadena, CA: Jet Propulsion Laboratory), 102–108.

Vane, G., Green, R.O., Chrien, T.G., Enmark, H.T., Hansen, E.G. and Porter, W.M., 1993, The Airborne Visible/Infrared Imaging Spectrometer (AVIRIS), *Remote Sensing of Environment*, **44**, 127–143.

Webster, R., 1985, Quantitative spatial analysis of soil in the field, *Advances in Soil Science*, **3**, 1–70.

Wrigley, R.C., Card, D.H., Hlavka, C.A., Hall, J.R., Mertz, F.C., Archwamety, C. and Schowengerdt, R.A., 1984, Thematic mapper image quality: Registration, noise and resolution, *IEEE Transactions on Geoscience and Remote Sensing*, **22**, 263–271.

6
Modelling and Efficient Mapping of Snow Cover in the UK for Remote Sensing Validation

Richard E.J. Kelly and Peter M. Atkinson

6.1 Introduction

Data on the spatial distribution of snow cover properties in the UK is required routinely by the Environment Agency (formally, the National Rivers Authority) and regional water authorities for general monitoring and flood prediction purposes (Hymas, 1993). Regional and local snow cover measurements are also potentially important inputs to global circulation models and weather forecasts because they can help to explain atmospheric processes (Peterson and Hoke, 1989). The most useful measures for characterizing the spatial distribution of snow cover are snow area extent and snow water equivalent. However, the available data are limited to point measurements of snow depth made at sparsely located meteorological and volunteer stations at 9:00 a.m. each morning. Further, snow data are not made available for some time after the snow event and because variation in snow depth across the UK is large over short spatial distances, point observations are difficult to interpret at the national or regional scale. These data are obtained and quality controlled by the Meteorological Office and until 1993 the data comprised the Snow Survey of Great Britain database. Since 1993 there has been a decline in the number of data on snow cover obtained by the Meteorological Office and so alternative ways of monitoring snow cover accurately are increasingly sought after.

Several projects have been reported which tackle the problem of providing near-real-time snow maps using satellite imaging sensors (Baumgartner, 1994; Carroll,

---

Advances in Remote Sensing and GIS Analysis. Edited by Peter M. Atkinson and Nicholas J. Tate.
© 1999 John Wiley & Sons Ltd.

1995). Initially, visible and infrared radiometers were used to provide high spatial resolution (Landsat TM) or frequent coverage (NOAA AVHRR) maps of snow area extent and, to a limited extent, snow depth (Dozier, 1989; Harrison and Lucas, 1989; Xu et al., 1993). For effective snow mapping in the UK, however, the occurrence of partial or complete cloud cover, which often persists during entire snow cover events, is a recurrent problem (Archer et al., 1994). During cloudy conditions visible and infrared sensors are rendered useless and resource managers need to look elsewhere for an operational snow mapping system.

Microwave sensors offer a potential solution to the problems caused by clouds. Active systems such as synthetic aperture radar (SAR) have been operational since 1991 through the ERS-1 and ERS-2 satellites. With their high spatial resolution and all-weather capability, the ERS SAR systems can be used to identify snow in the melting phase (e.g., Guneriussen et al., 1996; Kelly, 1996) which is critical for effective water resource management. However, ERS orbit configurations restrict temporal coverage over the UK to less than once in every two weeks. Since the duration of snow cover in the UK is often shorter than this period, the frequency of coverage provided by the ERS SAR systems is presently inadequate for snow monitoring in the UK. One alternative to available active microwave systems is provided by passive microwave (PM) systems and there is one such instrument that can be used operationally to monitor UK snowpacks.

The US Defense Meteorological Satellite Program can provide quantitative synoptic climate and surface data to environmental scientists from its Special Sensor Microwave Imager (SSM/I) sensor (Hollinger et al., 1987). SSM/I imagery can be used for snow mapping at regional to local scales (Fiore and Grody, 1992; Kelly, 1995; Grody and Basist, 1996). In the USA, PM imagery is used operationally to generate daily snow maps at spatial resolutions ranging from 40 km to 100 km, and in near real time (Armstrong and Hardman, 1991). In the UK, PM imagery can be used to estimate snow cover properties for the whole of the UK twice a day (Kelly, 1995). An operational snow monitoring system for the UK based on PM imagery must be calibrated and validated using accurate snow maps generated from non-SSM/I sources. In part, this is because the SSM/I spatial resolution is insufficient to resolve variation in snow depth in the UK.

One potential source of data for calibrating and validating PM snow algorithms is the Snow Survey of Great Britain. Although conventional ground observations of snow depth are not necessarily the most useful real or near-real-time indicators of snow cover, they can be used to calibrate and validate PM estimates of snow cover. Tait and Armstrong (1996) suggest that differences between PM satellite-derived snow depth estimates and ground measurements may arise as a result of the comparison between areal averages and point data. Hence, to reduce potential errors, point snow depth observations need to be 'scaled up' to give areal estimates. The general objective of this chapter, therefore, was to determine the most efficient technique for deriving snow depth maps from the Meteorological Office's Snow Survey data.

Given point measurements of snow depth for several dates, the aim was to produce accurate maps of snow depth for each date, co-registered with SSM/I imagery and defined on the same support (the size, geometry and orientation of the space over which an observation is defined). Using an optimum interpolation algorithm applied to SSM/I imagery developed by Poe (1990), we define our regular support as 2 km × 2 km. In addition, it is well known that topography (elevation, slope and aspect) affects the continual modification of a snowpack (McKay and Gray, 1981). With the availability of co-located digital elevation data (DEM) for the entire UK, the initial aim was extended to one of producing accurate maps from a combination of point snow depth data and elevation data. Furthermore, it was anticipated that in the future it will be possible to continually refine the algorithms using data from new snowfall events or to apply the method to snow cover data in other countries. Therefore, the method used to estimate snow depth should be generally applicable so that precise maps may be obtained routinely in the future under varying conditions.

There are many possible techniques for mapping snow depth on a raster grid of support size n km × m km. The objective was to decide which technique provides the most efficient solution. We examined techniques for mapping snow depth that fall into two broad categories: traditional interpolation and geostatistical interpolation. The former category includes a simple inverse distance weighting squared (IDWS) technique and a regression procedure based on the relation between snow depth and elevation. The latter group includes kriging and cokriging (based on the relation between snow depth and elevation).

The criterion by which the most accurate technique was selected is the root mean square error (RMSE) estimated using a cross-validation technique in which each sample observation is estimated in turn using the surrounding observations only. Differences between the original measured and estimated values were used to estimate RMSE values and from these the most efficient technique was chosen.

6.2 Sample Data

6.2.1 Source of Data

The Snow Survey of Great Britain primarily consisted of data obtained from a network of voluntary observers throughout the UK. These data were supplemented by measurements of snow depth obtained from operational weather stations to produce a complete data set of approximately 560 (1988) and 570 (1991) point measurements over the UK. Most observers in the Snow Survey network were associated with stations that submitted other meteorological information to the Meteorological Office. These particular data were usually observed at the same sites as the meteorological stations. During the winter season (October until May), observers sent their snow depth reports to the Meteorological Office at the end of each month. These reports were quality controlled (the procedures of which are not released), aggregated and eventually published in the following November as the Snow Survey of Great Britain.

Daily readings of snow depth and station altitudes are stored in large monthly files for all UK stations. It is possible to obtain data from the archive which are quality controlled but not aggregated and it is these data that were used here. Snow depth values were extracted from the monthly files along with station elevation and British National Grid (BNG) reference. The locations of these point observations of snow depth were mapped to the 2 km × 2 km grid derived for the SSM/I data such that the centre of a single 2 km × 2 km cell corresponded to each point measurement of snow depth. If stations did not return an observation for a selected date, it was assumed that the snow depth at this location was zero.

6.2.2 Snowfall Events and Data Sets

On the evening of 21 January 1988, a westerly frontal system, fed by cold polar air, passed over the southern counties of the UK depositing several centimetres of snow. By the following morning, the front turned northwards and deposited snow over the Midlands and northern England. At the end of the day, much of the UK was covered in snow with some places recording over a metre in depth. By the morning of 23 January, a new warm frontal system was approaching the UK from the west. By the time it had passed the UK in the evening, most of the lowland snow cover had melted and only patches remained on higher ground in Scotland.

Snow cover was present in parts of the UK for most of February 1991. The main period of snowfall producing extensive coverage was between 7 and 18 February. A sustained cold period accompanied by strong cyclonic activity resulted in snow being deposited from 7 to 12 February. After 12 February, no further significant snow was deposited and by 22 February most UK snow had melted with the exception of the Scottish snowpacks.

Point snow depth data for January 1988 and February 1991 were obtained from the Snow Survey of Great Britain database. The first set of data was taken from the 1988 archive and the analysis was undertaken on data for 22 and 23 January 1988 only. The second set of data was taken from the 1991 archive and the analysis was conducted on data for 12 and 16 February 1991 only, representing the time of maximum snow coverage and halfway through the melting period respectively.

6.3 Interpolation and Cross-validation

6.3.1 Traditional Approaches

There are several techniques for interpolation which fall into the 'traditional' category. These range from simple nearest neighbour fill techniques to complex high-order polynomial splines. An important criterion when choosing a technique for interpolating is whether the technique is global or local (Petrie and Kennie, 1990). Global techniques fit some function to all data such that each estimated point lies on or is as close as possible to the fitted surface. For widespread observations of snow depth over the UK on any date, this assumes that there is, for example, some relation between snow depth in the south of England and in Scotland. Clearly, this might not always be the case especially since snowpacks in the Scottish Highland may be permanent throughout winter and snow covers in lowland England may be short-lived. Local interpolators assume that there exists a local correlation which decreases with increasing distance from the location of the estimate. Thus, for snow cover, local techniques are the intuitive choice since snow depth is spatially dependent at a local scale.

The first local interpolator chosen was IDWS (Lancaster and Šalkauskas, 1986; Davis, 1973; Watson, 1992). The IDWS is used widely and routinely by geographers and spatial modellers so it provides an interesting benchmark for comparison with other techniques. The second traditional technique for interpolation is that based on a least squares regression model of the relation between snow depth and elevation. Regression was undertaken at two spatial scales: at the UK scale and at a regional scale for mountainous areas in Scotland, England and Wales. The following two subsections describe the implementation of the traditional interpolation techniques.

6.3.1.1 Inverse Distance Weighting Squared Interpolation

The IDWS technique is used widely for estimating unknown values at locations between sample observations (e.g., Watson, 1992). The technique is straightforward and consists of estimating a value $z(\mathbf{x})$ at a location $\mathbf{x}$ using N surrounding observations $z(\mathbf{x}_i)$ and their distances d_i from the point to be estimated

$$\hat{z}(\mathbf{x}) = \frac{\sum_{i=1}^{N} z(\mathbf{x}_i) d_i^{-p}}{\sum_{i=1}^{N} d_i^{-p}} \qquad (1)$$

where p is the power used in the weighting function. A formal expression of the weighting function w_i is

$$w_i = \frac{d_i^{-p}}{\sum_{i=1}^{N} d_i^{-p}} \quad i = 1,\ldots,N \qquad (2)$$

such that

$$\hat{z}(\mathbf{x}) = \sum_{i=1}^{N} w_i z(\mathbf{x}_i) \qquad (3)$$

The selected value of p determines the nature of the interpolation. If $p > 1$, the weighting function w_i will generate a 'local' interpolation effect in which local detail is preserved, but the interpolated map may have a somewhat 'blocky' appearance. If $p < 1$, the weighting function produces a more 'regional' smoothed interpolation. At $p = 1$, a non-power IDWS is generated where distance from the estimate is the prime weighting determinant.

Many researchers use IDWS with $p = 2$ and do not consider the implications of this choice further. Watson (1992, p. 115) notes that using inverse distance weighting,

the (interpolated) surface is confined to the range of the data; interpolated values cannot lie above or below the maximum and minimum measured values. For this reason the method cannot be fully responsive to local trends, and unsampled peaks cannot be inferred although they may be implicit in the data. Flat but tilted regions are interpolated poorly by inverse distance weighting.

In this chapter, we select a power function $p = 2$, and the minimum number of points

N required to perform each interpolation is set to 3 (within a maximum search radius of 100 km). In this way, we standardize the inverse distance weighting parameters.

6.3.1.2 Regression-based Interpolation

The relationship between snow depth and terrain elevation in the UK is complex. For selected meteorological conditions governing snowfall dynamics and subsequent snowpack development, snow cover distributions may or may not be related to terrain elevation. In high mountainous regions of the world, there is usually an implicit positive relationship between snow depth and elevation. However, in the UK as a whole, the relationship is not straightforward. For example, a cyclonic frontal system may bring deep snow to low-lying areas and completely miss mountainous areas. To complicate matters, a historic presence of mountainous snow covers combined with a new snowfall event in lowland areas may produce no relationship between snow depth and elevation at that time. Therefore, any relation, whether assumed or obtained empirically, between snow depth and elevation must be used with great care.

To simplify regression, we restricted our analyses to upland areas in the UK. A mountain/non-mountain mask of the UK was constructed from DEM data. The mask consisted of four mountainous regions (Scottish, English, Welsh and 'other' mountains). An image of the mask is shown in Figure 6.1.

For each mountainous region, paired observations of snow depth and elevation were extracted from the original snow depth data file. Correlation coefficients were calculated between snow depth and elevation to quantify the magnitude and direction of the relations. Then, a criterion was devised to identify when regression may or may not be used to estimate snow depth. The first step was to select a statistical significance level to test whether or not a correlation coefficient was significant (Norcliffe, 1977). Choosing a conservative level of 99.99% provided a strict measure of statistical confidence; if correlation coefficients were significant, then the next stage could be undertaken, otherwise IDWS was used as the interpolator. The second step was to identify a correlation coefficient threshold so that only when correlations were greater than or equal to this threshold would regression modelling be employed. We used a threshold of 0.6.

For completeness, the least squares regression model can be expressed as

$$\hat{z}(\mathbf{x}_i) = \beta_0 + \beta_1 y(\mathbf{x}_i) + \varepsilon_0 \tag{4}$$

where $z(\mathbf{x}_i)$ is the estimated snow depth, $y(\mathbf{x}_i)$ is the observed elevation and β_0, β_1 and ε_0 are coefficients of regression obtained from least squares fitting.

6.3.2 Geostatistical Approaches

The present objective was to estimate, for each study date, the mean snow depth Z over supports or blocks $\mathbf{V}$ of 2 km by 2 km set out on a 2 km by 2 km square grid of locations $\mathbf{x}_j$, $j = 1,2,3, \ldots, m$, from sparsely located point measurements, $z(\mathbf{x}_i)$, $i = 1,2,3, \ldots, n$. Theoretically, neither IDWS nor regression achieve this objective since in both of these

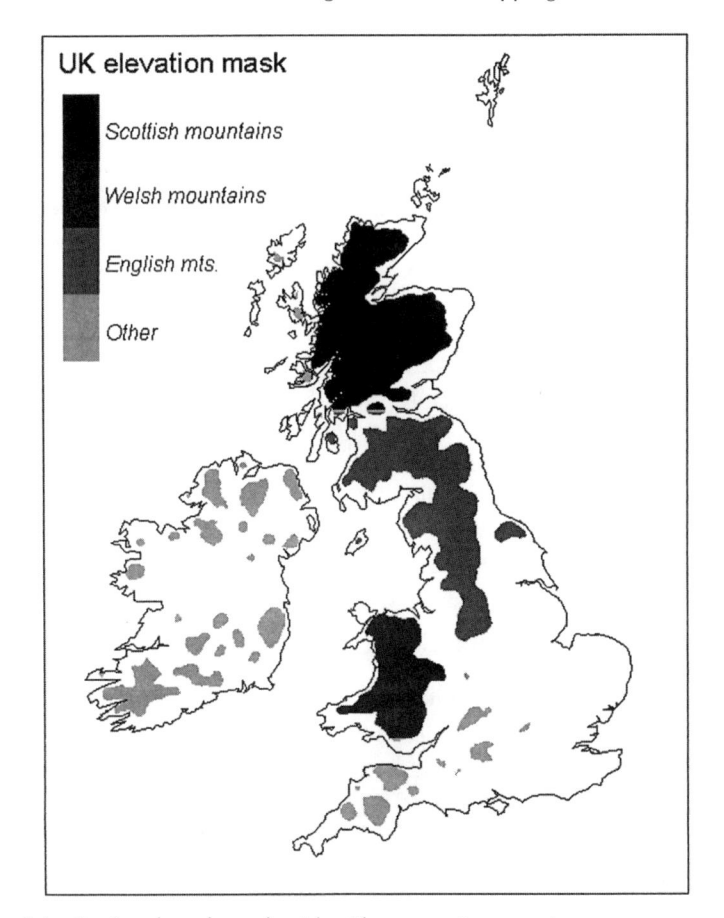

Figure 6.1 *Regional mask used to identify mountainous regions*

cases the support of the estimated values is a point. In this section, block kriging (using snow depth data alone) and block cokriging (using snow depth and elevation data) are described. These techniques allow one to estimate values for supports of any size.

6.3.2.1 *Kriging*

Kriging has been applied extensively in mining (Journel and Huijbregts, 1978), soil survey (e.g., Burgess and Webster, 1980a, 1980b), hydrology (e.g., Philips *et al.*, 1992) and in remote sensing (e.g., Atkinson, 1991). Kriging is similar to IDWS in that it may be used to interpolate from sparse sample data of a single variable. However, whereas IDWS is deterministic, kriging is based on a statistical model. More importantly, whereas IDWS assumes a model for the distance weighting function, kriging uses a function estimated from the data themselves to determine the most efficient weights. With kriging, therefore, there is no need to assume a power for the distance weighting function.

Kriging is an optimal technique for linear unbiased estimation. The estimated value is effectively a weighted average of observations surrounding the position of the estimate. Kriging estimates optimally by referring to the spatial dependence in the variable of interest represented by the variogram (or other structure function). Observations that are close to the estimate receive more weight than those that are further away. The exact weights are determined from the mathematical model fitted to the sample variogram, and they are chosen to ensure unbiasedness (the weights sum to one) and to minimize the estimation variance (via the kriging equations).

In statistics, observations of a given property are often modelled as realizations of a random variable, RV. A regionalized variable, ReV, is simply a realization of the spatial set of RVs covering the region of interest, known as a random function, RF (Matheron, 1965, 1971; Journel and Huijbregts, 1978; Isaaks and Srivastava, 1989; Deutsch and Journel, 1992). In geostatistics, a sample of a spatially varying property is commonly represented as a regionalized variable, ReV, that is, as a realization of a random function, RF. The semivariance may be defined as half the expected squared difference between the RFs $Z(\mathbf{x})$ and $Z(\mathbf{x} + \mathbf{h})$ at a particular lag $\mathbf{h}$. The variogram, defined as a parameter of the RF model, is then the function that relates semivariance to lag (Equation 5)

$$\gamma(\mathbf{h}) = 1/2 E[\{Z(\mathbf{x}) - Z(\mathbf{x} + \mathbf{h})\}^2] \qquad (5)$$

The sample variogram $\hat{\gamma}(\mathbf{h})$ can be estimated for $p(\mathbf{h})$ pairs of observations or realizations, $\{z(\mathbf{x}_l + \mathbf{h}), l = 1, 2, \ldots, p(\mathbf{h})\}$ by Equation 6

$$\hat{\gamma}(\mathbf{h}) = 1/2p(\mathbf{h}) \sum_{l=1}^{p(\mathbf{h})} \{z(\mathbf{x}_l) - z(\mathbf{x}_l - \mathbf{h})\}^2 \qquad (6)$$

To make use of the sample variogram in kriging it is usually necessary to fit a mathematical model to the estimated values. This is because the sample variogram is comprised of a set of sample semivariances at a set of discrete lags. To estimate the semivariances between the lags for which we have empirical estimates it is necessary to fit a continuous mathematical model. However, not just any model will do: the model fitted should ensure that negative variances cannot result from linear combinations of the RVs that make up the RF $Z(\mathbf{x})$. The model should be conditional negative semi-definite or CNSD. To ensure that the CNSD criterion is satisfied it is common practice to select from a set of CNSD models (Webster and Oliver, 1992). The selected model should then be fitted to the data, for example using weighted least squares approximation (McBratney and Webster, 1986).

In general, the variogram model is either unbounded (i.e., increases indefinitely with lag) or bounded (i.e., increases to a maximum value of semivariance, known as the sill, at a finite positive lag, known as the range a). The sill is equal to the *a priori* variance (that defined for an infinite region) of the RF while the range indicates the limit to spatial dependence, beyond which data are statistically uncorrelated. Often the model approaches and intercepts the ordinate at some positive value of semivariance known as the nugget variance c_0. The nugget variance results from measurement error (Atkinson, 1993), the uncertainty in estimating the variogram from a sample, the uncertainty in model fitting, and spatially dependent variation acting at scales finer than the sampling interval. The structured component of variation c_1 is then the sill minus the nugget variance (so that $c_0 + c_1 =$ sill).

The block kriging estimator $\hat{z}(\mathbf{V})$ is a linear weighted sum of the n observations (Equation 7)

$$\hat{z}(\mathbf{V}) = \sum_{i=1}^{n} \lambda_i z(\mathbf{x}_i) \qquad (7)$$

where the n sample observations $\{z(\mathbf{x}_i), i = 1, 2, \ldots, n\}$ are defined on quasi-point supports, $\mathbf{x}_i$. The n weights, λ_i, are then chosen such that the estimate is unbiased and has minimum estimation or kriging variance. For ordinary block kriging, unbiasedness is ensured when the n weights sum to one and the estimation variance is minimized by optimizing the weights using standard Lagrangian techniques. The result is a system of $(n + 1)$ linear equations in $(n + 1)$ unknowns, i.e., the kriging system (Equation 8)

$$\sum_{j=1}^{n} \lambda_j \gamma(\mathbf{x}_i, \mathbf{x}_j) + \psi = \bar{\gamma}(\mathbf{x}_i, \mathbf{V}) \qquad \text{for all } i = 1, \ldots, n$$

$$\sum_{j=1}^{n} \lambda_j = 1 \qquad (8)$$

where ψ is the Lagrange parameter. The kriging variance is then estimated by Equation 9

$$\sigma_k^2 = \sum_{i=1}^{n} \lambda_j \bar{\gamma}(\mathbf{x}_i, \mathbf{V}) + \psi - \bar{\gamma}(\mathbf{V}, \mathbf{V}) \qquad (9)$$

We will use the terms 'auto-variogram' and 'auto-kriging' to refer to the variograms and kriging methods described above, to contrast with the cross-variograms and cokriging using more than one variable described below.

6.3.2.2 Cokriging

Cokriging has the potential, at least in theory, to be more accurate than both auto-kriging (because it incorporates information from a second variable) and regression (because it simultaneously incorporates information on spatial dependence in both the primary (snow depth) and secondary (elevation) variables and also the cross-correlation between the two variables). The cokriging equations for estimating a primary variable from a coregionalized set of variables are simply extensions of those for auto-kriging. Cokriging is most profitable where the primary variable of interest is less densely sampled than the other(s). This is clearly likely to be the case where the secondary variable is provided by a DEM (which covers the region of interest completely). There are now many examples of cokriging in the environmental sciences, for example, Leenaers et al. (1989); Seo et al. (1990a, 1990b); Bhatti et al. (1991); Cressie (1991); Stein et al. (1991); Atkinson et al. (1992, 1994); Nash et al. (1992); Rossi et al. (1992); Zhang et al. (1992); Gohin and Langlois (1993).

For any two variables $k = u$ (snow depth) and $k = v$ (elevation) we assume that their auto-correlation over the lags of interest may be represented by the auto-variograms while their cross-correlation may be represented adequately by the cross-variogram (Equation 10)

$$\gamma_{uv}(\mathbf{h}) = \frac{1}{2}\mathrm{E}[\{Z_u(\mathbf{x}) - Z_u(\mathbf{x}+\mathbf{h})\}\{Z_v(\mathbf{x}) - Z_v(\mathbf{x}+\mathbf{h})\}] \qquad (10)$$

where $Z_u(\mathbf{x})$, $Z_v(\mathbf{x})$, $Z_u(\mathbf{x}+\mathbf{h})$ and $Z_v(\mathbf{x}+\mathbf{h})$ are the values of u and v at places $\mathbf{x}$ and $\mathbf{x}+\mathbf{h}$ respectively. Given data in the form $z_k(\mathbf{x}_i)$, $i = 1, 2, \ldots$, where the $z_k(\mathbf{x}_i)$ are the measured values of k and the $\mathbf{x}_i$ refer to places at which they were obtained, we can compute estimates of the cross-semivariances for each pair $p(\mathbf{h})$ of variables u and v using Equation 11

$$\hat{\gamma}_{uv}(\mathbf{h}) = \frac{1}{2p(\mathbf{h})} \sum_{i=1}^{p(\mathbf{h})} \{z_u(\mathbf{x}_i) - z_u(\mathbf{x}_i + \mathbf{h})\}\{z_v(\mathbf{x}_i) - z_v(\mathbf{x}_i + \mathbf{h})\} \qquad (11)$$

with the constraint that only pairs of places where both variables have been measured contribute.

As for the auto-variograms, it is necessary to select models that are CNSD. In addition, it is necessary to select models that in combination conform to the linear model of coregionalization. The principles are somewhat involved (Goulard and Voltz, 1992) and, therefore, for modelling and the cokriging equations themselves the reader is referred to Atkinson *et al.* (1992) where full details are given.

6.3.4 Cross-validation

A technique was required to test the accuracy of each interpolator. The technique needed to be a standard one so that it could be applied to compare both traditional and geostatistical approaches. In this chapter, we used cross-validation to evaluate accuracy.

The cross-validation technique requires that after interpolation has been undertaken and a map of snow depth generated, each sparse data point is re-estimated in turn from the surrounding sparse data only (i.e., omitting the datum to be estimated). Thus, estimates are made at the locations of the original data only. The RMSE is then computed between all re-estimated data and the original values. The RMSE indicates the accuracy of the technique; the smaller the RMSE, the more accurate the interpolation.

6.4 Analysis and Results

Analyses of January 1988 and February 1991 data were completed and the results are presented in the following sections. However, for illustrative purposes, graphical results are presented for 12 February 1991 only.

Table 6.1 Summary tables for accuracies of different interpolation techniques. Statistics shown are RMSE values of snow depth in centimetres

	Traditional interpolation		Geostatistical interpolation		
	IDWS	Regression+IDWS	Kriging	Cokriging	Hybrid (k+ck)
22 Jan. 1988	43.5	N/A	33.4	31.0	N/A
23 Jan. 1988	33.0	N/A	28.9	27.5	N/A
12 Feb. 1991	81.8	76.8	70.6	71.7	69.7
16 Feb. 1991	47.9	41.9	42.4	43.8	40.4

6.4.1 January 1988

6.4.1.1 *IDWS Interpolation*

Snow cover was more extensive in the morning of 22 January than 23 January 1988. The deepest snow was located in the Midlands and mid to north Wales. Cross-validation of data for 22 and 23 January gives RMSE values of 43.5 cm and 33.0 cm respectively (Table 6.1).

6.4.1.2 *Regression*

Correlation coefficients were calculated between snow depth and elevation for each day of January 1988. The correlation coefficients are shown in Figure 6.2 along with the number of paired observations used. Only stations that recorded positive snow depth were used for analysis. The four graphs show correlation coefficients for each day for all data in the UK (Figure 6.2a), Scottish mountains (Figure 6.2b), English mountains (Figure 6.2c), and Welsh mountains (Figure 6.2d) as defined by the elevation mask. For all graphs, the largest sample sizes were recorded between 19 and 25 January with maximum coverage on 22 and 23 January. When there were less than three data points, no correlation coefficient was calculated.

From Figure 6.2, there is a positive relationship between snow depth and elevation for all-winter snowpacks before and after snowfall 'events'. This is demonstrated by the large correlation coefficients (around 0.7) before 22 January and after 23 January for data recorded at Scottish mountain locations (Figure 6.2b). When snow cover is widespread (indicated by a large sample size of 100 or more) correlation coefficients approach zero, that is, there is little observable relationship between snow depth and station elevation and, hence, there is no justification for predicting snow depth using regression. Thus, for 22 and 23 January 1988, when correlation coefficients were less than 0.6, regression could not be used to predict snow depth in the UK. In Table 6.1, this is indicated by 'N/A' entered for the RMSE.

6.4.1.3 *Kriging*

Sample auto-variograms were computed for snow depth on 22 and 23 January and the variograms fitted with the spherical model with a range of 70 km, albeit with different

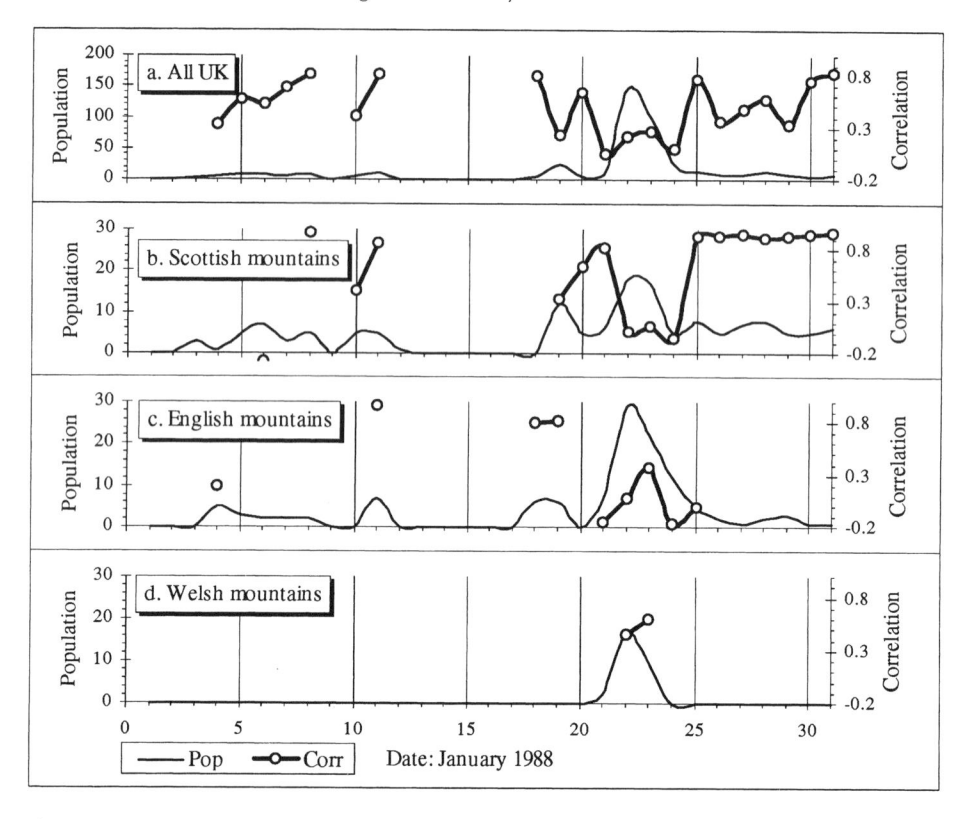

Figure 6.2 *Daily correlation coefficients between snow depth and elevation for January 1988. The heavy lines represent sequences of correlation coefficients and the accompanying thin lines represent the number of paired observations used to calculate each daily correlation. The four graphs labelled a–d represent correlations for (a) all UK observations, (b) Scottish, (c) English and (d) Welsh mountains observations*

nugget variances and structured components. The model type and its coefficients were input to the kriging equations to estimate snow depth for the entire UK. While kriging provides automatically an estimate of the minimum estimation or kriging variance, it was necessary to estimate the kriging RMSE to provide a standard comparison with IDWS. The RMSE values obtained for kriging are shown in Table 6.1. Kriging is more accurate than IDWS in both cases (22 and 23 January).

6.4.1.4 Cokriging

In addition to the auto-variograms for snow depth and elevation, cross-variograms between the two were computed for both 22 and 23 January 1988. These functions were also fitted with the spherical model with a range of 70 km. The common range is necessary to satisfy the constraints of the linear model of coregionalization (Journel and Huijbregts, 1978). A further constraint is that the matrix of coefficients of the models is positive definite, and this was found to be the case. The common range of 70 km implies that both snow depth and elevation are spatially dependent up to a lag of 70 km, beyond which data are uncorrelated. It implies (i) that patches of snow are up to 70 km in diameter, and (ii) that mountainous regions are up to 70 km in diameter. Since the cross-variogram shows evidence of structure akin to that in the two auto-variograms it is likely that the snow patches correspond to mountainous regions which are up to 70 km across.

The models and their coefficients were entered into the cokriging equations to estimate snow depth over the whole of the UK. The RMSE values estimated for cokriging were in all cases smaller than for kriging, albeit by a small amount (Table 6.1).

6.4.2 February 1991

6.4.2.1 *IDWS Interpolation*

IDWS snow depth maps were produced for 12 and 16 February 1991. Figure 6.3a shows interpolated snow depth map for 12 February 1991. Using IDWS, RMSE values of 81.1 cm and 47.9 cm were obtained for 12 and 16 February respectively (Table 6.1).

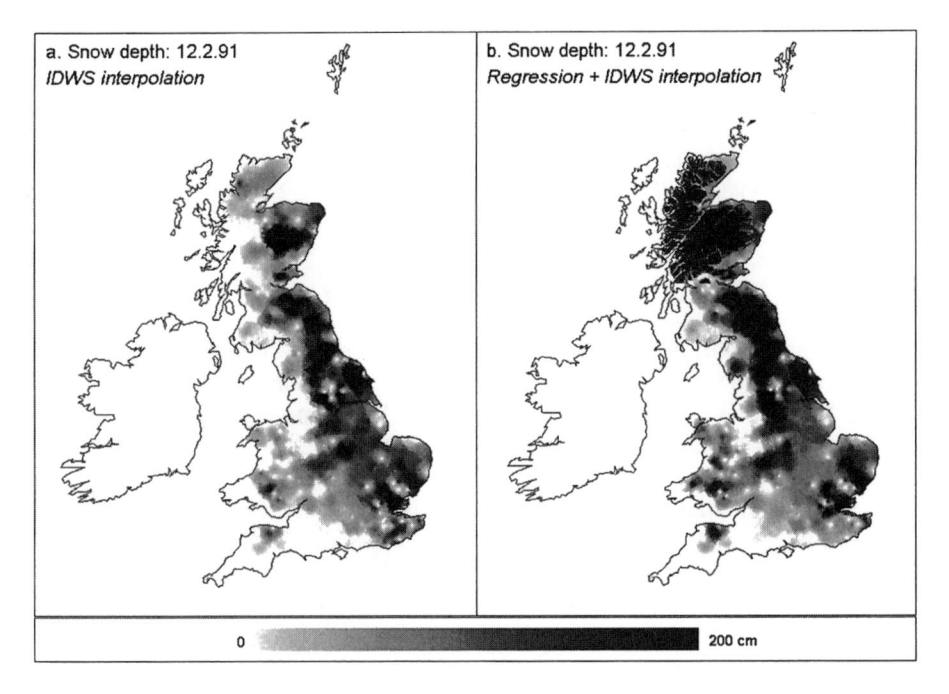

Figure 6.3 Snow depth maps for 12 February 1991 produced by (a) IDWS and (b) regression for Scottish mountains combined with IDWS for all other locations

88   Advances in Remote Sensing and GIS Analysis

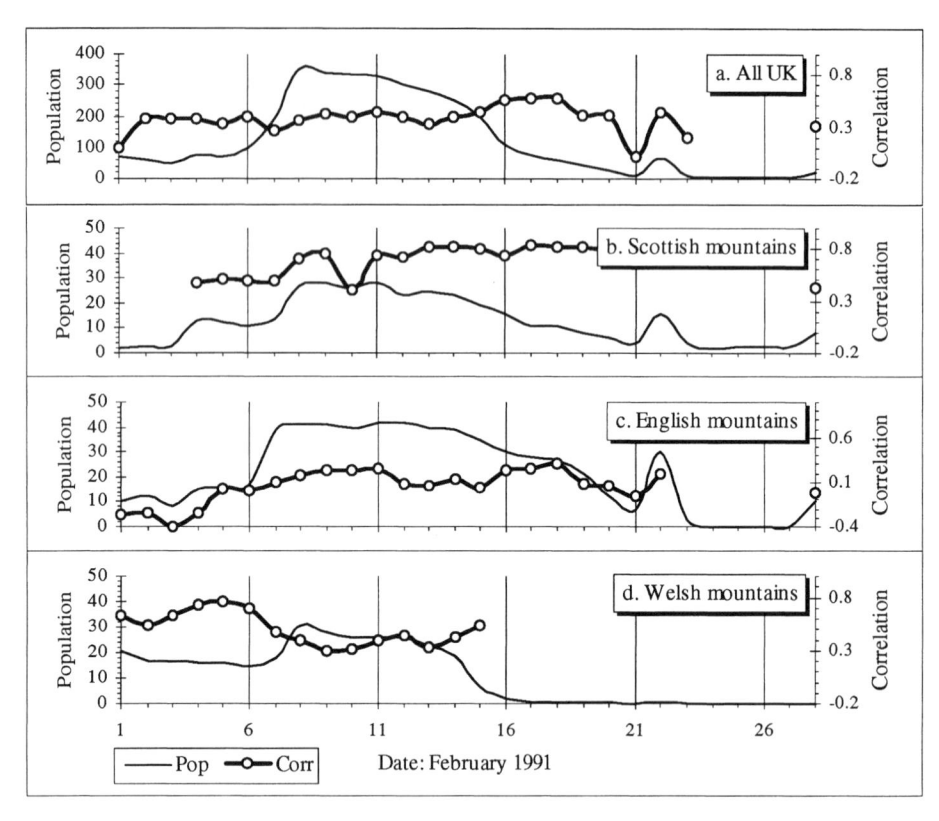

Figure 6.4 *Daily correlation coefficients between snow depth and elevation for February 1991. The heavy lines represent sequences of correlation coefficients and the accompanying thin lines represent the number of paired observations used to calculate each daily correlation. The four graphs labelled a–d represent correlations for (a) all UK observations, (b) Scottish, (c) English and (d) Welsh mountains observations*

6.4.2.2 Regression

Figure 6.4 shows daily correlation coefficients between snow depth and elevation for February 1991. The four different graphs relate data for the entire UK, and for the Scottish, English and Welsh mountains. Again, when a snowfall event occurs in the UK, the correlation between snow depth and station elevation decreases. As snow-covered areas contract in size, the correlation between depth of snow and elevation increases. Thus, in Figure 6.4a–d, the pattern is one of increasing correlation with decreasing areal snow coverage (as indicated by a reduction in sample size). This feature is demonstrated best by the correlation coefficients for the Scottish and Welsh mountains (Figure 6.4b and d), although the remaining plots also show the essence of this inverse relationship.

Using the regression approach outlined in Section 6.3.1.2, snow depth in the UK for 12 and 16 February was estimated using a combination of regression modelling and IDWS. For the Scottish mountains region, correlation coefficients were significant and greater than 0.6 so a regression relation between snow depth and elevation was derived and applied. However, outside the Scottish mountains region, correlation coefficients were not significant and snow depth was estimated using unconstrained IDWS interpolation.

Cross-validation gave RMSE values of 76.8 cm for 12 February and 41.9 cm for 16 February. The map for 12 February 1991 is shown in Figure 6.3b.

6.4.2.3 Kriging

For 12 and 16 February 1991 sample auto-variograms were estimated and fitted with the spherical model with a range of 70 km as for the 1988 data set. For 12 February the scatter of estimated values of semivariance around the fitted model was less than for 16 February. This was due to the smaller sample size for 16 February. The variogram for 12 February is shown with its fitted model in Figure 6.5a.

The models and their coefficients were entered into the kriging equations and snow depth mapped for the whole of the UK for both 12 and 16 February 1991. The RMSE values obtained for 12 and 16 February are shown in Table 6.1. As for the January 1988 data, kriging is once again more accurate than IDWS and, in this case, it is also more accurate than the hybrid regression and IDWS approach. A map of kriged snow depth for 12 February 1991 is shown in Figure 6.5b.

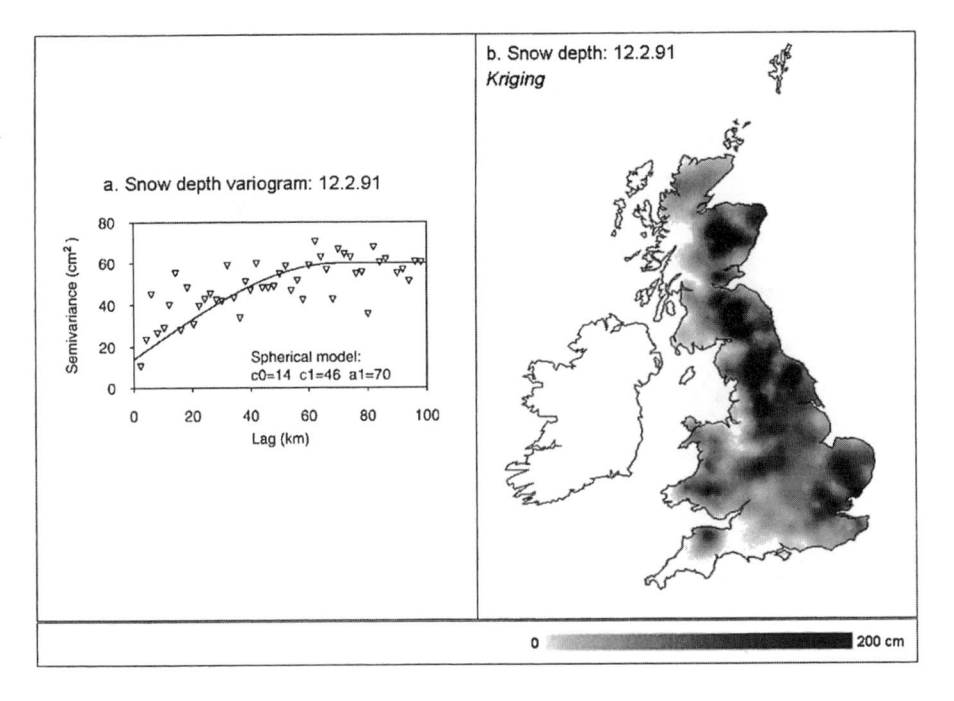

Figure 6.5 *Variogram of snow depth observations and kriged snow depth map for 12 February 1991. Figure 6.5a shows the variogram fitted with a spherical model and Figure 6.5b shows the resulting kriged snow depth map*

6.4.2.4 Cokriging

The auto-variograms for elevation and the cross-variograms between snow depth and elevation for 12 and 16 February 1991 were estimated and their functions fitted with the spherical model with a range of 70 km. The fit of the models to the variograms for elevation appears acceptable, but somewhat less so for the cross-variograms. For both 12 and 16 February the fitted models represent a compromise which was necessary to conform to the constraints of the linear model of coregionalization. The elevation and cross-variograms are shown for 12 February in Figures 6.6a and 6.6b respectively.

The models and their coefficients were entered into the cokriging equations to estimate snow depth for the whole of the UK for 12 and 16 February 1991. The cokriged map of snow depth for 12 February is shown in Figure 6.6c and embodies more local-scale variation in relief than the kriged map in which the local variation of snow depth is smoothed. However, on both 12 and 16 February the RMSE values from cokriging were found to be larger than those for kriging (Table 6.1). Therefore, it is not automatic that cokriging will lead to an increase in accuracy above that obtainable with kriging. It depends on the cross-correlation between the primary and secondary variables. For both 12 and 16 February, the correlation coefficients between snow depth and elevation were small (less than 0.6) and so one might not expect cokriging to be profitable.

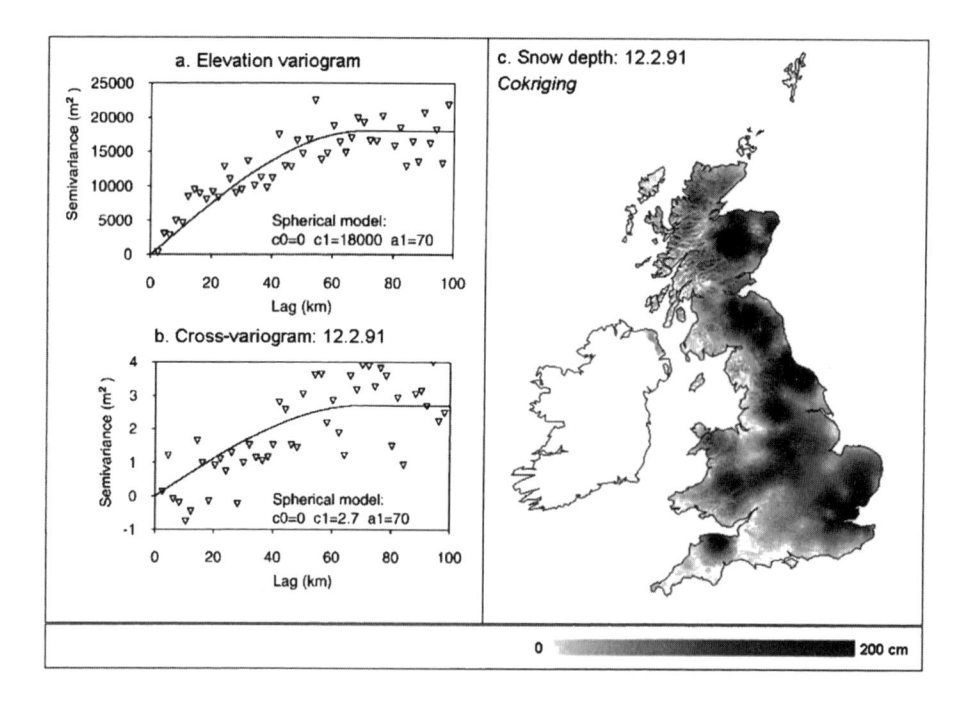

Figure 6.6 *Variogram, cross-variogram and cokriged snow depth map for 12 February 1991. Figures 6.6a and b show the auto-variogram for elevation and cross-variogram of elevation and snow depth. Both figures are fitted with spherical models. Figure 6.5c shows the cokriged map of snow depth*

A possible explanation for the similar accuracy is that the models of coregionalization describe inadequately the spatial variation in snow depth and elevation, either through inaccuracy in the sample variograms or non-stationarity of the variograms across the UK. Non-stationarity in the relation between snow depth and elevation might exist in some localities because the cyclonic nature of snowfall produces snow cover whose depth varies irrespective of elevation; ambient temperatures are cold enough for deep snow covers to persist at both low and high altitudes. A better model of the coregionalization of snow depth and elevation, therefore, might be one which gives more weight to snow depth when estimating at lower elevations and less weight when estimating at higher elevations. After the initial snow deposition event, however, snow cover at low elevation and outside mountain areas may be subject to rapid melt as ambient temperatures increase more quickly than at locations at higher elevations. The UK snow cover distribution is then comprised of all-winter snowpacks and so the usual model of coregionalization in which snow depth varies with elevation will apply.

To model the above non-stationarity and to compare more accurately the results of the hybrid regression and IDWS technique with geostatistical techniques, we combined cokriging for the Scottish mountains with kriging for the remaining UK for both dates in 1991. The RMSE was 69.7 cm and 40.4 cm for 12 and 16 February respectively. These values were the most accurate for mapping snow depth when compared with all other techniques used in this chapter. The result of applying this hybrid technique to data for 12 February is shown in Figure 6.7.

6.5 Discussion

The results from this chapter indicate that techniques for geostatistical interpolation produce more accurate snow depth maps than those derived from traditional approaches. Table 6.2 shows the average RMSE values for 1988 and 1991 data and that kriging and cokriging are consistently more accurate than traditional forms of interpolation. If all RMSE data are averaged (Table 6.3), cokriging is shown to have the greatest interpolation accuracy when compared with other interpolation techniques (it was not possible to combine kriging and cokriging for the 1988 data so this example is not included in the overall averages). However, this table masks the fact that both kriging and cokriging have similar accuracies when confidence levels are attached to their RMSE values.

For the 1991 case studies selected, the most accurate technique for interpolation is the hybrid combination of cokriging for the Scottish mountain data and kriging for the remainder of the UK. This approach reflects the nature of the complex relationship between snow depth and elevation. In the UK, when snow cover is not widespread but is restricted to all-winter snowpacks in mountainous terrain, strong associations exist between snow depth and elevation. This relationship is illustrated comprehensively by the correlation plots for the Scottish mountains snowpacks in both the 1988 and the 1991 data sets (Figures 6.2 and 6.4). Signs of this relationship are also revealed by the 1991 Welsh mountain data. Conversely, when temperatures are zero and sub-zero over

92 Advances in Remote Sensing and GIS Analysis

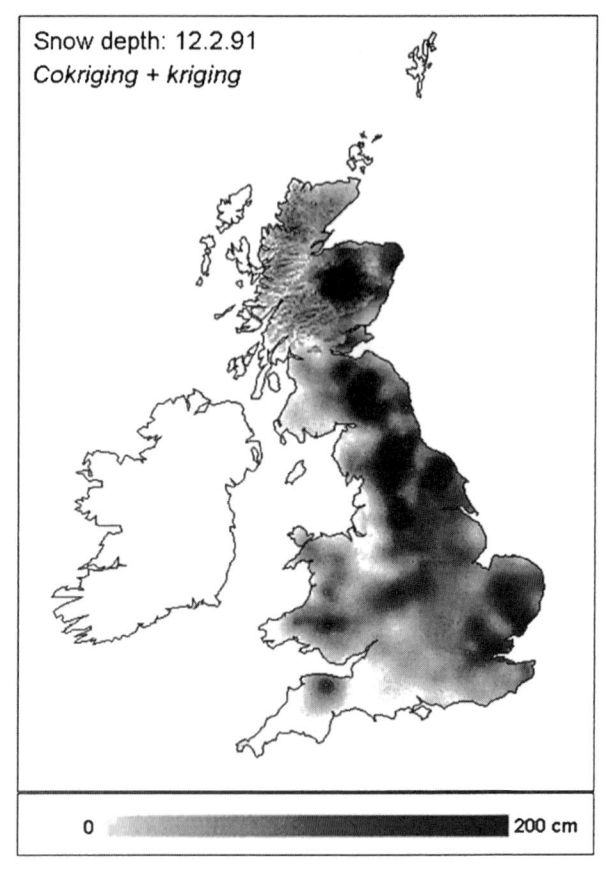

Figure 6.7 *Map of snow depth derived from a combination of cokriging and kriging. Cokriging was applied to the Scottish mountains and kriging was applied to all other locations*

a large area, snow cover may be widespread. In such cases, deposition of snow cover varies irrespective of relief.

The advantages of the geostatistical approaches are that interpolations *de facto* are optimal and unbiased in their estimation of snow depth. In addition, kriging and cokriging can be applied over any support desired by the user. The disadvantage with geostatistical interpolation (as with any interpolation), is that data are smoothed by the interpolator. This is a problem that has been addressed by Atkinson and Kelly (1997) where they quantify the variance in the original snow depth data and 'add' this variance back to the final interpolated map. By accounting for the smoothing of the ground data in this way, calibration and validation of SSM/I snow depth estimates are unbiased.

6.6 Conclusions

In all snow cover events, geostatistical interpolation was found to be more accurate

Table 6.2 Average RMSE accuracies of different interpolation techniques for 1988 and 1991. Statistics shown are RMSE values of snow depth in centimetres

	Traditional interpolation		Geostatistical interpolation		
	IDWS	Regression + IDWS	Kriging	Cokriging	Hybrid (k + ck)
Jan. 1988	38.3	N/A	31.2	29.3	N/A
Feb. 1991	64.9	59.4	56.5	57.8	55.1

Table 6.3 Overall average RMSE accuracies of different interpolation techniques. Statistics shown are RMSE values of snow depth in centimetres

	Traditional interpolation		Geostatistical interpolation		
	IDWS	Regression + IDWS	Kriging	Cokriging	Hybrid (k + ck)
Ave. RMS	51.6	N/A	43.8	43.6	N/A

than traditional interpolation. For the 1988 case study, regression modelling could not be applied because of a lack of statistical confidence in the correlation between snow depth and station elevation. Nevertheless, both kriging and cokriging were found to be more accurate than IDWS interpolation. Between kriging and cokriging, there was little difference in RMSE.

For the 1991 case study, kriging and cokriging were consistently more accurate than IDWS interpolation. Compared with the regression and IDWS interpolation technique, kriging and cokriging once again had smaller RMSE values and the hybrid combination of kriging and cokriging produced the best results. Thus, for all data sets analysed, it was shown that geostatistical techniques for interpolation are consistently more accurate than traditional techniques.

Acknowledgements

The authors acknowledge the financial support of the Leverhulme Trust and the Natural Environmental Research Council in the preparation of this chapter.

References

Archer, D.R., Bailey, J.O., Barrett, E.C. and Greenhill, D., 1994, The potential of satellite remote sensing of snow over Great Britain in relation to cloud cover, *Nordic Hydrology*, **25**, 39–52.

Armstrong, R. and Hardman, M., 1991, Monitoring global snow cover, *Proceedings of IGARSS 1991* (New York: IEEE), 1947–1949.

Atkinson, P.M., 1991, Optimal ground-based sampling for remote sensing, *International Journal of Remote Sensing*, **12**, 559–567.

Atkinson, P.M., 1993, The effect of spatial resolution on the experimental variogram of airborne MSS imagery, *International Journal of Remote Sensing*, **14**, 1005–1011.

Atkinson, P.M. and Kelly, R.E.J., 1997, Scaling-up point snow depth data in the UK for comparison with SSM/I imagery, *International Journal of Remote Sensing*, **18**, 437–443.

Atkinson, P.M., Webster, R. and Curran, P.J., 1992, Cokriging with ground-based radiometry, *Remote Sensing of Environment*, **41**, 45–60.

Atkinson, P.M., Webster, R. and Curran, P.J., 1994, Cokriging with airborne MSS imagery, *Remote Sensing of Environment*, **50**, 335–345.

Baumgartner, M.F., 1994, Towards an integrated geographic analysis system with remote sensing, GIS and consecutive modeling for snow cover monitoring, *International Journal of Remote Sensing*, **15**, 1507–1517.

Bhatti, A.U., Mulla, D.J. and Frazier, B.E., 1991, Estimation of soil properties and wheat yields on complex eroded hills using geostatistics and thematic mapper images, *Remote Sensing of Environment*, **37**, 181–191.

Burgess, T.M. and Webster, R., 1980a, Optimal interpolation and isarithmic mapping of soil properties I. The semi-variogram and punctual kriging, *Journal of Soil Science*, **31**, 315–331.

Burgess, T.M. and Webster, R., 1980b, Optimal interpolation and isarithmic mapping of soil properties II. Block kriging, *Journal of Soil Science*, **31**, 333–341.

Carroll, T.R., 1995, GIS used to derive operational hydrologic products from in situ and remotely sensed snow data, in A. Carrara and F. Guzzetti (eds), *Geographical Information Systems in Assessing Natural Hazards* (London: Taylor & Francis), 335–342.

Cressie, N.A.C., 1991, *Statistics for Spatial Data* (New York: Wiley).

Davis, J.C., 1973, *Statistics and Data Analysis in Geology* (New York: Wiley).

Deutsch, C.V. and Journel, A.G., 1992, *GSLIB Geostatistical Software Library User's Guide*, (Oxford: Oxford University Press).

Dozier, J., 1989, Spectral signature of alpine snow cover from Landsat Thematic Mapper, *Remote Sensing of Environment*, **28**, 9–22.

Fiore, Jr, J.V. and Grody, N.C., 1992, Classification of snow cover and precipitation using SSM/I measurements: case studies, *International Journal of Remote Sensing*, **13**, 3349–3361.

Gohin, F. and Langlois, G., 1993, Using geostatistics to merge in situ measurements and remotely sensed observations of sea surface temperature, *International Journal of Remote Sensing*, **14**, 9–19.

Goulard, M. and Voltz, M., 1992, Linear coregionalization model: Tools for estimation and choice of cross variogram matrix, *Mathematical Geology*, **24**, 264–286.

Grody, N.C. and Basist, A.N., 1996, Global identification of snowcover using SSM/I measurements, *IEEE Transactions on Geoscience and Remote Sensing*, **34**, 237–249.

Guneriussen, T., Johnsen, H. and Sand, K., 1996, DEM corrected ERS-1 SAR data for snow monitoring, *International Journal of Remote Sensing*, **17**, 181–195.

Harrison, A.R. and Lucas, R.M., 1989, Multi-spectral classification of snow using NOAA AVHRR imagery, *International Journal of Remote Sensing*, **10**, 907–916.

Hollinger, J., Lo., R., Poe, G., Savage, R. and Pearce, J., 1987, *Special Sensor Microwave/Imager User's Guide* (Washington, DC: Naval Research Laboratory).

Hymas, K., 1993, The Meteorological Office National Severe Weather Warning System (MSWWS), *The Meteorological Magazine*, **122**, 53–61.

Isaaks, E.H. and Srivastava, R.M., 1989, *Applied Geostatistics* (Oxford: Oxford University Press).

Journel, A.G. and Huijbregts, C.J., 1978, *Mining Geostatistics* (London: Academic Press).

Kelly, R.E.J., 1995, Snow monitoring in the UK using passive microwave imagery, *RSS'95 Remote Sensing in Action* (Nottingham: Remote Sensing Society), 1223–1230.

Kelly, R.E.J., 1996, Snow monitoring in the UK using active microwave data, in E. Parlow (ed.), *Progress in Remote Sensing Research and Applications, Proceedings of the 15th EARSeL Symposium, Basel, 4–6 September, 1995* (Rotterdam: Balkema), 253–259.

Lancaster, P. and Šalkauskas, K., 1986, *Curve and Surface Fitting: An Introduction* (London: Academic Press).

Leenaers, H., Burrough, P.A. and Okx, J.P., 1989, Efficient mapping of heavy metal pollution on floodplains by cokriging from elevation data, in J.F. Raper (ed.), *Three Dimensional Applications in Geographical Information Systems* (London: Taylor & Francis), 37–50.

McBratney, A.B. and Webster, R., 1986, Choosing functions for semi-variograms of soil properties and fitting them to sampling estimates, *Journal of Soil Science*, **37**, 617–639.

McKay, G.A. and Gray, D.M., 1981, The distribution of snowcover, in D.M. Gray and D.H. Male (eds), *Handbook of Snow* (Toronto: Pergamon Press), 153–190.

Matheron, G., 1965, *Les Variables Regionalisées et Leur Estimation* (Paris: Masson).

Matheron, G., 1971, *The Theory of Regionalized Variables and its Applications* (Fontainebleau: Centre de Morphologie Mathématique, Ecole des Mines de Paris).

Nash, M., Toorman, A. and Wierenga, P., 1992, Estimation of vegetation curves in an arid rangeland based on soil moisture using cokriging, *Soil Science*, **154**, 25–36.

Norcliffe, G.B., 1977, *Inferential Statistics for Geographers* (London: Hutchinson).

Peterson, R.A. and Hoke, J.E., 1989, The effect of snow cover on the regional analysis and forecast system (RAFS) low level forecasts, *Weather Forecaster*, **4**, 253–257.

Petrie, G. and Kennie, T.J.M., 1990, *Terrain Modelling in Surveying and Civil Engineering* (Caithness: Whittles).

Philips, D.L., Dolph, J. and Marks, D., 1992, A comparison of geostatistical procedures for spatial analysis of precipitation in mountainous terrain, *Agricultural and Forest Meteorology*, **58**, 119–141.

Poe, G.A., 1990, Optimum interpolation of imaging microwave radiometer data, *IEEE Transactions on Geoscience and Remote Sensing*, **28**, 800–810.

Rossi, R.E., Mulla, D.J., Journel, A.G. and Franz, E.H., 1992, Geostatistical tools for modeling and interpreting ecological spatial dependence, *Ecological Monographs*, **62**, 277–314.

Seo, D-J., Krajewski, W.F. and Bowles, D.S., 1990a, Stochastic interpolation of rainfall data from rain gauges and radar using cokriging 1. Design of experiments, *Water Resources Research*, **26**, 469–477.

Seo, D-J., Krajewski, W.F., Azimi-Zonooz, A. and Bowles, D.S., 1990b, Stochastic interpolation of rainfall data from rain gauges and radar using cokriging 2. Results, *Water Resources Research*, **26**, 915–924.

Stein, A., Staritsky, I.G., Bouma, J., Van Eijnsbergen, A.C. and Bregt, A.K., 1991, Simulation of moisture deficits and areal interpolation by universal cokriging, *Water Resources Research*, **27**, 1963–1973.

Tait, A. and Armstrong, R., 1996, Evaluation of SMMR satellite-derived snow depth using ground-based measurements, *International Journal of Remote Sensing*, **17**, 657–665.

Watson, D.F., 1992, *Contouring: A Guide to the Analysis and Display of Spatial Data* (Oxford: Pergamon Press).

Webster, R. and Oliver, M.A., 1992, Sample adequately to estimate variograms of soil properties, *Journal of Soil Science*, **43**, 177–192.

Xu, H., Bailey, J.O., Barrett, E.C. and Kelly, R.E.J., 1993, Monitoring snow area and depth with integration of remote sensing and GIS, *International Journal of Remote Sensing*, **14**, 3259–3268.

Zhang, R., Warrick, A.W. and Myers, D.E., 1992, Improvement of soil textural estimates using spectral properties, *Geoderma*, **52**, 223–234.

7
Using Variograms to Evaluate a Model for the Spatial Prediction of Minimum Air Temperature

Dan Cornford

7.1 Why Map Minimum Air Temperatures?

Air temperature (more correctly screen temperature) is measured 1.25 m above the Earth's surface in a standard exposure screen (Meteorological Office, 1956; Linacre, 1992, p. 29) at approximately 570 disparate locations across Great Britain (Figure 7.1). Daily measurements are made at 09.00 Greenwich Mean Time of the previous 24 hours' minimum and maximum temperatures. These stations comprise the 'climatological network', where recordings are undertaken by organizations and volunteers, the data being collated by the Meteorological Office (Meteorological Office, 1991). Each station is representative of a small (undefined?) area. Information on air temperature is, however, often required at unsampled locations (Linacre, 1992) which obligates spatial prediction by interpolation from the observations.

The focus of this work was to predict air temperature in the context of road weather studies (Cornford and Thornes, 1996a, 1996b; Johns, 1996) using data from the climate network to establish the relation between 'climate' and winter road maintenance expenditure (the amount of money spent on gritting and snow clearance from roads). The techniques and results could be equally applicable to areas such as agriculture (Avissar and Mahrer, 1988a, 1988b), ecology (Levin, 1992; Aspinall and Matthews, 1994) and building design or construction (Shaviv, 1984; Pulpitlova *et al.*, 1991), or to any other situation in which minimum air temperature is important. This study

---

Advances in Remote Sensing and GIS Analysis. Edited by Peter M. Atkinson and Nicholas J. Tate.
© 1999 John Wiley & Sons Ltd.

98 Advances in Remote Sensing and GIS Analysis

Figure 7.1 *The distribution of climate stations in Great Britain*

demands the distinction between spatial interpolation of minimum temperatures and temporal prediction (forecasting), the former being the subject of discussion.

7.2 A Model to Map Minimum Air Temperatures

Knowledge of the behaviour of the atmosphere derived from the governing physical laws, previous observations (expert knowledge) and exploratory data analysis suggests that the spatial variation in daily minimum air temperatures could be attributed to three factors:

- large-scale, airmass derived, temperature variation,
- interaction of the airflow with the local terrain, and
- microclimate effects.

Thus a model (summarized in Figure 7.3) was designed to incorporate all of these aspects (Cornford, 1996). When mapping daily minimum air temperatures over Great Britain terrain information is crucial (Tabony, 1985; Lennon and Turner, 1995). Terrain information was included in the model using five terrain variables. The five terrain variables were chosen on the basis of their power to explain the distribution of minimum air temperatures and their mutual independence (Cornford, 1996). They were:

ln(CD) logarithm of distance to the nearest coast (km)
ln(DD) logarithm of distance to the nearest drainage feature (km)
ln(UD) logarithm of distance to the nearest urban area (km)
CP   a land cover derived index of 'cooling potential' (unitless)
Z   elevation from a Digital Elevation Model with a 500 m horizontal resolution (m)

The variables were derived from GIS coverages of Great Britain. The distance to coast, drainage feature and urban area were computed from the corresponding Bartholomew's vector data sets, for every 500 × 500 m square in Great Britain within the ARC/INFO GIS (ESRI, 1992; Cornford, 1996). The (natural) logarithm of distance to coast, drainage feature and urban areas was used since the effect of these variables on minimum air temperatures was observed to be non-linear in distance space but linear in log distance space.

The 'cooling potential' index was constructed from land cover information on a kilometre square grid derived from LANDSAT scenes over the whole of Great Britain. Each class in the 26-class classification (Fuller et al., 1994) was assigned a subjective 'cooling potential' (Cornford, 1996). A score of 100 was assigned to the land class likely to cool most rapidly (i.e., short pasture) and a score of zero was assigned to the land class felt to cool least rapidly (i.e., urban areas). The 'cooling potential' (CP) index was defined as the weighted sum of the land class 'cooling potentials' within each grid square. The weights were proportional to the percentage of each class observed within each one kilometre square. The elevation data were obtained from a 100 m resolution Digital Elevation Model (DEM) re-sampled to 500 m resolution in Arc-Info. The terrain variables at each climate station were assigned the value of the cell in which the station is located.

The model used linear regression on the terrain variables, together with a first-order polynomial in easting (X) and northing (Y) to model the large-scale spatial variation. A first-order polynomial was used to enable simple visualization of the results, the coefficients being readily interpretable. The large-scale and terrain effects model was thus given by

$$T_{\min} = \alpha_0 + \alpha_1 X + \alpha_2 Y + \alpha_3 \ln(CD) + \alpha_4 \ln(DD) + \alpha_5 \ln(UD) + \alpha_6 CP + \alpha_7 Z + \varepsilon \quad (1)$$

where α_i are the regression coefficients and ε is the residual (error) from the model. This multiple regression was solved using an ordinary least squares estimator (Draper and Smith, 1981).

Several factors are neglected in the model as proposed above. First, no explicit account is taken of the fact that temperature might be spatially correlated (Kawashima and Ishida, 1992; Hudson and Wackernagel, 1994). In addition, several potentially important terrain variables, such as the specific heat capacity of the soil, are not included. The first-order polynomial may also be too inflexible to represent all the large-scale variation. To account for these weaknesses, the regression model was used to predict the minimum air temperature for each station. This produced a temperature surface that took into account local terrain and large-scale variation. These values were subtracted from the observations to produce residuals (ε) at each station. These

100 *Advances in Remote Sensing and GIS Analysis*

residuals were then analysed in a geostatistical context. Figure 7.2 provides an overview of the model.

7.2.1 Geostatistics and the Variogram

The residuals from the regression model were observed to be spatially correlated with each other. Mathematically this can be expressed using the concept of a random variable as defined by Journel and Huijbregts (1978, p. 29): 'A random variable ... is a variable which takes a certain number of numerical values according to a certain probability distribution.'

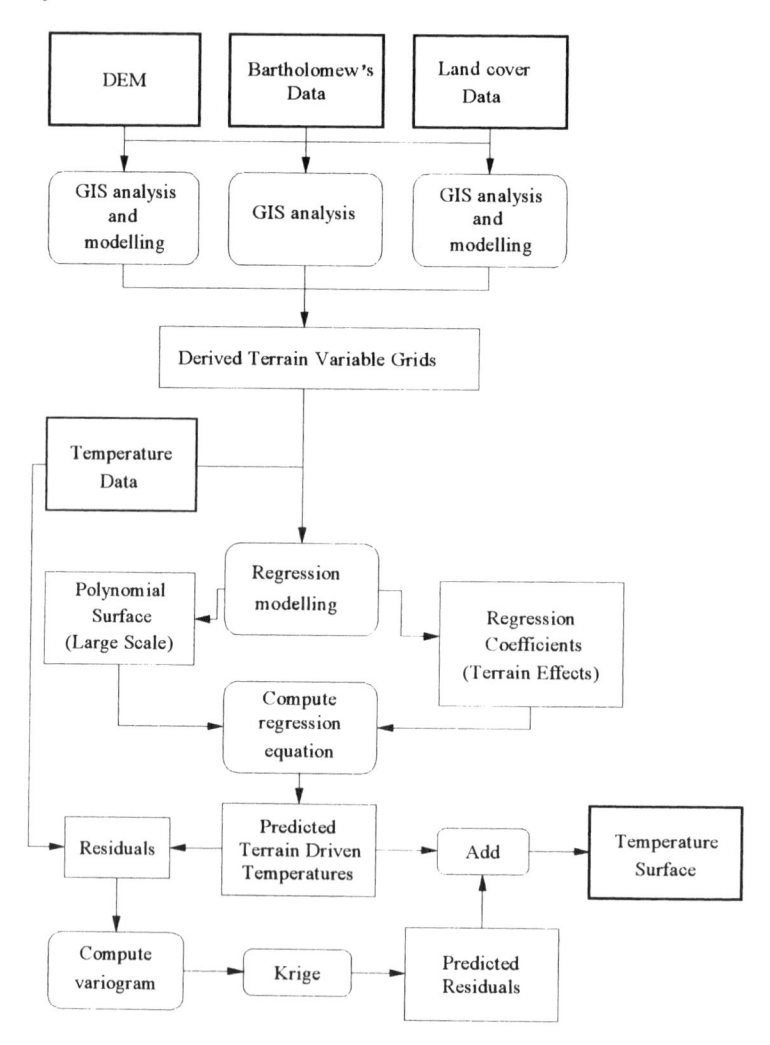

Figure 7.2 *A flowchart summarizing the model used to spatially interpolate minimum air temperatures. (DEM = Digital Elevation Model, GIS = Geographic Information System)*

A set of random variables for all points in a study area is considered a random function (or field). Suppose the variable of interest, t (e.g., minimum air temperature) varies spatially: let **x** represent the spatial location **x** = (x,y). The trend or mean value of the random function is represented by $m(\mathbf{x})$, and the autocorrelated component by ε so that

$$t(\mathbf{x}) = m(\mathbf{x}) + \varepsilon \qquad (2)$$

If the above model of spatial variation (of t) defines the behaviour correctly, then traditional ordinary least squares estimators of the trend function, $m(\mathbf{x})$, will be biased and will not make optimal use of the information in the data (Cressie, 1986; Lamorey and Jacobson, 1995). The solution is to use an iterative generalized least squares algorithm (Cressie, 1993). However, in this work the computationally less demanding least squares algorithm was used.

Theoretically, to use the autocorrelation between two locations requires a complete knowledge of the autocorrelation structure of the field between all observed and unobserved locations. In general, this can never be achieved and further assumptions are necessary. It is assumed that the random function of the residuals is intrinsically stationary (Cressie, 1993).

The variogram (γ) of the random function ε is defined to be

$$\gamma(\mathbf{h}) = \tfrac{1}{2}\mathrm{E}[(\varepsilon(\mathbf{x} + \mathbf{h}) - \varepsilon(\mathbf{x}))^2] \qquad (3)$$

where **h** denotes the separation vector between two points and E[ ] the expectation. In much practical work the weaker assumption of intrinsic stationarity is made, this being that $\gamma(\mathbf{h})$ exists and is finite for all vectors **h** and that the mean of the differences is zero (Wackernagel, 1995).

7.2.1.1 Computation of the Variogram

The experimental variogram was computed for the residuals, ε (Equation 1), using a method of moments (classical) estimator (Cressie, 1993)

$$\hat{\gamma}(\mathbf{h}) = \frac{1}{2N(\mathbf{h})} \sum_{i=1}^{N(\mathbf{h})} [\varepsilon(\mathbf{x}_i + \mathbf{h}) - \varepsilon(\mathbf{x}_i)]^2 \qquad (4)$$

where the summation is over all $N(\mathbf{h})$ variable pairs such that $h_1 < \mathbf{h} < h_2$. In this way **h**, the separation distance, is divided into intervals (lag classes) and the variogram is computed as the mean of the summed semivariances for each lag interval: the mean lag, h_i, is attributed to the ith lag class. By computing the semivariances for a series of lags, the experimental variogram is determined. Variograms tend to have small semivariances at small separation distances. Theoretically, the variogram is zero at separation distance zero, but if the variogram is extrapolated to zero separation distance it usually is observed to have a positive intercept, called the nugget variance. As the separation distance (lag) increases the semivariance also increases, but often there is a certain separation distance (usually referred to as the range) beyond which the semivariance remains at a constant value equating to the nugget plus sill variance. This often corresponds to the sample variance.

The experimental variogram is discontinuous and does not necessarily meet the conditionally non-negative definite criteria, which ensures the variance is non-negative (Cressie, 1993). To achieve a continuous representation of the variogram, a functional form is used. There are several potential variogram functions although not all are admissible (Christakos, 1984). The model that matched the minimum air temperature data best (in a weighted least squares sense) was the exponential variogram. The other models assessed were the squared exponential model (often called the Gaussian model), the linear model and the spherical model (Cressie, 1993).

The exponential variogram is given by

$$\gamma(\mathbf{h}) = c_0 + c\left(1 - e^{\left(\frac{-\mathbf{h}}{r}\right)}\right) \tag{5}$$

where c_0 is the nugget variance, c the sill variance and r the distance parameter (a working range is $3r$). By fixing the functional form in advance, the variogram model was fitted to the experimental variogram by optimizing three parameters (c_0, c and r). The function to be minimized is non-linear in r and hence could not be fitted using linear ordinary least squares techniques. A conjugate gradient algorithm (Bishop, 1995) minimizes the weighted residual sum of squares given by

$$\sum_{k=1}^{n_l} |N(\mathbf{h})| \left(\frac{\hat{\gamma}(\mathbf{h}_k)}{\gamma(\mathbf{h}_k)} - 1\right)^2 \tag{6}$$

where the experimental variogram is in the denominator, the fitted variogram model is in the numerator and n_l is the number of lag classes. Figure 7.3 shows an example of an experimental variogram and the fitted variogram model.

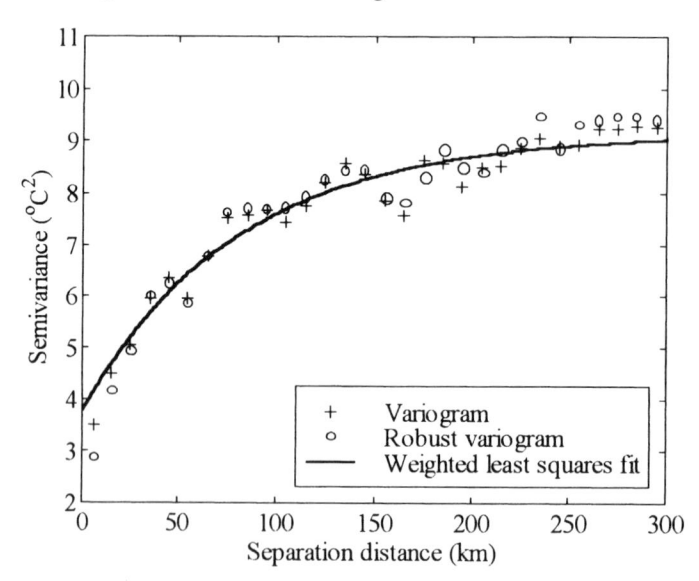

Figure 7.3 An example of an experimental variogram (+) and a robust experimental variogram (o) together with the weighted least squares fit of the exponential model to the experimental variogram, for the first training set of day 42

The experimental variogram is known to be sensitive to outliers in the data set, since it is calculated as the sum of squared differences (Krige and Magri, 1982). Earlier work had highlighted the possible existence of outliers in the data set. These were removed prior to the analysis described in this chapter. In order to verify that the experimental variograms were not affected by outliers, the robust ('fourth root') variogram of Cressie and Hawkins (1980) was computed. This can be seen to be very similar to the 'usual' experimental variogram (Figure 7.3), suggesting that outliers are not a major problem (all days were checked visually for this condition). Strictly **h** is a vector; however, examination of the variograms along different directions (two examples are shown in Figure 7.4) established that the residuals were sufficiently isotropic to use a scalar h.

Having computed the experimental variograms and fitted model parameters, the residuals were interpolated using ordinary kriging (Deutsch and Journel, 1992; Cressie, 1993). The interpolated values of the residuals were then added to the regression-based temperature surface produced earlier (Equation 1) to yield the final model predictions (see Figure 7.2).

In order to analyse the performance of the model, a test set of approximately 70 observations was removed from the full data set of 570 observations, leaving approximately 500 observations in the training set. The aim was to test the predictions of the model on unseen data (i.e., the test set). Eight unique, randomly selected test sets were sequentially extracted from the full data set. The data remaining after the removal of the ith test set was assigned to the ith training set. Thus the model was trained (i.e., the parameters are estimated) eight times with 'overlapping' training sets. By examining the model parameters computed from the eight different training sets model stability can be assessed.

7.3 Performance of the Model over a Short Study Period

To understand the behaviour of the model (Equation 1), particularly with reference to the impact of the synoptic situation, it was studied over a short period. This period, from 26 November to 23 December 1991, was chosen since it exhibited a wide range of synoptic conditions. In this chapter the synoptic situation refers to the position of the cells of mid-latitude anticyclones and cyclones (with their attendant fronts) that govern the weather experienced by Great Britain. It also contained the day (13 December – day 44) for which the model performed worst in terms of the root mean square prediction error over the winter 1991–92. Before any analysis was undertaken, the period was divided into similar synoptic types, based on a subjective, expert view of the synoptic charts and the synopsis of the Monthly Weather Logs for November and December 1991. This resulted in the identification of five periods of distinct weather:

1. *Days 26 to 29*. Largely south-westerly flow, with fronts crossing the country from the west; tropical maritime or continental airmasses.
2. *Days 30 to 35*. Largely anticyclonic, dry and cloudy; tropical continental airmass.
3. *Days 36 to 42*. Anticyclonic with light easterly winds and clear skies (some freezing fog – especially in the south); polar continental airmass.
4. *Days 43 to 46*. Anticyclonic over southern Great Britain but cloudy/windy over

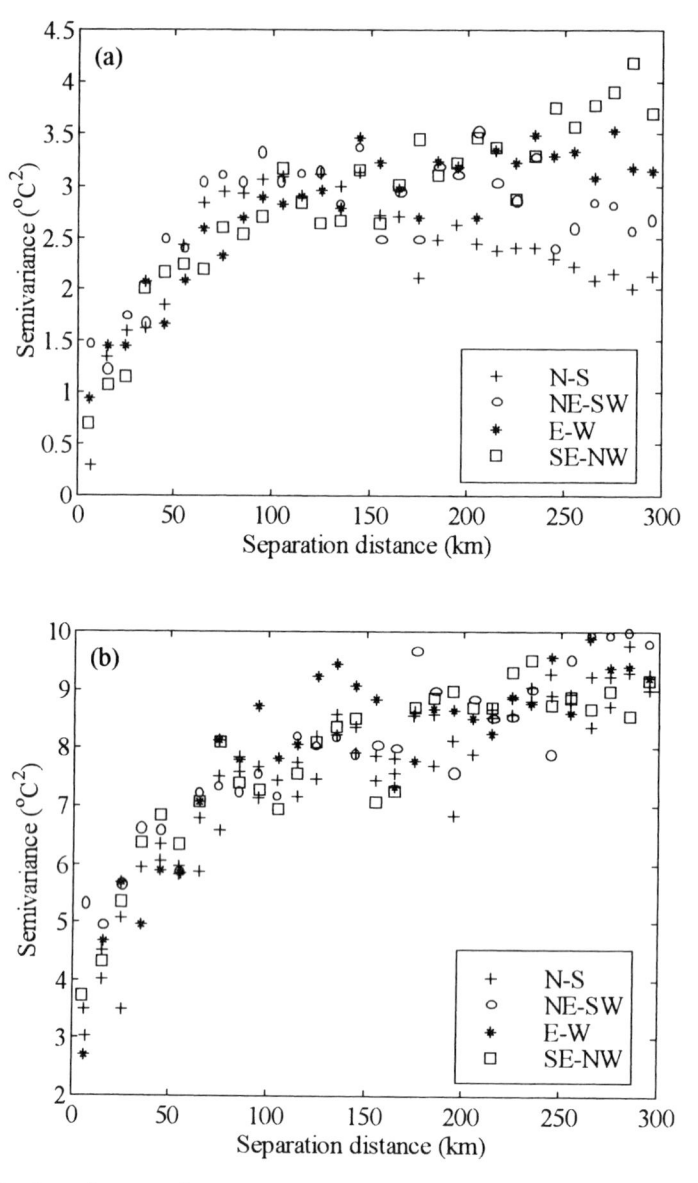

Figure 7.4 Four directional, experimental variograms for (a) day 31 and (b) day 42. The first training set is used

northern regions; largely polar continental airmass.
5. *Days 47 to 53.* Unsettled westerlies with frequent frontal passages and strong winds; alternating polar maritime and tropical maritime airmasses.

Figure 7.5 illustrates the variation of the estimated model parameters over the short time period. The five synoptic periods identified prior to the analysis are demarcated by solid vertical lines and each day has eight estimates from the different (overlapping) training sets. Also plotted (solid line) is the daily median value of the model parameters, computed from the eight estimates.

Figure 7.5a shows the regression constant over the period examined. This can be interpreted as the temperature that might be expected at the coast by a river in a town at sea level with zero cooling potential and with a location west of the Isles of Scilly ($X = 0$, $Y = 0$). It was not surprising that the minimum air temperatures for this constant were rather high – up to 15 °C in mid-December. For each different training set very similar values for the regression constant were estimated. There was considerable temporal variation and the different synoptic periods did not seem to be well defined (i.e., there were no marked jumps in the regression coefficient associated with changes in synoptic type).

Figures 7.5b and c show the variation of the regression coefficients of easting (X) and northing (Y). The estimation of the variation of minimum air temperature with Y was particularly stable, and generally negative, although days 41 to 45 show positive slopes, as might be expected with an anticyclone dominating over southern Great Britain. This can be explained by the existence of cold polar continental air over southern Britain, while warmer tropical maritime air is advected over north-western Britain (e.g., Meteorological Office, 1978). The impact of X was generally negative, notably so during the cold anticyclonic periods 3 and 4. In the more cyclonic periods, there was a small eastward cooling but this was temporally variable.

Figure 7.5d shows the variation of the regression coefficient in relation to the (logarithm of) distance to coast (CD). This was approximately zero for periods 1, 2 and 5 but strongly negative for periods 3 and 4. This implies that coastal locations were likely to have experienced warmer nights than inland regions, as expected. The relation between minimum air temperatures and (the logarithm of) the distance to drainage features shows the impact of the synoptic situation (Figure 7.5e), being generally zero but positive in the clear anticyclonic period 3. The estimation of the coefficients was rather variable across training sets within each day. It was expected that the estimated regression coefficient would be positive during anticyclonic weather (periods 3 and 4) since under these conditions one might expect katabatic (gravity) flows (Ye *et al.*, 1990). These would cause cooler valley bottoms, which will typically be near drainage features. Thus the distance to drainage variable may be acting as a simple surrogate for a model of katabatic flow.

Figure 7.5f shows the impact of the logarithm of the distance to urban areas on minimum air temperatures and presents a rather complicated picture. In general the impacts were positive, which is not what was expected: urban areas were believed to be generally warmer than rural areas *a priori*. Only days 31, 36 and 44 show the strong negative coefficients expected. On all days the estimated coefficients varied across the different training sets.

The 'cooling potential' (Figure 7.5g) has a more readily interpreted relation with minimum air temperatures: the impact was small for periods 1 and 5 (although day 27 has a negative coefficient), becoming more negative in period 2 and strongly negative for period 3. This coincides with *a priori* expectations that areas believed to have a higher 'cooling potential' would be cooler, particularly under clear sky conditions. It should be borne in mind that the 'cooling potential' also contains information on land

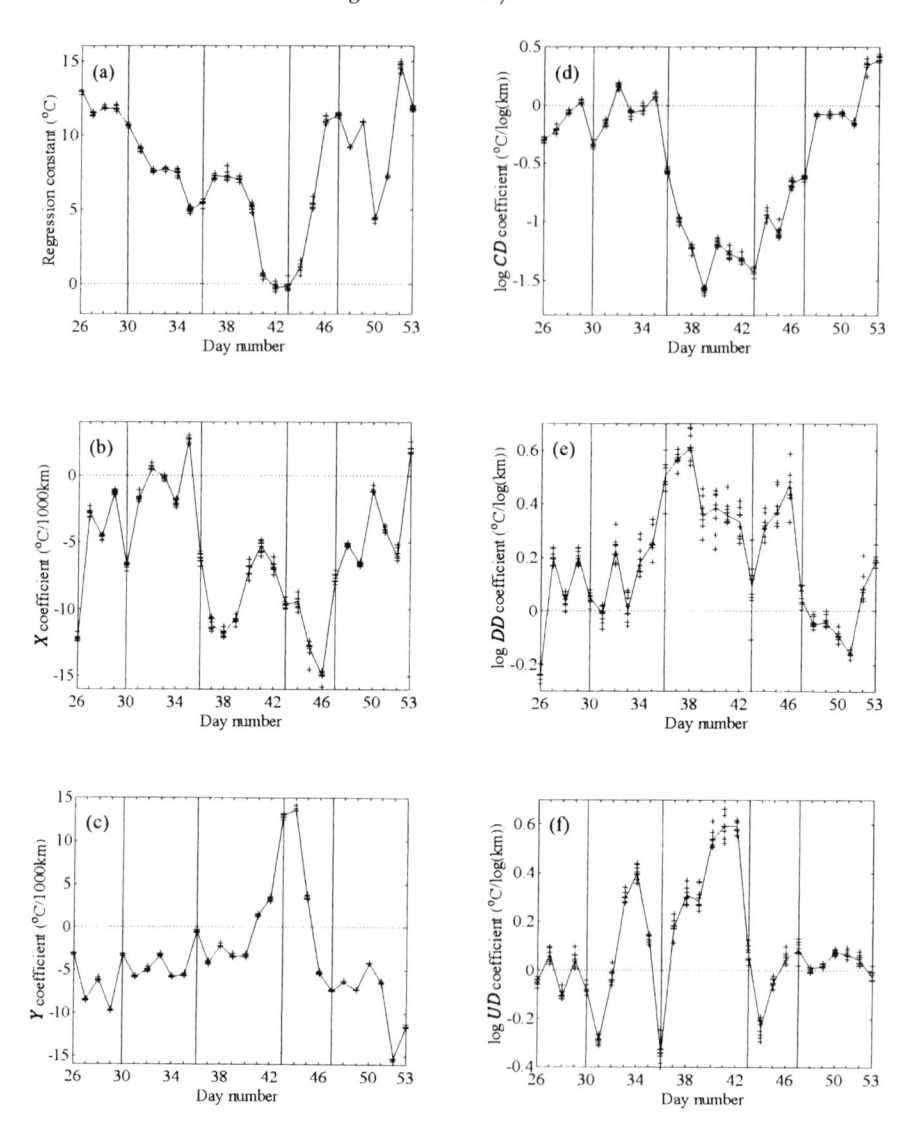

Figure 7.5 Temporal variation of the model parameters for the eight training sets (+) and their daily median (solid line): (a) regression constant, (b) X coefficient, (c) Y coefficient, (d) ln(CD) coefficient, (e) ln(DD) coefficient, (f) ln(UD) coefficient, (g) CP coefficient, (h) Z coefficient, (i) regression R^2, (j) variogram nugget variance, (k) variogram sill variance and (l) variogram distance parameter

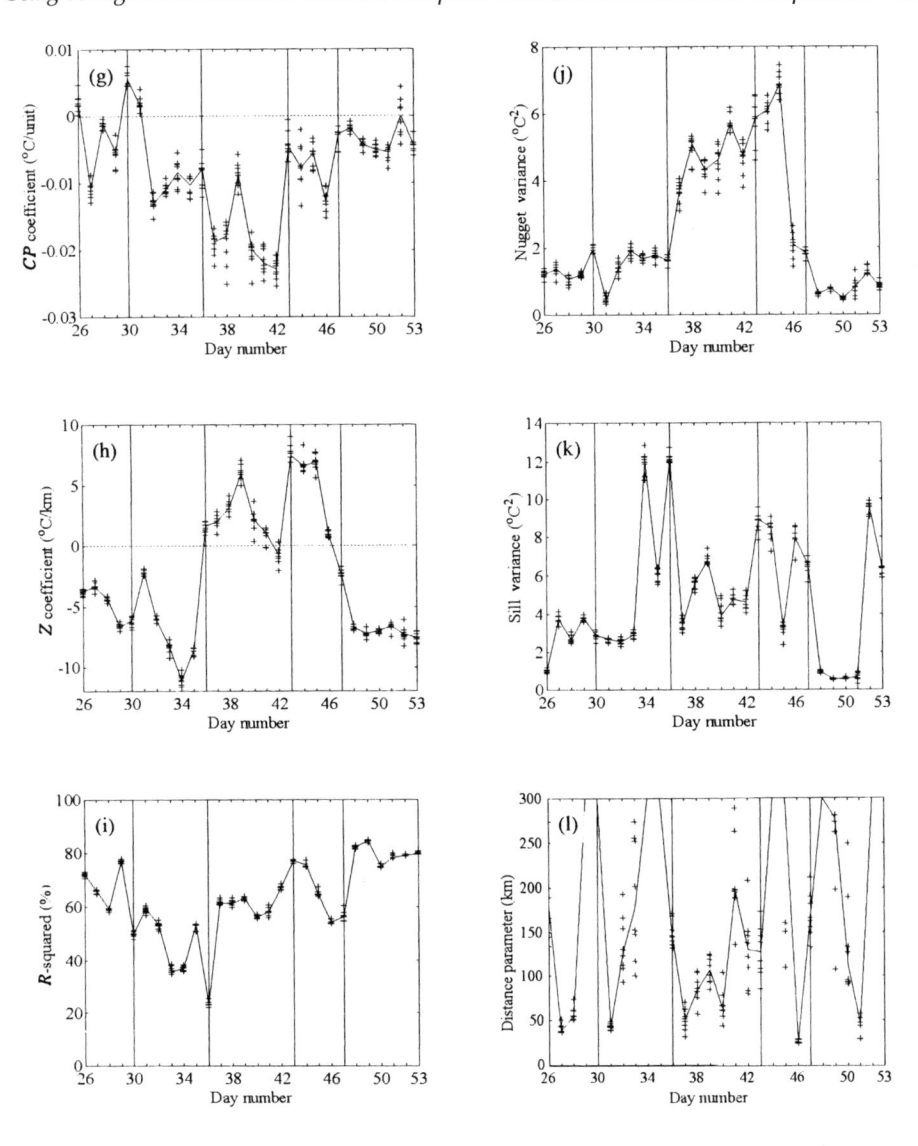

Using Variograms to Evaluate a Model for the Spatial Prediction of Minimum Air Temperature   107

cover and hence the distribution of (and thus possibly some measure of distance to) urban areas.

The impact of elevation is shown in Figure 7.5h. This also separates out the different synoptic periods: periods 1, 2 and 5 had negative slopes (equivalent to lapse rates of -4 to $-9°C/km$), while elevation had a small positive impact on minimum air temperatures during the clear periods 3 and 4. This is what might be expected. During cloudy, windy weather the atmospheric boundary layer is well mixed and the environmental lapse rate is typically $-7.5°C/km$ (Meteorological Office, 1978). During anticyclonic weather a surface inversion is formed and local radiative cooling can dominate, thus reducing the impact of elevation.

Figure 7.5i displays the performance of the regression model (Equation 1), in terms of R^2, showing that approximately 60% of the variance in the observed minimum air temperatures was explained by the regression model. The percentage variance explained ranged from below 30% on day 36 to over 70% on several days, especially in period 5. Day 36 produced an especially small value; however there was no apparent reason for this, although period 2 generally seems to have the smallest R^2 values.

Figures 7.5j, k and l show the variogram parameters for the short study period. The nugget variance (Figure 7.5j) was smaller in periods 1, 2 and 5 and larger in periods 3 and 4. The estimation of the nugget variance was fairly consistent across the different training sets. The sill variance demonstrates similar behaviour (Figure 7.5k), but the elevated values of sill variance on days 52 and 53 are difficult to explain. These two days both saw positive influences of the distance to coast but were otherwise unremarkable in the impact of the terrain variables.

Figure 7.5l shows the variation in the distance parameter of the variograms. This parameter displayed the greatest variation of all the model parameters estimated, with generally long ranges in all periods except 3. These shorter ranges, in period 3, corresponded to variograms showing a more defined spatial structure. Some care must be taken in interpreting these figures of the variogram parameters, since on several days the variograms (which were computed for separation distances up to 300 km) were almost linear in the range zero to 300 km. Since the exponential model can approximate a linear variogram by setting very large (some up to 10^{16} km) distance parameters (and thus also rather large sill variances), the plotted sill variance in Figure 7.5k is the semivariance attained at 100 km (minus the nugget variance) when the distance parameter is greater than 100 km (i.e., the range is greater than 300 km). The distance parameter is also constrained to have a maximum value of 300 km in Figure 7.5l. This convention is only used for the display of results rather than in any model computations. Figure 7.6 shows that different variogram parameters can produce rather similar variograms over the interval zero to 300 km. The eight model and experimental variograms (computed from the eight different training sets) are shown for day 42, which can be seen to have large variability in estimated variogram parameters.

Figure 7.7 shows the final model root mean square error for each day after computing both the regression and kriging components on the eight unseen, non-overlapping, test sets. The average root mean square error is plotted as a solid line and is in the range 0.5–2.2°C, which is small compared to previous studies (Avissar and Mahrer, 1988b; Laughlin and Kalma, 1990; Ishida and Kawashima, 1993) which showed errors of more than 2.0°C on most occasions.

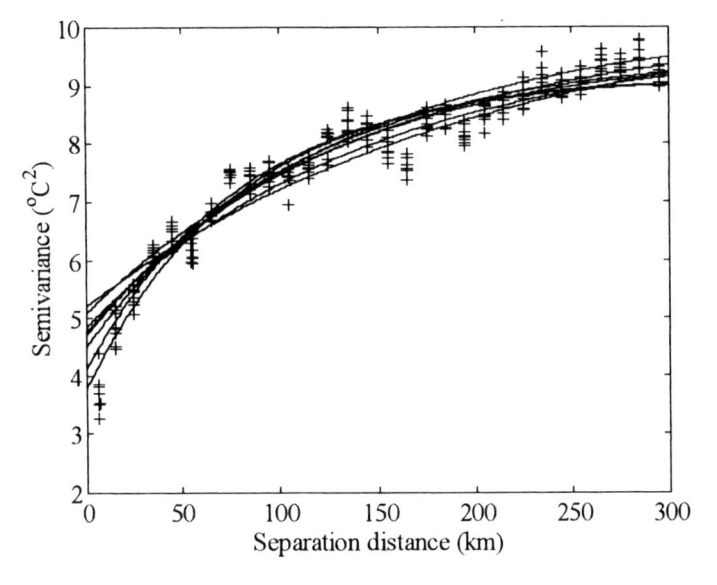

Figure 7.6 The eight experimental variograms and their fitted models for all training sets of day 42

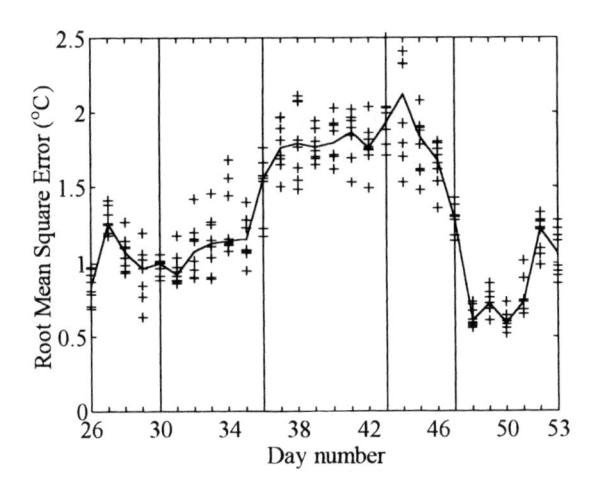

Figure 7.7 The temporal evolution of the root mean square (prediction) errors on the test sets (+) and their daily averages (solid line)

7.4 Analysis of the Model Performance over a Short Study Period

The results in Figure 7.5 indicate that the model made consistent estimates of the regression parameters using different training sets. Some regression coefficients are estimated more consistently across the different training sets than others. Figures 7.5b and c show that the X coefficient has a greater variation (within a given day) than the Y

coefficient. This might be partially attributed to the shape of Great Britain. Both coefficients are of similar magnitude (10°C/1000 km), but the Y variable covers a greater range of values since Britain is about twice as large in the Y (north–south) direction as in the X (east–west) direction. Thus the Y regression coefficient can be more reliably estimated.

The temporal variation of the impact of the terrain variables was identified well, with some apparent anomalies. Figure 7.5e shows the impact of the distance to the nearest drainage feature. The large positive impact in period 3 was expected, since there was a large, cloudless anticyclone over Great Britain. Thus the presence of cold air in valley bottoms (i.e., near to drainage features) has a sensible physical interpretation. More puzzling are the large positive ln(DD) coefficients in period 4 when the anticyclone was moving south. For a large bulk of the country this coefficient should have been positive, but over Scotland there ought to have been no effect, causing the estimated impact to be reduced. However, this was not the case, possibly because of the better defined drainage network in Scotland (it has fewer, larger rivers, on the Bartholomew's rivers coverage used) which gives generally larger distances to drainage features in (the warmer) Scotland. This is always a danger with regression models where there may be latent (unobserved) variables causing the variation in minimum air temperatures for which our input (terrain) variables are acting as proxies (Draper and Smith, 1981, p. 295).

The nugget variances (Figure 7.5j) showed the expected pattern of temporal variation, being generally less than $1°C^2$ during periods 1, 2 and 5 and of the order of $5°C^2$ during periods 3 and 4. The difference between the two groups of periods relates two factors: the impact of each climate station's microclimate and the ability of the regression model to account for the spatial variation in temperatures. The impact of a station's microclimate on observations of minimum air temperatures has never been quantified explicitly, although a reasonable guess might be $\pm 2°C$ which would imply a microclimate effect of around $1°C^2$. Thus on many occasions the nugget variance may well be approximating the microclimate impact (plus errors in the data). The fact that the microclimate varies with synoptic type is to be expected, since the impact of a location's microclimate, which is driven by the local energy balance, is greatest during calm, clear sky conditions and least during windy, cloudy conditions. The high nugget variances observed in periods 3 and 4 suggest that microclimate and very local influences can have a large effect on the ability of the climate station network to resolve the spatial variation of minimum air temperatures (given the terrain variables used in this study). It does not seem possible that the large nugget variances could be attributed to enhanced observational errors in the minimum air temperature data since this is thought to be temporally stable and thus less than the minimum observed nugget variance of $0.5°C^2$.

The sill variances reflected a similar picture, with the largest variances occurring during anticyclonic conditions (periods 2, 3 and 4). Period 3 is also the only one with clearly defined distance parameters (and hence ranges). Thus during period 3 the modelled variograms implied considerable spatial structure in the residuals from the regression model suggesting that some (terrain) factor, which was not included in the regression model, was causing this variation. During periods 1 and 5, in particular, the sill variances are small, indicating that there is little spatial structure in the regression

Using Variograms to Evaluate a Model for the Spatial Prediction of Minimum Air Temperature 111

residuals, and thus that the regression model has largely captured the spatial variation of minimum air temperatures.

Figure 7.8 shows some examples of the experimental and fitted variogram models for selected days. Both Figures 7.8a (day 28) and 7.8f (day 50) are characteristic of periods 1 and 5 (westerlies). The exponential model fits well and the variograms are consistently estimated across the different training sets. The nugget variance is generally small, indicating microclimate effects on minimum air temperatures are small.

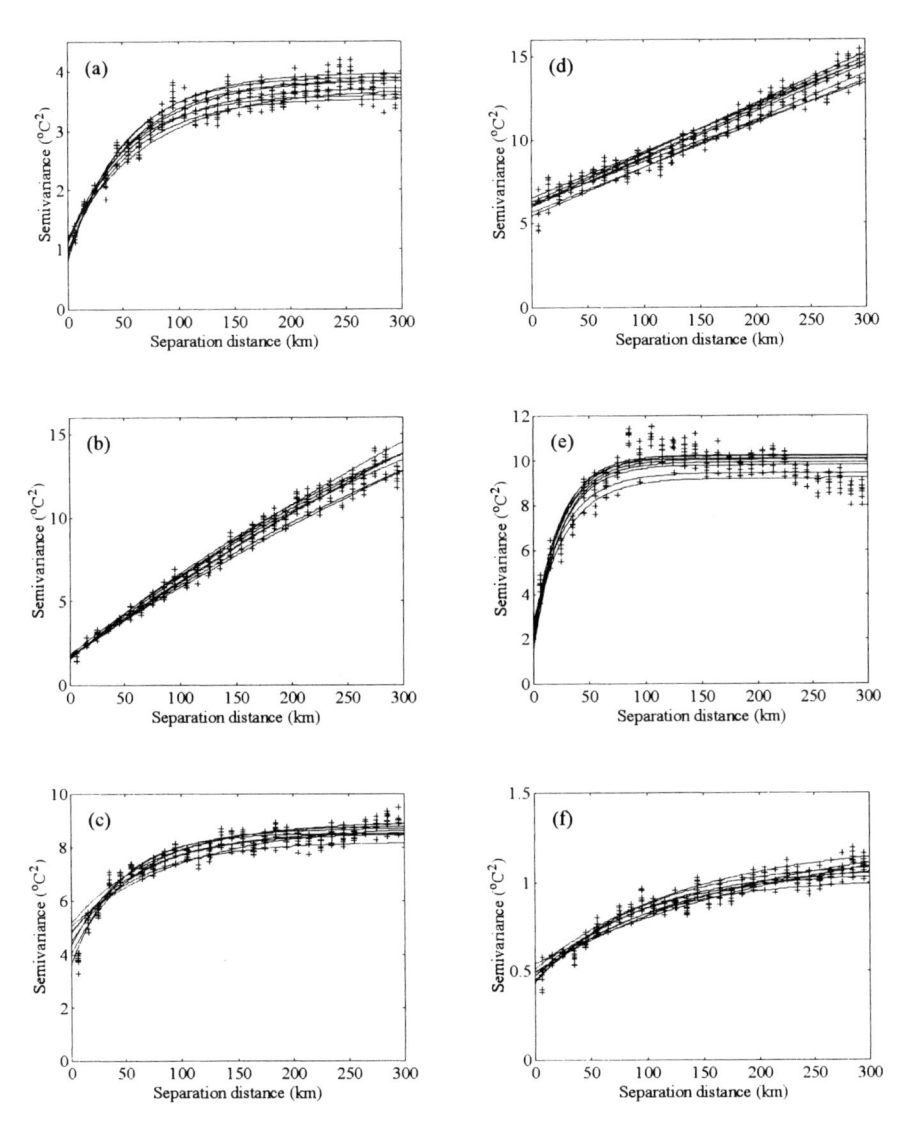

Figure 7.8 *Experimental and modelled variograms for the eight training sets on: (a) day 28, (b) day 34, (c) day 40, (d) day 44, (e) day 46 and (f) day 50*

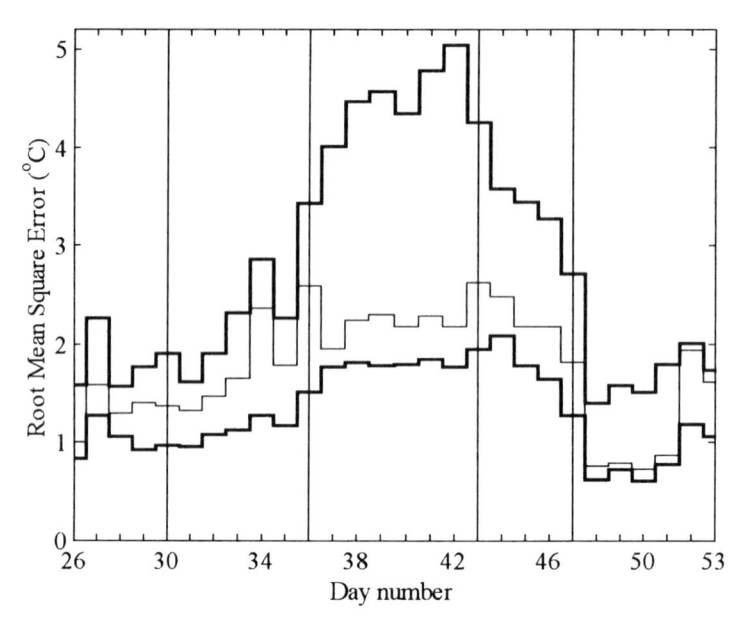

Figure 7.9 *The temporal evolution of the daily average root mean square (prediction) error, of the model components, on the test sets. Upper solid line: error after applying the first-order polynomial in X and Y, middle (thin) line: error after applying the regression model (Equation 1), and, lower solid line: error after applying the full model (Equation 1 and kriging the residuals)*

The sill variance is also rather small (especially on day 50) indicating that the regression model is accounting for much of the spatial variation in minimum air temperatures.

The effect of fitting the exponential model to variograms that exhibit linear behaviour over the lag intervals examined can be seen in Figures 7.8b (day 34) and 7.8d (day 44). The variogram models fit the experimental variograms well. It is thought that if the experimental variogram were estimated over lags greater than 300 km a sill would be reached (there may be evidence of this in Figure 7.8b where the final three lags at 270 to 300 km appear to be reaching a sill). Thus the exponential model is still theoretically sensible since the semivariance is bounded. These variograms suggest a large degree of spatial autocorrelation in the residuals from the regression model. Figures 7.8c (day 40) and 7.8e (day 46) are characteristic of what might be expected during anticyclonic weather. Both exhibit large sill variances, with pronounced spatial structure at ranges of 150 to 300 km (which correspond to distance parameters of 50 to 100 km). This suggests some relatively small (~ 200 km) scale physical process (which may represent the impact of some unmeasured terrain variable) is generating this autocorrelation.

7.5 Theoretical Implications

Several hypotheses could be formulated by examining the performance of the minimum air temperature model over this short period. If the synoptic periods showed marked contrasts in the regression parameters then it might well be possible to forecast (temporally) the spatial distribution of minimum air temperatures using only the regression model, where the regression coefficients were determined by the forecast synoptic type. Unfortunately, the variation within similar synoptic types suggested that forecasting (temporally) would require a great deal of further work.

More usefully, several observations can be made about the geostatistical analysis. The geostatistical part of the model accounts for two features of the temperature field: the microclimate and the spatial autocorrelation in the regression residuals. In this work the *a priori* assumption was that the spatial distribution of daily minimum air temperatures could be modelled using a first-order polynomial in location to represent large-scale 'airmass' variations of temperature and a linear regression model in terrain variables to represent the interaction of the atmosphere with the local terrain, together these being referred to as the regression model (Equation 1).

The geostatistical analysis of the residuals from this regression model has two possible interpretations. It could be that the geostatistical analysis is regarded as a 'structural analysis' where the basic aim is the identification of the spatial structure of the autocorrelated variable, generally using variograms. The determined structure (variogram) is then used in spatial prediction (kriging), although this type of analysis is usually applied directly to the variable of interest rather than residuals. This work not only applies knowledge of the variogram when kriging the residuals, but also interprets the autocorrelation structure to enhance our understanding of the performance of the regression model. The spatial autocorrelation, especially in periods 3 and 4, suggests that there is a physical process not represented in the regression model, which influences the spatial distribution of minimum air temperatures.

Identifying terrain variables that exert an influence on minimum air temperatures is arguably as important as the interpolation algorithm used. The kriging step improved the root mean square (prediction) error of the model (applied to the unseen test sets) by between 0.1 and 1.0°C (Figure 7.9). Figure 7.9 shows the root mean square errors associated with the different parts of the model when applied to the test sets. The upper solid line shows the prediction errors that might be expected if only a first-order polynomial regression model (in X and Y) were used to predict minimum air temperatures. In periods 1, 2 and 5 this very simple model leads to prediction errors which are generally less than 2°C; however, in periods 3 and 4 very large errors of up to 5°C result. If the terrain regression is included in the model, prediction errors shown by the thin middle line are obtained. This can be seen to greatly reduce the prediction errors, notably during periods 3 and 4. The lower thick line gives the prediction error of the full model, which includes the regression model and the final kriging step. As expected, this always produces smaller prediction errors than the regression model alone; that is kriging always improves the prediction accuracy. The distance between the middle thin line and the lower thick line gives the increases in prediction accuracy which can be obtained by kriging the residuals.

The interpretation of trend (regression model) and autocorrelation in this work is similar to that expressed by Cliff and Ord (1981) where the emphasis is on regression, the residuals from this being analysed to assess the degree of autocorrelation. Indeed, Ripley (1981) states:

The main thrust of research [into the application of auto-regressive models] has been into investigating tests of no correlation among the observations or among residuals from a regression of the observations on explanatory variables. Indeed, the philosophy adopted seems to have been that if 'spatial autocorrelation' is found more explanatory variables should be introduced until it disappears!

It is interesting that the geographically driven 'autocorrelation' texts (of which Cliff and Ord (1981) is a classic example) and the geostatistical texts (e.g., Journel and Huijbregts, 1978) present two different interpretations of similar problems. The difference stems from the phenomena that interested geostatistical and geographical workers. Geostatistics was largely geologically initiated (Upton and Fingleton, 1985; Cressie, 1990), being ultimately interested in mapping the grade of ore given samples from boreholes. Depending on the genesis of the ore body, there were generally few covariates, thus most research effort went into predicting the spatial distribution of a variable from one realization. The processes which were associated with the formation of the ore body would often be diffusive and hence well modelled using the random function concept.

The geographical studies were typically interested in assessing measures of human wealth/comfort and relating these to census variables as a tool for evaluating social policy (Cliff and Ord, 1981; Ripley, 1981). Thus, their primary interest was identifying those (census) variables that were linked to a particular phenomenon (and hence producing models of social conditions). Correlation in the residuals was therefore taken to imply that their model had omitted variables and that additional variables might produce a better model.

Thus when mapping minimum air temperatures it appears that an interpretation along the lines of the geographical studies is plausible, in that the degree of spatial autocorrelation in the residuals from the regression model varies with synoptic type. This suggests that there is little intrinsic correlation in the minimum air temperature field above the minimum variogram lag separation distance of ~ 8 km. The autocorrelation observed through the variograms can be attributed to the impact of terrain variables not included in the model. These terrain variables are especially important during clear, anticyclonic conditions. In the atmosphere, however, turbulent and diffusive mixing means that over small separation distances the random function model for minimum temperatures will be appropriate.

7.6 Conclusions

The model (Equation 1) has been shown to predict daily minimum air temperatures over Great Britain with root mean square errors of 0.5 to 2.0°C on independent test sets. The prediction accuracy can be seen to vary with synoptic weather type. During cloudy, windy weather, altitude and location dominated the spatial distribution of minimum air temperatures. Clear, anticyclonic conditions caused more complex spatial distributions of minimum air temperature, with terrain variables (notably the logarithm of distance to coast, distance to drainage feature and 'raw' land cover derived 'cooling potential') becoming more important.

During cloudy, windy weather the variograms of the residuals from the regression model had small nugget and sill variances. This implies that the regression model was encompassing most of the variation in minimum air temperatures. The remaining variance can be largely explained by temperature variation at distances less than the sampling interval (microclimate) and errors in the data (which are possibly large – especially in the terrain data).

During anticyclonic weather the variograms generally had a well-defined structure with larger nugget variances. This leads to the interpretation that the nugget variance is effectively representing the microclimate plus errors in the data. After fitting the regression model, the spatial structure remaining in the minimum air temperature field is attributed to terrain factors missed in the regression model. It is suggested that if a complete set of terrain variables were used then, much as during the cloudy, windy conditions, the variograms would tend to pure nugget (i.e., a variogram which has a constant value at all lags). This interpretation is more akin to the geographical studies of autocorrelation (Cliff and Ord, 1981) than classical geostatistics (Journel and Huijbregts, 1978). This conclusion is appropriate to the spatial separation distances of the 'climate station' network and is consistent with a stochastic interpretation of minimum air temperature as being a random function.

The model could be greatly improved in many ways. Ordinary least squares estimation of the regression coefficients followed by a method of moments estimation of the variogram is known to bias the model parameter estimates (Lamorey and Jacobson, 1995). Model parameters could be estimated using a generalized least squares algorithm, together with a parametric form for the covariance (Cressie, 1993), and the parameters could be estimated using iterative techniques (e.g., Upton and Fingleton, 1985, p. 367). The accuracy of the predictions might be further improved by considering the temporal dimension in the model, since the current model treats each day as being totally independent. Further work could seek to determine additional terrain variables, such as soil information or the spatial distribution of cloudiness, which may improve the spatial prediction of minimum air temperatures. The conclusions about the importance of various terrain variables could be strengthened by performing planned experiments over small, intensively monitored study areas.

Acknowledgements

The author would like to thank the anonymous reviewers and Dr Peter Atkinson for the helpful comments and suggestions which greatly improved the chapter. The work was supported by a studentship from the School of Geography at the University of Birmingham. I would like to thank my supervisor, Dr John E. Thornes, for his input and support.

References

Aspinall, R.J. and Matthews, K.B., 1994, Climate-change impact on distribution and abundance of wildlife species – an analytical approach using GIS, *Environmental Pollution*, **86**, 217–223.
Avissar, R. and Mahrer, T., 1988a, Mapping frost-sensitive areas with a three-dimensional

local-scale numerical model. Part I: Physical and numerical aspects, *Journal of Applied Meteorology*, **27**, 400–413.

Avissar, R. and Mahrer, Y., 1988b, Mapping frost-sensitive areas with a three-dimensional local-scale numerical model. Part II: Comparison with observations, *Journal of Applied Meteorology*, **27**, 414–426.

Bishop, C.M., 1995, *Neural Networks for Pattern Recognition* (Oxford: Oxford University Press).

Christakos, G., 1984, On the problem of permissible covariance and variogram models, *Water Resources Research*, **20**, 251–265.

Cliff, A.D. and Ord, J.K., 1981, *Spatial Processes: Models and Applications* (London: Pion).

Cornford, D., 1996, The development and application of techniques for mapping daily minimum air temperatures, PhD thesis, University of Birmingham, UK.

Cornford, D. and Thornes, J.E., 1996a, A comparison between spatial winter indices and expenditure on winter road maintenance in Scotland, *International Journal of Climatology*, **16**, 339–357.

Cornford, D. and Thornes, J.E., 1996b, Winter severity indices, geostatistics and GIS, *Proceedings of the 8th International Road Weather Conference* (Birmingham: University of Birmingham), 27–36.

Cressie, N., 1986, Kriging nonstationary data, *Journal of the American Statistical Association*, **81**, 625–634.

Cressie, N., 1990, The origins of kriging, *Mathematical Geology*, **22**, 239–252.

Cressie, N., 1993, *Statistics for Spatial Data* (New York: Wiley).

Cressie, N. and Hawkins, D.M., 1980, Robust estimation of the variogram, *Mathematical Geology*, **12**, 115–125.

Deutsch, C.V. and Journel, A.G., 1992, *GSLIB – Geostatistical Software Library and User's Guide* (Oxford: Oxford University Press).

Draper, N.R. and Smith, H., 1981, *Applied Regression Analysis* (New York: Wiley).

ESRI, 1992, *GRID Command References* (Redlands, CA: Environmental Systems Research Institute Inc.).

Fuller, R.M., Groom, G.B. and Jones, A.R., 1994, The land-cover map of Great Britain – an automated classification of LANDSAT Thematic Mapper data, *Photogrammetric Engineering and Remote Sensing*, **60**, 553–562.

Hudson, G. and Wackernagel, H., 1994, Mapping temperature using kriging with external drift: theory and an example from Scotland, *International Journal of Climatology*, **14**, 77–92.

Ishida, T. and Kawashima, S., 1993, Use of cokriging to estimate surface air temperature from elevation, *Theoretical and Applied Climatology*, **47**, 147–157.

Johns, D., 1996, The MOORI. A new winter index for winter road maintenance, *Proceedings of the 8th International Road Weather Conference* (Birmingham: University of Birmingham), 22–26.

Journel, A.G. and Huijbregts, C.J., 1978, *Mining Geostatistics* (London: Academic Press).

Kawashima, S. and Ishida, T., 1992, Effects of regional temperature – wind speed and soil wetness on spatial structure of surface air temperature, *Theoretical and Applied Climatology*, **46**, 153–161.

Kitanidis, P.K., 1993, Generalized covariance functions in estimation, *Mathematical Geology*, **25**, 525–540.

Krige, D.G. and Magri, E.J., 1982, Studies of the effects of outliers and data transformation on variogram estimates for a base-metal and a gold ore-body, *Mathematical Geology*, **14**, 557–564.

Lamorey, G. and Jacobson, E., 1995, Estimation of semivariogram parameters and evaluation of the effects of data sparsity, *Mathematical Geology*, **27**, 327–358.

Laughlin, G.P. and Kalma, J.D., 1990, Frost risk mapping for landscape planning: a methodology, *Theoretical and Applied Climatology*, **42**, 41–51.

Lennon, J.J. and Turner, J.R.G., 1995, Predicting the spatial distribution of climate: temperature in Great Britain, *Journal of Animal Ecology*, **64**, 370–392.

Levin, S.A., 1992, The problem of pattern and scale in ecology, *Ecology*, **73**, 1943–1967.

Linacre, E., 1992, *Climate Data and Resources* (London: Routledge).

Matheron, G., 1973, The intrinsic random functions and their application, *Advances in Applied Probability*, **5**, 439–468.
Meteorological Office, 1956, *Observer's Handbook* (London: HMSO).
Meteorological Office, 1978, *A Course in Elementary Meteorology* (London: HMSO).
Meteorological Office, 1991, *January 1991, Monthly Weather Report* (London: HMSO).
Pulpitlova, J., Matiasovsky, P., Nakamura, H. and Darula, S., 1991, The issues of urban and building climatology for building science, *Energy and Buildings*, **15**, 399–405.
Ripley, B.D., 1981, *Spatial Statistics* (New York: Wiley).
Shaviv, E., 1984, Climate and building design – tradition, research and design tools, *Energy and Buildings*, **7**, 55–69.
Smith, K., 1992, *Environmental Hazards: Assessing Risk and Reducing Disaster* (London: Routledge).
Tabony, R.C., 1985, Relations between minimum temperature and topography in Great Britain, *Journal of Climatology*, **5**, 503–520.
Upton, G.J.G. and Fingleton, B., 1985, *Spatial Data Analysis by Example. Volume 1: Point Pattern and Quantitative Data* (New York: Wiley).
Wackernagel, H., 1995, *Multivariate Geostatistics* (Berlin: Springer-Verlag).
Ye, Z.J., Garratt, J.R., Segal, M. and Pielke, R.A., 1990, On the impact of atmospheric thermal stability on the characteristics of nocturnal downslope flows, *Boundary-Layer Meteorology*, **51**, 77–97.

8
Modelling the Distribution of Cover Fraction of a Geophysical Field

John B. Collins and Curtis E. Woodcock

8.1 Introduction

A recurring problem in remote sensing is identification of the extent of a geophysical field within an image or study area. The term 'field' refers to any entity which exists over some spatial domain. The problem addressed here is the separation of a 'target field' from a contrasting 'background field' corresponding to all other scene elements. A typical approach to the problem is the application of a threshold to remotely sensed data or some quantity derived from it, for example a vegetation index such as NDVI. The pixels which are identified as being covered by the field are those whose value exceeds the chosen threshold.

Examples of this type of analysis include the identification of clouds from remotely sensed images (Hutchison and Hardy, 1995; Shin *et al.*, 1996), usually using imagery with relatively coarse spatial resolutions. Key *et al.* (1994) also discuss the issue in the context of identifying the extent of fractures in sea ice, which has important implications for estimates of sensible and latent heat fluxes in polar regions. Thesholding of Landsat Thematic Mapper (TM) and AVHRR images has been used in arid zones to quantify the extent of ephemeral water bodies (Verdin, 1996), with important implications for ecological modelling. Potential applications in forest ecosystems are plentiful. For example, estimation of canopy closure, or the proportion of an area covered by tree crowns, is important because of its relevance to forest management and its relationship to deforestation and habitat fragmentation, especially in tropical zones (Malingreau *et al.*, 1989). Frequently, change detection is undertaken by thesholding

---

Advances in Remote Sensing and GIS Analysis. Edited by Peter M. Atkinson and Nicholas J. Tate.
© 1999 John Wiley & Sons Ltd.

an image of a 'change indicator' derived from multitemporal imagery (Coppin and Bauer, 1994).

Certainly the spectral properties of the target and background are relevant to any thresholding operation. The spectral waveband or vegetation index chosen for the thresholding should be one in which the target field is easily distinguished from other scene elements. Besides the spectral qualities of the scene, the other main consideration in a thresholding operation is the spatial resolution of the sensor. The sensor Field of View (FOV) interacts in a complex way with the spatial pattern of the target field. When the FOV is small enough that pixels tend to be homogeneous with respect to the presence or absence of the field, results of a thresholding operation are straightforward. However, when the FOV gets larger, pixels tend to contain some mixture of target and background. Any thresholding operation which identifies pixels as either 'covered' or 'not covered' will be imprecise.

The problem of applying a threshold to a coarse spatial resolution image is that pixels are not characterized by complete presence or absence of a target field, but instead contain a 'cover fraction', or proportional covered area. When the cover fraction is high, a pixel will be more spectrally similar to a 'pure' example of the target field than will a pixel with a low cover fraction. Thus selection of a threshold on the image data corresponds to selection of some critical cover fraction above which a pixel will be identified as covered and below which the pixel will be identified as not covered. The calculated extent of the target field in the image corresponds to the number of pixels whose cover fraction exceeds the chosen threshold.

The question arises as to the accuracy of the area estimates made by such methods. To answer this question, Key (1994) constructed a mathematical model relating the spatial pattern of a geophysical field, pixel size and the distribution of the cover fraction. Since the factors influencing cover fraction are not generally known *a priori*, a stochastic model was chosen to describe its spatial variability. Thus the model takes the form of a probability distribution function (pdf). A critical parameter of this pdf is its variance. The paper by Key (1994) modelled spatial structure of the target field using an exponential autocovariance function, and used this measure to estimate the indicator variance as a function of FOV size. This chapter presents the same type of analysis, but uses variograms of arbitrary forms to characterize spatial structure. Such a mathematical model can be used to test the sensitivity of the area estimates as a function of the threshold employed and the sensor spatial resolution. This chapter outlines the theory and methods of such a modelling approach, and presents an application to the problem of estimating the area covered by snow in a boreal forest.

8.2 Mathematical Modelling of Cover Fraction Images

Mathematically, a scene can be considered a set of points in an area $D \subset R^2$, a subset of two-dimensional space. The individual points composing D will be denoted **s**, representing spatial position vectors. A number of variables can be defined on the collection of points $\mathbf{s} \in D$, including reflected radiance in specific wavelength bands, vegetation indices, etc. Any quantity which varies as a function of spatial position within the domain set D is called a *regionalized variable* (Matheron, 1963; Journel and

Huijbregts, 1978; Cressie, 1993). For analysis of a geophysical field, a useful type of regionalized variable is an *indicator function* $I(\mathbf{s})$ which takes on a value of 1 if a particular field is present at $\mathbf{s}$ and takes on a value of 0 otherwise. Thus the mathematical object we are dealing with is the regionalized variable denoted

$$\{I(\mathbf{s}): \mathbf{s} \in D\} \qquad (1)$$

The field whose extent is to be quantified is defined as all points for which $I(\mathbf{s}) = 1$.

This simple model is not adequate to describe a remotely sensed image taken over D since image values are not available for the individual points $\mathbf{s}$. Rather, data are available over finite fields of view (FOV), denoted w. The FOV is closely associated with the idea of the measurement *support* used in geostatistics. The term *regularization* describes the process by which values of a regionalized variable are integrated to produce a single value for a support. To describe the regularization of a geophysical field, the support can be modelled as a disk centred at the origin of an arbitrary coordinate system, with a diameter equivalent to the sensor spatial resolution. The regularized indicator function over support w centred at the point $\mathbf{s}$ is denoted $I(\mathbf{s};w)$, where

$$I(\mathbf{s};w) = \frac{1}{|w|} \int_w I(\mathbf{s} + \mathbf{t}) d\mathbf{t} \qquad (2)$$

The variable of integration $\mathbf{t}$ sweeps through all points within the support w, whose area is given by $|w|$. So Equation (2) represents the integrated indicator function over a pixel divided by the pixel's area, which is equivalent to the proportion of the pixel for which $I(\mathbf{s}) = 1$. That is, the quantity $I(\mathbf{s};w)$ represents the cover fraction.

According to a stochastic model, the cover fraction follows some pdf among the pixels in an image. Taking a parametric approach, a specific model for this distribution must be chosen in order to further analyse the properties of the regionalized variable. The beta distribution is a useful choice. For an arbitrary regionalized variable X, let the function $f_X(x)$ describe the associated probability density. The beta distribution is defined as

$$f_X(x) = \frac{\Gamma(\alpha + \beta)}{\Gamma(\alpha)\Gamma(\beta)} x^{\alpha-1}(1-x)^{\beta-1}, 0 < x < 1 \qquad (3)$$

where α and β are parameters, and Γ represents the gamma function

$$\Gamma(z) = \int_0^\infty y^{z-1} e^{-y} dy \qquad (4)$$

The two parameters are related to the mean (μ) and variance (σ^2) of the distribution as follows:

$$\beta = \frac{1-\mu}{\sigma^2} [\mu(1-\mu) - \sigma^2] \qquad (5)$$

$$\alpha = \frac{\mu\beta}{1-\mu} \qquad (6)$$

The beta distribution is well suited to describe proportions, since the random variable is constrained to the interval (0, 1). Additionally, the behaviour of the distribution as a function of its variance accords with what one would expect of distributions of cover fractions under changes of support. This can be seen by considering two extreme cases for the distribution of cover fraction. The first occurs when the support is taken to be the entire image domain. In this case, there is only one 'pixel', and its cover fraction is equal to μ, the mean cover fraction over the scene. Thus the 'probability' of observing this mean value is 1, and the probability of observing anything else is zero. So $f_X(x)$ is a degenerate probability mass function, written as

$$X_X(x) = \begin{cases} 1, & x = \mu \\ 0, & \text{otherwise} \end{cases} \quad (7)$$

Such a distribution has a mean of μ and a variance of $\sigma^2 = 0$.

The other extreme case occurs when supports are infinitesimal. In this case, the only cover fractions which can be observed at a point are 0 (when the target field is absent) and 1 (when the target field is present). The probability of the latter is to equal to the mean cover fraction μ. Such a random variable has a Bernoulli distribution.

$$f_X(x) = \begin{cases} \mu, & x = 1 \\ 1 - \mu, & x = 0 \\ 0, & \text{otherwise} \end{cases} \quad (8)$$

The mean of a Bernoulli distribution is μ, and the variance is $\sigma^2 = \mu(1 - \mu)$.

This discussion points to the fact that over all support sizes, the mean of the cover fraction distribution is unchanging while its variance falls in the interval $(0, \mu(1 - \mu))$. An important characteristic of the beta distribution is that its variance is constrained to the same interval. Furthermore, as the variance of the beta distribution approaches zero, it approximates the degenerate case above. As the variance approaches $\mu(1 - \mu)$, the distribution begins to resemble that of a Bernoulli random variable (Figure 8.1). Thus the beta distribution is at least capable of appropriately describing the extreme cases.

Having adopted the beta model, the precise distribution of cover fraction depends primarily on its variance. The variance, in turn, depends on the relationship between the support size and the spatial pattern of the target field. For example, if the target field contains large homogeneous patches, then the cover fraction variance will decrease slowly as the spatial resolution becomes more coarse. Alternatively, if the target field is such that its indicator function varies over small spatial frequencies, then spatial detail (and hence variance) will decline more quickly as the spatial resolution degrades.

Geostatistical analysis can be used to determine precise relationships between resolution and spatial structure. Specifically, the spatial structure of a regionalized variable can be described by its *variogram*. The variogram $\gamma(\mathbf{h})$ is a function of a vector lag $\mathbf{h}$, and is equal to half of the expected value of the squared difference between observations separated by that lag (Matheron, 1963)

$$\gamma_I(\mathbf{h}) = \frac{1}{2} E\left\{[I(\mathbf{s}) - I(\mathbf{s} + \mathbf{h})]^2\right\} \quad (9)$$

If the spatial structure is assumed to be isotropic, the variogram is a function only of the length h of the lag vector $\mathbf{h}$. In practice, theoretical models are fitted to observed

Modelling the Distribution of Cover Fraction of a Geophysical Field

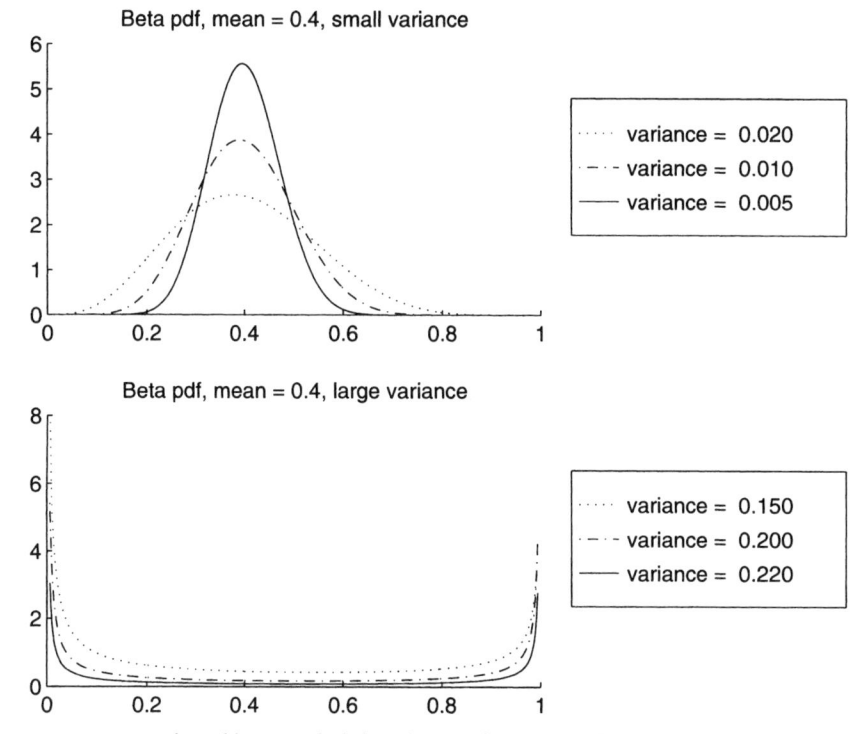

Figure 8.1 Examples of beta probability density functions. The mean of all distributions is $\mu = 0.4$, and the variance is as indicated. The distribution is unimodal when the variance is near zero, but becomes bimodal as the variance approaches its maximum possible value of 0.24

variograms. One such model is the exponential type, which can be written as

$$\gamma_I(h) = c(1 - e^{-h/a}) \tag{10}$$

The parameter c is the *sill* of the variogram, and represents the value that the variogram approaches at large lags. The parameter a is the *range* of the model, and its magnitude is related to the separation distance at which observations are essentially independent. Larger values of a indicate a persistence of autocorrelation at longer lags. The simple exponential model often provides a reasonable fit to data. But several other permissible models exist, such as spherical, Gaussian, and linear. Linear combinations of these simpler models are referred to as 'nested' models, and often provide a better fit to observed data when spatial structure is more complex.

Questions concerning relationships between variance of a regionalized variable and spatial structure can often be answered using the idea of an *average variogram* between sets (Rendu, 1978). Consider two arbitrary subsets, $A, B \subseteq D$ of the domain on which a regionalized variable is defined. Two instances of an FOV could be an example of such sets. The average variogram between A and B is defined as

$$\bar{\gamma}_I(A,B) = \frac{1}{|A\|B|} \int_A \int_B \gamma_I(\mathbf{s} - \mathbf{t}) d\mathbf{t} d\mathbf{s} \tag{11}$$

The average variogram can be used to determine the variance of cover fraction for

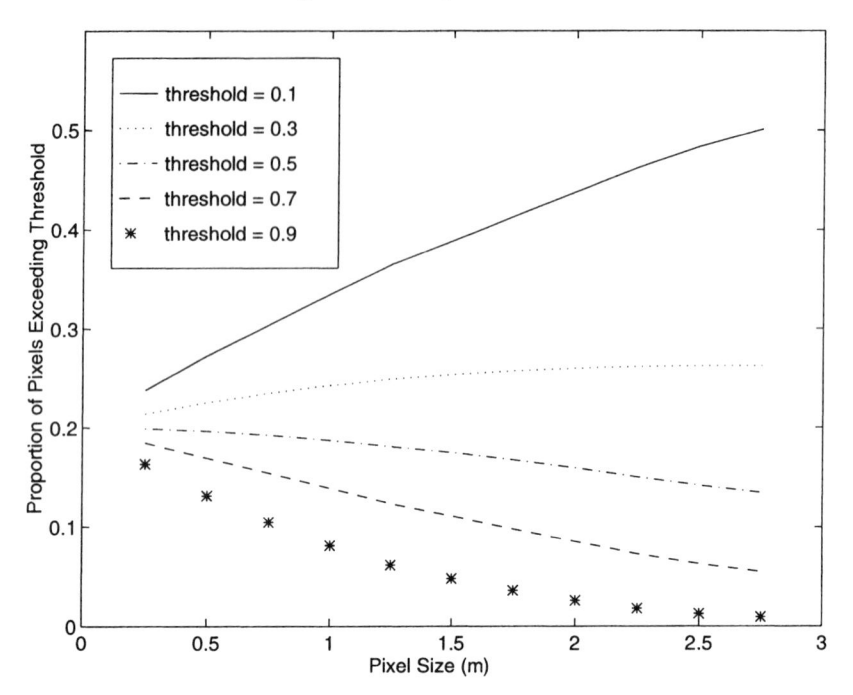

Figure 8.2 Modelled relationships between the proportion of image pixels exceeding a given threshold as a function of pixel size, for several thresholds. At small pixel sizes, all thresholds produce answers close to the true area coverage of 0.2

arbitrary spatial resolutions. Specifically, when a regionalized variable is observed by taking averages over supports w, the variance of these values is known as the *dispersion variance* of w in D, denoted $\sigma^2(w, D)$ (Journel and Huijbregts, 1978). It can be calculated as

$$\sigma^2(w,D) = \bar{\gamma}_I(D,D) - \bar{\gamma}_I(w, w) \qquad (12)$$

The dispersion variance of the indicator function of a target field is equal to the average variogram over the entire domain of the scene, minus the average variogram within a single support. Since the variogram is typically a non-decreasing function of the lag, the term $\bar{\gamma}_I(w,w)$ becomes larger as the field of view represented by w becomes larger, causing dispersion variance to decrease. This equation agrees with the intuitive notion that the variance observed using a coarse spatial resolution sensor is less than would be observed with a fine spatial resolution sensor. But Equation (12) makes this relationship more precise by including the variogram, a measure of spatial structure.

The result of these developments is that, given a variogram for an indicator function and a description of a support, one can determine the variance of the cover fraction. This variance can be used to parameterize a beta distribution, making it a simple matter to determine the proportion of pixels in an image whose cover fraction exceeds a given threshold.

A simple example of this type of analysis was undertaken for a theoretical target field with a mean of $\mu = 0.2$ (and hence a variance of $\sigma^2 = \mu(1 - \mu) = 0.16$). An isotropic exponential model was chosen as a punctual variogram

$$\gamma(h) = 0.16(1 - e^{-h/a}) \tag{13}$$

This implies that the sill is equal to the variance of the data. This is not strictly true, but provides a good approximation in most cases (Barnes, 1991). The range a and the lag h both have units of length. For this simple example, the range is chosen to be $a = 1.0$ m.

Figure 8.2 shows, for several thresholds, how area coverage estimates vary as pixel size changes. When the FOV is small, varying the threshold does not produce large variations in the estimated area coverage. Estimated coverages are all close to the correct value of 0.2 at fine spatial resolutions. As the spatial resolution coarsens, results begin to depend more strongly on threshold. For the coarser spatial resolutions shown here, large errors can result from applying arbitrary thresholds.

Similar data are shown in Figure 8.3. Here the abscissa indicates threshold values, while each curve corresponds to a single pixel size. When the diameter of the FOV is small relative to the range of spatial structure, the curve is nearly flat, indicating that any choice of threshold produces an acceptable answer. As resolution becomes more coarse, the curves become progressively steeper, indicating a greater sensitivity of the results to the selected threshold.

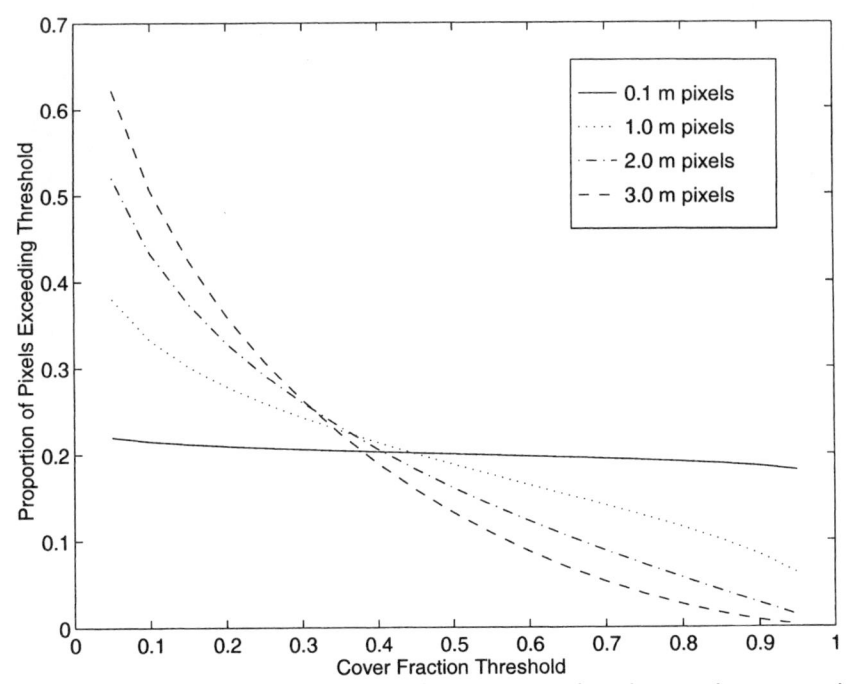

Figure 8.3 Modelled relationships between the proportion of pixels exceeding a given threshold, for several pixel sizes. As pixel size increases, estimated area coverage becomes more sensitive to the chosen threshold

8.3 Analysis of Cover Fraction Images

The example given in the preceding section illustrates some of the problems of deriving area coverage estimates from thresholding operations. This section describes a similar analysis based on a real data set.

One geophysical field of interest in many applications is snow cover. Figure 8.4 shows an image in which pixel values represent the proportion of the pixel for which snow is visible from the viewing position. The dimensions of the image are approximately 25 km × 30 km. The image was created at a spatial resolution of 30 m using spectral mixture analysis based on data from the Landsat TM. The darkest values represent areas which have no snow cover, while the brightest values indicate snow cover of approximately 50%. The mean cover fraction for a pixel (and hence for the whole image) is 0.1675, with an associated variance of 0.015. In terms of the mathematical model developed in the last section, this image represents the variable $I(\mathbf{s};w)$, where w is a 30 m field of view. That is, the image represents a regularized indicator function.

A variogram for this image was calculated from a random sample of approximately 10% of the image pixels, using the classical variogram estimator

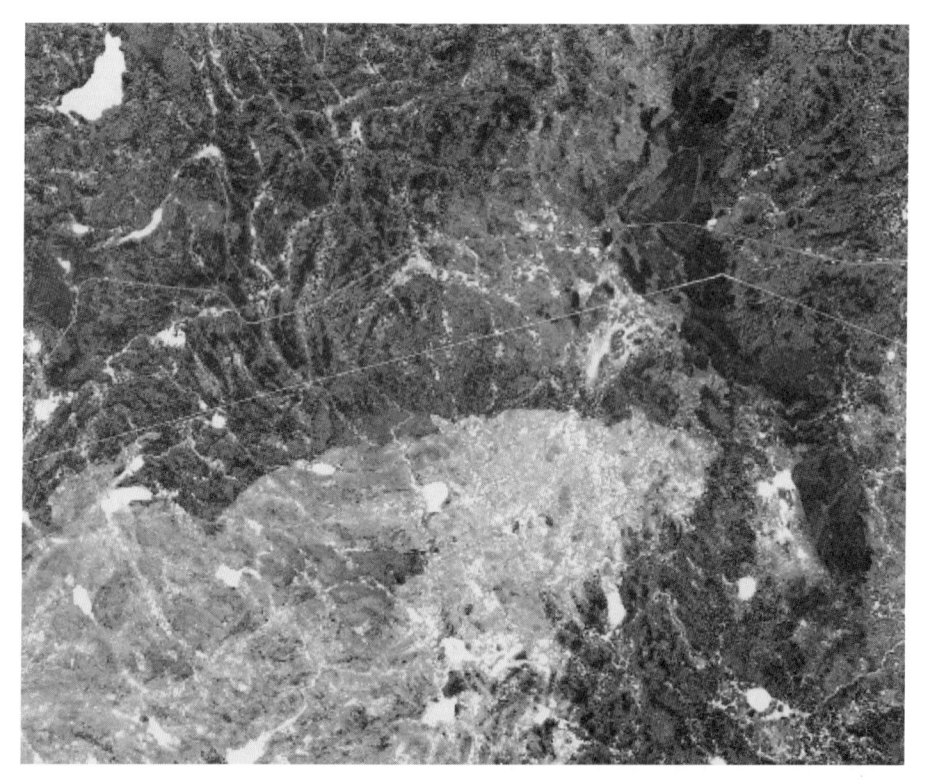

Figure 8.4 A snow cover image. Pixel intensities are proportional to snow cover fraction. The image measures approximately 25 km × 30 km

Modelling the Distribution of Cover Fraction of a Geophysical Field

$$\gamma_I(h;w) = \frac{1}{2N(h)} \sum_{i=1}^{N(h)} [I(\mathbf{s}_i;w) - I(\mathbf{s}_i + \mathbf{h};w)]^2 \tag{14}$$

where $N(h)$ is the number of pairs of points separated by h, represented by $\mathbf{s}_i$ and $\mathbf{s}_i - \mathbf{h}$. It was assumed that the spatial structure in the scene can be modelled as isotropic. Thus the variogram is a function of only the length of the lag vector. Note that since the data values are regularized over the support w, the estimated quantity is itself regularized. The inclusion of w in $\gamma_I(h;w)$ indicates this fact. Figure 8.5 shows a plot of the variogram for lags ranging from zero to 3000 m. A model was fit to the data using a weighted least squares algorithm proposed by Cressie (1985). The best fit was obtained using a nested model consisting of the sum of two exponential functions, and is given by

$$\gamma_I(h;w) = 0.0007 + 0.0050\, e^{-h/1517} + 0.0056\, e^{-h/131} \tag{15}$$

where the lag h is in metres.

The equations determining dispersion variance of cover fraction depend not on the regularized variogram, but on the *punctual* variogram derived from point-scale measurements. Since only regularized observations are available, these equations do not apply directly. The problem of regularization can be dealt with in two steps. First, a punctual variogram must be estimated based on the observed data. Second, the relationships between support size and dispersion variance must themselves be modified slightly.

The main complication arising when dealing with regularized data is the existence of a 'nugget effect', or an apparent non-zero intercept in the regularized variogram. The

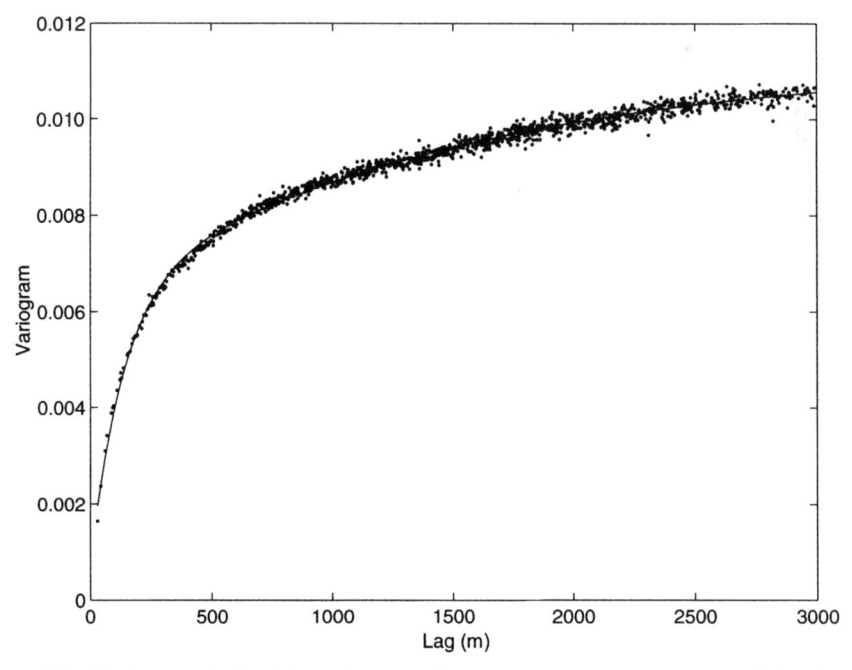

Figure 8.5 Variogram derived from the cover fraction image in Figure 8.4, with fitted nested model

nugget effect can be modelled as arising from either random measurement error, or from the existence of 'microstructures' in the data which are too small to be detected given the support size. The term 0.0007 in Equation (15) represents the nugget effect. The main problem associated with the nugget effect is that it behaves differently under changes of support than does the spatially dependent component of variation. For this reason, it is helpful to separate the observed variogram into two components as follows

$$\gamma_I(h;w) = N(w) + \gamma_I^*(h;w) \tag{16}$$

where $N(w)$ denotes the nugget effect (a function only of the support w and not of the lag h), and $\gamma_I^*(h;w)$ represents a regularized variogram whose nugget effect has been subtracted.

The nugget effect behaves as a 'white noise' process (Rendu, 1978), and at supports W larger than w, its magnitude can be calculated as

$$N(W) = \frac{|w|}{|W|} N(w) \tag{17}$$

It is important to note that this relationship does not hold when $|W| < |w|$. That is, the nugget effect does *not* behave as a white noise process for supports smaller than the one for which the data were originally observed. This is due to the fact that $N(w)$ may have arisen via regularization of spatially dependent microstructures, whose exact nature cannot be determined from coarse-scale observations. This has the important implication that esimation of variance at resolutions finer than that of the original data may not be valid.

A second component of the punctual variogram contributes to the spatially dependent component of variation via the regularization process. If its structure is represented by $\gamma_I^*(h;w)$, it is possible to estimate the punctual component associated with this variation; call it $\gamma_i^*(h)$. The relationship between a punctual variogram and its regularized counterpart for support w is

$$\gamma_I^*(h;w) = \bar{\gamma}_i^*(w,w_h) - \bar{\gamma}_i^*(w,w) \tag{18}$$

The term w_h represents a translation of the support w by a distance of h. Thus the regularized variogram at lag h is the average variogram between two supports separated by distance h, less the average variogram within a single support. While this relationship cannot be inverted to find $\bar{\gamma}_i^*(w,w_h)$, it can serve as the basis of a numerical estimation technique. The method used here is based on one discussed in Atkinson and Curran (1995). The strategy is to assume that the punctual variogram is equal to the regularized variogram, plus some to-be-determined fine-scale variogram. Parameters are initially guessed for this fine-scale term, and the resulting estimated punctual variogram is regularized by Equation (18). The process iterates, adjusting the sill of the fine-scale term, until the re-regularized variogram matches the observed data to an acceptable tolerance.

The separation of the nugget effect and the estimation of the punctual zero-nugget variogram resulted in the following estimates:

$$\begin{aligned} N(w) &= 0.0007 \\ \gamma^*(h) &= 0.0050\, e^{-h/1517} + 0.0056\, e^{-h/131} + 0.0036\, e^{-h/5.6} \end{aligned} \tag{19}$$

The term $0.0036\ e^{-h/5.6}$ represents the fine-scale variogram estimated from the de-regularization process.

When a nugget effect is present in an observed variogram, Equation (12) must be modified slightly. To find the dispersion variance for supports W larger than w, one may use (Zhang et al., 1990)

$$\sigma^2(W,D) = \left[\frac{|w|}{|W|} - \frac{|w|}{|D|}\right] N(w) + \bar{\gamma}^*(D,D) - \bar{\gamma}^*(W,W) \tag{20}$$

The first term represents the contribution of the nugget effect to the overall variance for supports of W, and the second term represents the dispersion variance arising from regularization of the zero-nugget punctual variogram.

Thus it is possible to estimate variance of cover fraction for arbitrary supports. Given a constant mean cover fraction, it is then possible to parameterize a beta pdf describing the cover fraction. From this distribution, it is possible to determine the proportion of pixels exceeding any threshold for any size of support greater than the original 30 m resolution of the data.

8.4 Results and Discussion

Figure 8.6 shows modelled distributions of cover fractions for several spatial resolutions. The plot corresponding to 30 m resolution includes a normalized histo-gram of the original image, which should theoretically match the associated pdf. This plot provides a useful assessment of the chosen statistical model. The relationship between the histogram and its model is far from ideal. Although the model tends to underestimate the probability of finding sparsely covered pixels, it captures the general trend of the distribution. So this model was retained, but one should be cautious when applying the results. The approximate nature of the fit between the model and reality indicates that further analyses may give only general indications about the thresholding process.

The progression of cover fraction distributions shows some expected patterns. As sensor FOV increases, one would expect less variance of cover fraction. The increasing kurtosis of the distributions corroborates this expectation. But the graphs in Figure 8.6 are also revealing for characteristics that they lack. As was mentioned above, when the variance of the beta distribution approaches its maximum, the distribution itself becomes bimodal (Figure 8.1). This corresponds to the case of decreasing support sizes. When the distribution takes on its bimodal form, the results of a thresholding operation are far less sensitive to the chosen threshold. Consider the bimodal curves in Figure 8.1. The proportion of the distribution exceeding a cover fraction of 0.5 does not differ greatly from the proportion exceeding 0.6 for example. Heuristically, this indicates that when support is small enough, results are not strongly dependent on the chosen threshold.

The opposite case is found for the distributions shown in Figure 8.6. Over the range of resolutions examined here, all distributions are unimodal. The observed variance of the 30 m data is too small to allow the bimodal nature of the indicator function to

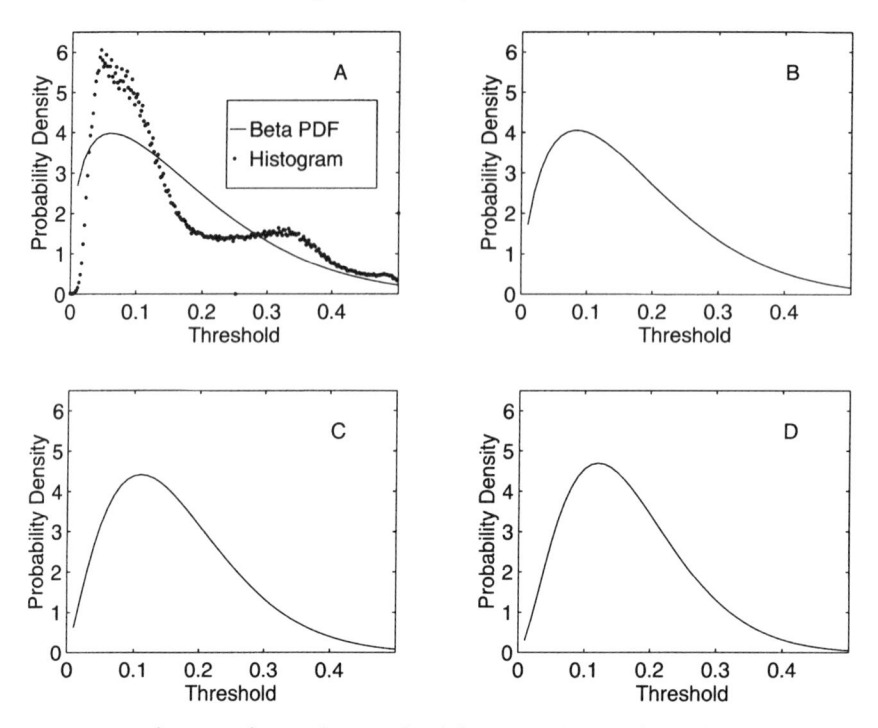

Figure 8.6 Distributions of cover fraction for differing pixel sizes, derived using the variogram shown in Figure 8.5. A. 30 m resolution. B. 100 m resolution. C. 500 m resolution. D. 1000 m resolution. Figure 8.6A also includes a histogram of the cover fraction image. Modelled distributions become more peaked as resolution becomes more coarse

manifest itself. The fact that much of the variance of the data is lost could be inferred from the image variance. The variance is 0.015. However, the mean of the image is $\mu = 0.1675$, which implies that the point-scale variance is $\mu(1 - \mu) = 0.1394$, much higher than that observed. So regularization by a 30 m FOV has a drastic effect on the resulting variance of this image. There clearly exist fine-scale structures with autocorrelation ranges considerably smaller than 30 m which constitute a significant portion of the overall point-scale variance. The nugget effect is likely a manifestation of the regularization of these structures. Implications of this fact are illustrated in Figure 8.7. This figure shows the same information as that contained in Figure 8.3 and Figure 8.2, but for the data derived from the actual image instead of from an idealized example. Each line in the upper plot corresponds to a single support size, and indicates the proportion of an image identified as covered using a particular threshold. Also shown is the 'true' cover fraction for the test image, corresponding to 0.1675. For all resolutions, the necessary threshold for identifying the correct cover fraction is approximately 0.3. This indicates the inadequacy of using an arbitrary threshold of 0.5, for example, to differentiate covered from uncovered areas. As is obvious in this case, very few pixels contain greater than 50% snow cover, so results of such an operation would be incorrect.

An interesting characteristic of these plots, which is shown more clearly in the lower

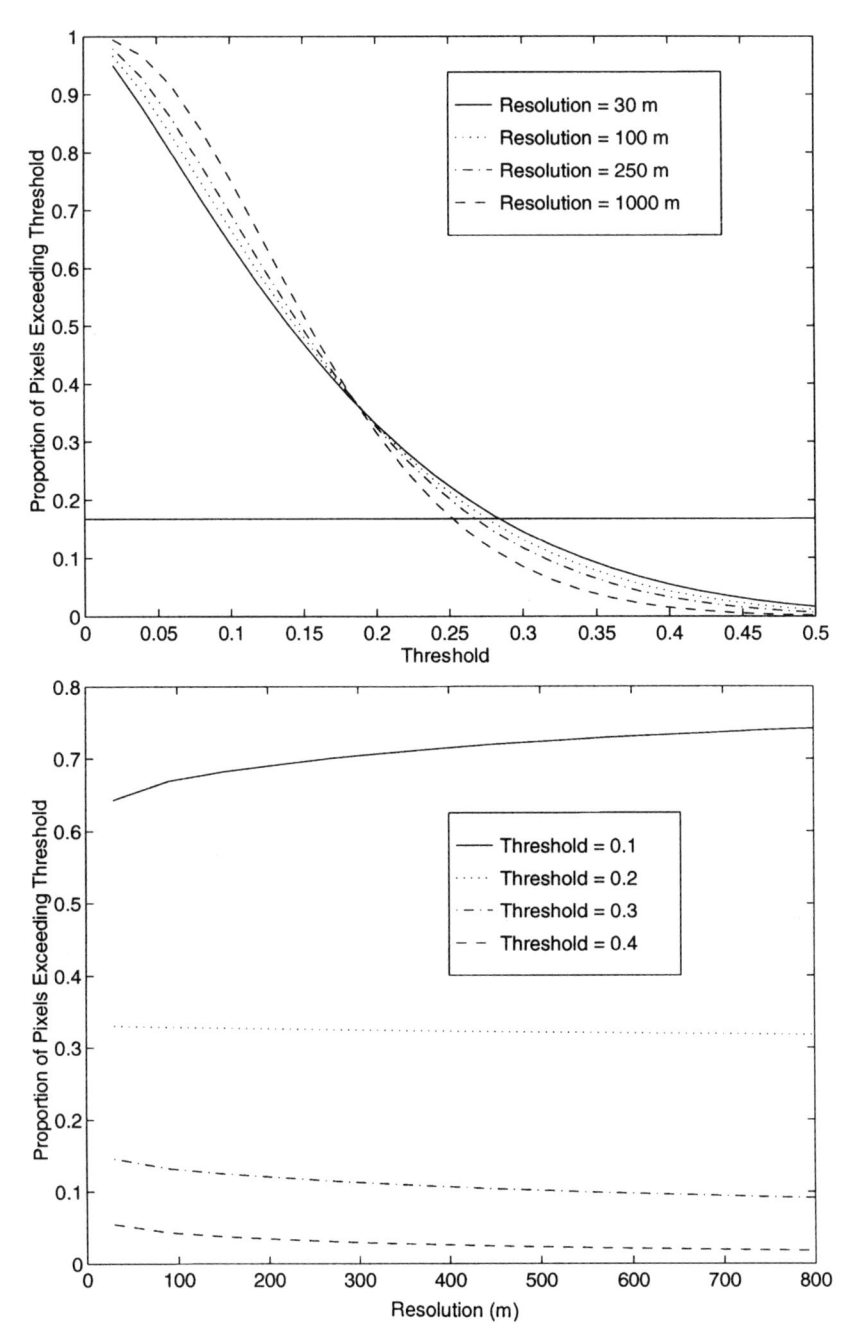

Modelling the Distribution of Cover Fraction of a Geophysical Field 131

Figure 8.7 Relationships among thresholds, resolutions, and cover fractions for the snow cover image. Over the range of resolutions investigated here, estimated area coverage depends strongly on the chosen threshold

plot of Figure 8.7, is the relative insensitivity to resolution over the range of resolutions modelled here. In this plot, each line corresponds to a particular threshold, and indicates the proportion of a scene identified as covered using that threshold as resolution varies. The lowest resolution plotted in this figure is the original 30 m. One would expect that at very fine resolutions (i.e., as the lines of this plot approach an abscissa of zero) the chosen threshold would not matter, and all of these lines would have an ordinate close to the true proportion of 0.1675. But as was discussed above, it is not possible to make inferences about supports smaller than the one used for the original data. However, all of the lines in Figure 8.7 begin to show such a tendency at the finest spatial resolutions. This figure is useful because it gives an indication as to the speed with which the relationship between threshold and cover becomes sensitive to the chosen threshold. When spatial resolution reaches a relatively fine 30 m, enough of the scene variation is subsumed that the choice of an appropriate threshold is a non-trivial matter.

These conclusions have important implications for remote sensing. For all scenes with spatial properties similar to the one on which this study is based, great care must be used when choosing a threshold for identifying covered and non-covered areas. The range of simulated spatial resolutions covers that of most of the sensors currently used for monitoring land surface conditions. For the range of cases examined here, spatial resolution may not have a strong influence on the results, but the choice of a threshold does.

8.5 Conclusions

This chapter has examined relationships among cover fraction of geophysical fields and sensor spatial resolution, with emphasis on thresholding operations designed to quantify a target field within an image domain. A reasonable model for the distribution of cover fraction is a beta pdf. Its parameters can be related to the variance of cover fraction, which in turn can be related to sensor field of view via the variogram of the regularized indicator function. Analyses provide support for some intuitive notions of how the regularization process affects the distribution of cover fraction. Specifically, the distribution is shown to become more peaked as support size increases. This corresponds to a situation in which the identified cover fraction becomes highly sensitive to the threshold chosen to quantify it. For images with similar spatial structure to the one used here, the choice of threshold is far more important than the choice of spatial resolution in determining overall cover fraction.

References

Atkinson, P.M. and Curran, P.J., 1995, Defining an optimal size of support for remote sensing investigations, *IEEE Transactions on Geoscience and Remote Sensing*, **33**, 768–776.

Barnes, R.J., 1991, The variogram sill and the sample variance, *Mathematical Geology*, **23**, 673–678.

Coppin, P.R. and Bauer, M.E., 1994, Processing of multitemporal Landsat TM imagery to optimize extraction of forest cover change features, *IEEE Transactions on Geoscience and Remote Sensing*, **32**, 918–927.

Cressie, N.A.C., 1985, Fitting variogram models by weighted least squares, *Mathematical Geology*, **17**, 563–586.

Cressie, N.A.C., 1993, *Statistics for Spatial Data* (New York: Wiley).

Hutchison, K.D. and Hardy, K.R., 1995, Threshold functions for automated cloud analyses of global meteorological satellite imagery, *International Journal of Remote Sensing*, **16**, 3665–3680.

Journel, A.G. and Huijbregts, C.J., 1978, *Mining Geostatistics* (London: Academic Press).

Key, J.R., 1994, The area coverage of geophysical fields as a function of sensor field-of-view, *Remote Sensing of Environment*, **48**, 339–346.

Key, J.R., Maslanik, J.A. and Ellefsen, E., 1994, The effects of sensor field-of-view on the geometrical characteristics of sea ice leads and implications for large-area heat flux estimates, *Remote Sensing of Environment*, **48**, 347–357.

Malingreau, J.P., Tucker, C.J. and Laporte, N., 1989, AVHRR for monitoring global tropical deforestation, *International Journal of Remote Sensing*, **10**, 855–867.

Matheron, G., 1963, Principles of geostatistics, *Economic Geology*, **58**, 1246–1266.

Rendu, J.M., 1978, *An Introduction to Geostatistical Methods of Mineral Evaluation* (Johannesburg: South African Institute of Mining and Metallurgy).

Shin, D., Pollard, J.K. and Muller, J-P., 1996, Cloud detection from thermal infrared images using a segmentation technique, *International Journal of Remote Sensing*, **17**, 2845–2856.

Verdin, J.P., 1996, Remote sensing of ephemeral water bodies in western Niger, *International Journal of Remote Sensing*, **17**, 733–748.

Zhang, R., Warrick, A.W. and Meyers, D.E., 1990, Variance as a function of sample support size, *Mathematical Geology*, **22**, 107–121.

9
Classification of Digital Image Texture Using Variograms

James R. Carr

9.1 Introduction

Digital image classification can be performed using both pixel-specific and neighbourhood algorithms (Schowengerdt, 1983). Maximum likelihood (Bayesian) and minimum-distance-to-mean algorithms are two examples of pixel-specific classification algorithms for classifying the spectral content of digital images. Textural classification can additionally be performed using a spatial co-occurrence matrix (Haralick *et al.*, 1973; Haralick, 1979) and is an example of a neighbourhood algorithm. In fact, texture information is contained in the average spatial relationship exhibited among grey levels within a particular class (Haralick *et al.*, 1973). Therefore, when attempting the classification of image texture, a neighbourhood classification scheme is necessary and is the focus herein.

Rather than using a spatial co-occurrence matrix approach for textural classification, the variogram function (Matheron, 1963) is instead used to capture average grey-level spatial dependence. Variograms have been used for some time when analysing and describing digital images (e.g., Serra, 1982; Carr and Myers, 1984; Curran, 1988; Woodcock *et al.*, 1988). The use of this function for digital image classification is a more recent application (Miranda *et al.*, 1992, 1996; Miranda and Carr, 1994). The variogram is shown to be advantageous for digital image classification with respect to algorithm accuracy and efficiency.

Advances in Remote Sensing and GIS Analysis. Edited by Peter M. Atkinson and Nicholas J. Tate.
© 1999 John Wiley & Sons Ltd.

9.2 Variogram Textural Classification Algorithm

9.2.1 Application to Single Waveband Imagery

Let the grey-levels comprising a given digital image be represented as $G(x,y)$, in which x represents pixel position (horizontal position) and y represents row, or line position (vertical position). Then, the variogram for these grey-levels is written (adapted from Matheron, 1963)

$$2\gamma(h) = \frac{1}{2}\int_x \int_y [G(x,y) - G(x',y')]^2 dy dx \qquad (1)$$

in which h is the Euclidean distance (also known as lag distance, and is a vector) between the pixel location, (x,y), and the pixel location, (x',y'), and G represents pixel value, or grey-level as is noted above. In practice, this integral is approximated as

$$\gamma(h) = \frac{1}{2N} \sum_{i=1}^{N} [G(x,y) - G(x',y')]^2 \qquad (2)$$

in which N is the total number of pairs of pixel values, and $G(x,y)$ and $G(x',y')$ represent a pair of pixel values that are separated by a vector distance, **h**; note that this accommodates the compression from a double integral to a single summation.

As is mentioned above, calculation of the variogram can be constrained to particular spatial directions, hence implying a vector calculation. The following four examples show E–W, N–S, NE–SW, and NW–SE calculations respectively:

E–W:

$$\gamma(h) = \frac{1}{2N} \sum_{i=1}^{N} [G(x,y) - G(x+h,y)]^2 \qquad (3)$$

N–S:

$$\gamma(h) = \frac{1}{2N} \sum_{i=1}^{N} [G(x,y) - G(x,y+h)]^2 \qquad (4)$$

NE–SW:

$$\gamma(h) = \frac{1}{2N} \sum_{i=1}^{N} [G(x+h,y) - G(x,y+h)]^2 \qquad (5)$$

NW–SE:

$$\gamma(h) = \frac{1}{2N} \sum_{i=1}^{N} [G(x,y) - G(x+h, y+h)]^2 \qquad (6)$$

In each of these equations, N is the total number of pairs of pixel values separated by a distance, h, in a particular spatial direction. Furthermore, h is not shown in bold in these equations because it is a scalar quantity (only representing distance; the four different equations accommodate differing directions).

Directional variograms are useful for analysing, or classifying texture that shows a particular directional character (such as parallel geologic structure: folding, fracturing, or faulting). For texture that is the same in all spatial directions, an omni-directional variogram is useful. The term, omni-directional, herein represents a variogram that is an average over all spatial directions. One way to obtain such a variogram is to average the four directional calculations presented earlier.

Regardless of the approach used, directional or omni-directional, computational efficiency must be considered when computing the variogram. In application to image processing, the variogram is simplified using the absolute value, rather than the square, of pixel difference (Carr, 1996):

$$\gamma(h) = \frac{1}{2N} \sum_{i=1}^{N} ABS[G(x,y) - G(x',y')] \qquad (7)$$

When applied to image processing, the variogram function is obtained by starting at $h = 1$ (a 1 pixel offset), then incrementing h by 1 through a maximum of 20 increments. A variogram, either directional or omni-directional depending on the nature of the texture, is computed for each class using training sites of size $M \times M$ pixels. A variogram is also computed for the $M \times M$ sized region around each pixel to be classified. A minimum-distance metric is then used to determine the similarity of variograms

$$\text{distance} = \sum_{i=1}^{K} ABS[\gamma_t(i) - \gamma_p(i)] \qquad (8)$$

wherein K is the number of increments of h allowable given the constraint of the window size, M, the subscripts, t and p, represent the training site and pixel neighborhood variograms, respectively, and i represents a particular class. In practice, K is equal to M, but is designated as a unique symbol in the equation for clarity and emphasis. A pixel is assigned to the class, i, for which the value, distance, is a minimum.

9.2.2 Extension to Multispectral Imagery

Classifying texture in multispectral digital imagery is, in practice, computationally more rigorous than for monospectral image data, but the theory behind this classification is rather simple to explain.

Given W spectral wavebands, the textural classification algorithm is as follows: (i) compute W variograms, one for each spectral waveband, as described previously; then, (ii) compute T cross-variograms, one for each possible combination of two spectral wavebands; that is, $T = (W(W-1))/2$ total two-waveband combinations. Because this application involves multispectral imagery, texture information is assumed to consist of the average spatial relationship between pixels for a particular class, within a particular waveband as well as between spectral wavebands. The cross-variogram is used to capture the average spatial relationship of pixels for a particular class across spectral wavebands. In this case, the cross-variogram is actually a paired-sum variogram (Myers, 1982; Carr, 1996). For example, suppose the cross-variogram is wanted between wavebands, b and c, for a particular class. First, pixel values are simply summed for wavebands b and c to yield a new 'waveband' $G_{b+c} = G_b + G_c$ (this uses notation introduced earlier, by which G represents the total collection of pixel values for a particular waveband). Then, a variogram is simply computed for G_{b+c} using the procedure described earlier for single waveband imagery. Such a procedure is used for each possible two-waveband combination from a collection of W wavebands.

To understand why cross-variograms are required in addition to variograms when classifying texture within multispectral data, recall the procedure for spectral classification of multispectral imagery using a maximum likelihood (Bayesian) algorithm (cf. Schowengerdt, 1983). In this algorithm, a variance/covariance matrix is developed for each class. Diagonal entries in this matrix are the variances for the separate spectral wavebands for a class. Off-diagonal entries represent covariances for all possible two-waveband combinations for a class. That is, spectral classification of multispectral imagery is a function of both within and between waveband (spectral) correlation. This is analogous to what is described herein for textural classification. Texture is classified using variograms and cross-variograms representing, respectively, within and between waveband (spatial) correlation.

9.2.3 Classification Implementation

Computer software for classifying image texture using variograms (and cross-variograms) is described by Carr (1996). Supervised classification schemes are used that require training sites when defining classes. The procedure for selecting training sites is identical to that which is described in Carr (1995); readers are referred to this text for a complete discussion of training site selection and a tutorial. This computer software is capable of performing digital image classification based solely on spectral information, or solely on variogram information, or a combined spectral/variogram classification. Relying solely on spectral information is useful when the spectral signature of classes is of paramount importance. On the other hand, using spatial information (the variogram) for image classification is useful when texture is of paramount concern. Often, classes represent unique combinations of spectral and textural information. In this case, a combined spectral/variogram classification is useful. Sometimes, all three approaches are necessarily experimented with to obtain the best classification results.

Two different approaches may be used when implementing the combined classification. One approach first involves a minimum-distance-to-mean spectral classification. Then, the image is reclassified using only the variogram function. Pixels classified in the first step remain classified only if the variogram surrounding them is closest to the variogram for the class to which they were originally assigned. Otherwise, the pixel is assigned to class 0 (threshold). This is a very strict classification protocol, but pixels that remain classified have a much higher probability of belonging to a class than when classified using only spectral information.

The second approach combines spectral information in the distance calculation

$$DIST_a = \frac{1}{K+1}\left[\left(\sum_{i=1}^{K} ABS(\gamma_t(i) - \gamma_p(i))\right) + ABS(G - \overline{G_a})\right] \quad (9)$$

where G is the value of the pixel to be classified, $\overline{G_a}$ is the mean pixel value for class a, and K is the number of increments used in the variogram computation.

The foregoing distance calculation is designed for classifying single waveband (monospectral) images. In multispectral classification, this algorithm is expanded as follows. Training sites are first selected for which variograms and cross-variograms are computed. Once these functions are computed for each training site (class), classification of an image proceeds, pixel by pixel, as follows (a three-waveband image composite is assumed):

1. An $N \times N$ pixel region surrounding each pixel to be classified is extracted from the image; in this instance, N is odd and is the same size as is used for the training sites.
2. Three variograms are computed for the extracted region, one for each spectral waveband; depending on user preference, this calculation is performed for one of the four spatial directions, or an omni-directional (average) calculation is used; a visual inspection of the image necessarily determines if texture to be classified has any obvious directional character; if so, a directional calculation is used for the variogram. Users may find that a directional variogram calculation orthogonal to apparent texture direction is better for classification because the variogram will show an obvious oscillation, a unique signature not easily confused with variograms for classes associated with isotropic textural characteristics.
3. Three cross-variograms are computed for this local region surrounding the pixel using the three possible paired-sum waveband combinations: red + green, red + blue, and green + blue; as with variogram computation, either an omni-directional or directional computation may be implemented.
4. A 'distance' is computed between the variogram for the extracted region and the variogram for each of the training classes as follows

$$DIST_a = \frac{1}{6K}\left[\sum_{i=1}^{K} ABS(\gamma_{t,L}(i) - \gamma_{p,L}(i)) + \sum_{i=1}^{K} ABS(C_{t,s}(i) - C_{p,s}(i))\right] \quad (10)$$

in which L represents one of the three spectral wavebands, S represents one of the three spectral waveband paired combinations, $C_{p,s}$ is the cross-variogram for one of three paired sums for the extracted image region, and $C_{t,s}$ is the cross-variogram for the analogous paired sum for class a. Because L and S both range from 1 to 3, for six total variograms and cross-variograms, the last calculation is a division by $6K$ to obtain an average distance value over K increments and six variograms. $DIST_a$ is unique to each class. A pixel is classified to the class for which $DIST_a$ is the smallest.

The two approaches to a combined spectral/textural classification mentioned previously for monospectral classification are also used for multispectral classification. One approach allows a user to first perform a spectral classification using a minimum-distance-to-mean algorithm. This classification is then modified by using variogram and cross-variogram information for a second classification. The second approach combines spectral information in the distance calculation shown earlier

$$DIST_{final,a} = DIST_a + \frac{1}{3}\sum_{i=1}^{3} ABS(G_i - \overline{G_{a,i}}) \quad (11)$$

where G_i is the value of the pixel to be classified in the ith spectral waveband, $\overline{G_{a,i}}$ is the mean pixel value for class a, spectral waveband i, and $DIST_a$ is the distance value that is shown above, computed as a function of variogram and cross-variogram information.

9.3 Example Classifications

Perhaps the most difficult aspect of textural classification is understanding what constitutes texture in a digital image. As it is defined earlier, texture is the spatial relationship exhibited by grey-levels in a digital image. This may be visually evidenced by the patterns made by pixels, or it may be the more subtle variation in grey levels from one pixel position to another. Texture may be, for instance, the pattern of agricultural fields (shapes: square, rectangular, circular); or it may be the pattern of joints in rock, cutting across many spectral classes, such as vegetation and exposed soil. The challenge is to properly classify texture, however the image analyst wishes to define it. Often, classification maps of texture appear significantly different from spectral classifications because texture may cross spectral boundaries. A few examples illustrate this concept.

9.3.1 Single Waveband Classification of Texture: the Example of Topography

A mid-infrared (waveband 5) Landsat TM image of Lake Tahoe, Nevada, is shown in Figure 9.1. The north-east portion of the lake is shown in the lower, left portion of this image. Incline Village, Nevada, and surrounding topography are shown north and east of Lake Tahoe in this image. Topography creates the dominant texture in this scene. The experiment is to use the variogram to classify this texture, in comparison to a simple, minimum-distance-to-mean spectral classification.

A single training site, 31 × 31 pixels in size, is used to represent topography. The upper, left corner of this training site is at row 41, pixel number 281 (the entire image size shown in Figure 9.1 is 400 rows by 400 pixels per row). A training site of this size may be too large for spectral classification because of the increased chance of spectral mixing. With textural classification, however, larger training sites are preferred if the texture being classified is rather large in spatial extent. The training site must be large enough to capture the essence of the texture. In this case, a 31 × 31 pixel training site is the minimum size thought capable of capturing the texture of topography in this image. In this example, topography is an information class and is the only class used.

Classification results (Table 9.1) illustrate the difference between spectral and textural classification for this particular image. Confusion matrices show that only 33% of training site pixels, and 36% of test site pixels, are correctly classified using spectral classification. On the other hand, textural classification correctly classifies 100% of the training pixels, and 75% of the test site pixels. The variogram function is more capable of representing the texture of topography than is spectral classification alone. Visual results of spectral and textural classification are shown in Figure 9.2.

At first, this comparison seems unfair. Results from the spectral classification are logically poor due to smoothing over a rather large training site; consequently, many

Figure 9.1 A Landsat TM, band 5 image of Lake Tahoe, Nevada. Image size is 400 rows by 400 pixels per row. A portion of Lake Tahoe is shown in the lower left of this image. Topography is a dominant texture

Table 9.1 Confusion matrices: Lake Tahoe image

Spectral classification	Textural classification
Training site:	Training site:
Threshold = 67%	Threshold = 0%
Class 1 = 33%	Class 1 = 100%
Test site:	Test site:
Threshold = 64%	Threshold = 25%
Class 1 = 36%	Class 1 = 75%
Test site (upper, left corner) located at row 101, pixel 281	
Total number of test site pixels = 2500 (50 × 50)	

pixels are not classified due to thresholding. This is precisely the anticipated result. Topographic texture in this image is spatially pervasive. A larger training site is necessary to represent the essential spatial character of this texture; moreover, the variogram is able to mathematically capture this spatial signature. Obviously, this

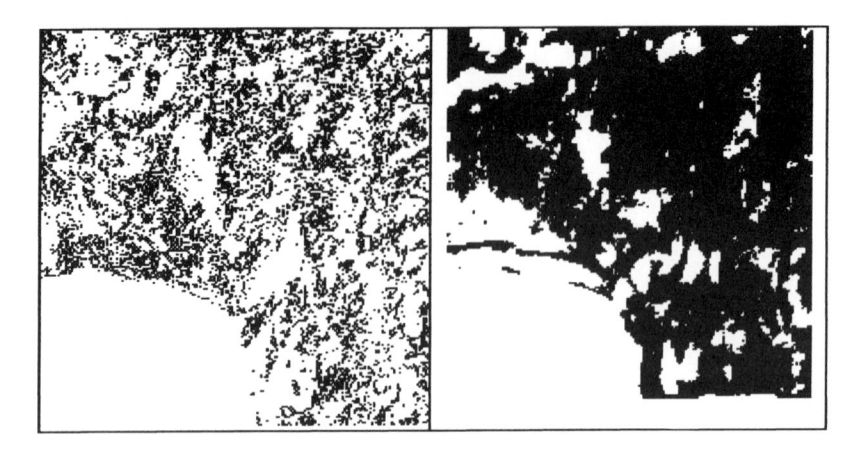

Figure 9.2 A mosaic of two classification results for the Lake Tahoe image. The left-hand classification is a minimum-distance-to-mean spectral classification. The right-hand classification is based solely on the variogram. Note that the variogram captures much more of the topographic texture than does pure spectral classification. This texture represents a highly variable spectral content. Classified pixels show as black in both maps

texture is highly spectrally variable, thus spectral classification alone cannot correctly associate any given pixel with this texture to a degree of certainty that exceeds a chosen threshold.

9.3.2 Multispectral Classification of Texture: the Example of the Fjords of Norway

A HRV SPOT image of the geographic region near Oslo, Norway, is shown in Figure 9.3. Numerous fjords are evident, as is a distinctive, two-directional network of joints in exposed rocks (associated with and highlighted by healthy vegetation). One training site, 11 × 11 pixels in size, is used to represent the texture of coastlines (the upper, left corner of this training site is located at row 96, pixel number 26). More specifically, the texture of interest is the contact between land and sea. The objective of this example is the correct classification of the unique shoreline that consists of numerous fjords.

Confusion matrices (Table 9.2) show the difference between spectral and textural classification for this application. Because of the spectral mixing within the training site, spectral classification correctly identifies only 14% of the training pixels, and 18% of the test site pixels. Textural classification, in contrast, correctly identifies 81% of the training pixels, and 74% of the test site pixels. In this application, the combined use of variograms and cross-variograms represents the texture of coastlines better than does spectral classification alone. Visual results of spectral and textural classification are shown in Figure 9.4.

An explanation is warranted for the smaller training site size used in this example in contrast to that used for the Lake Tahoe image. Textural classification of a single waveband image was employed for the Lake Tahoe image, in part as a demonstration.

Classification of Digital Image Texture Using Variograms 143

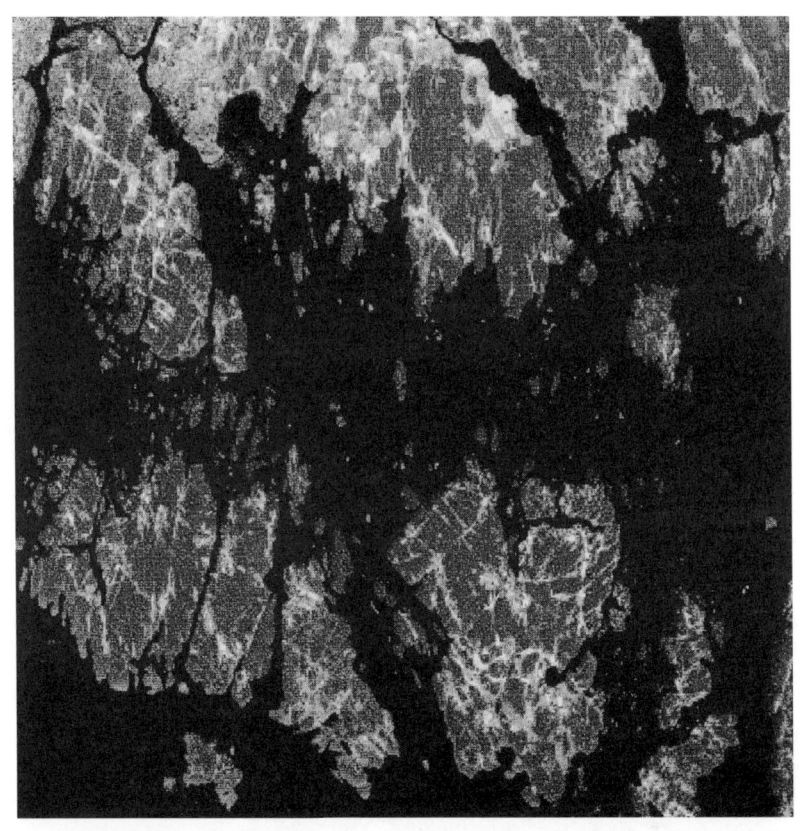

Figure 9.3 One of the three wavebands (infrared, waveband 3) for the SPOT HRV image of Oslo, Norway. The land–sea contact, associated with numerous fjords, is a distinctive textural feature. The image size is 400 rows by 400 pixels per row. © CNES/SPOT Image Corp., 1998

Table 9.2 Confusion matrices: SPOT Oslo, Norway image

Spectral classification	Textural classification
Training site:	Training site:
Threshold = 86%	Threshold = 19%
Class 1 = 14%	Class 1 = 81%
Test site:	Test site:
Threshold = 82%	Threshold = 26%
Class 1 = 18%	Class 1 = 74%
Test site (upper, left corner) located at row 216, pixel 26	
Total number of test site pixels = 121	

No cross-variograms are computed when classifying the texture within a single waveband. When classifying texture in multispectral imagery, however, cross-variograms are computed. Consequently, smaller training sites are necessarily used to increase classification speed.

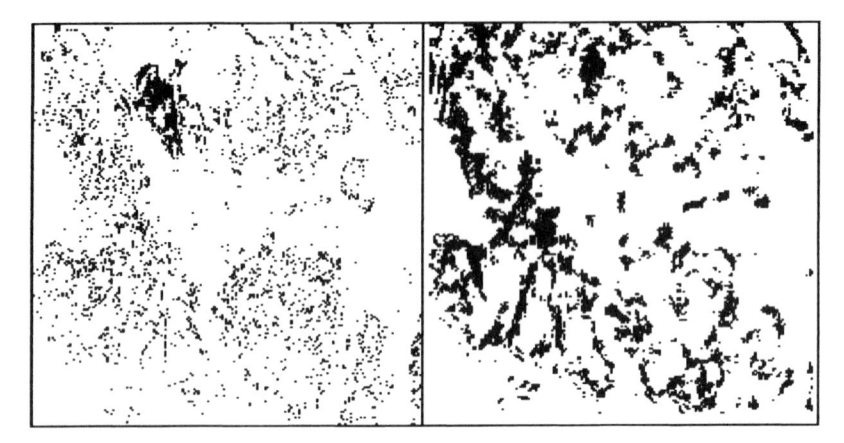

Figure 9.4 A mosaic of two classification results for the Oslo image. The left-hand result was obtained using spectral classification and a minimum-distance-to mean algorithm. The right-hand result was obtained using a pure textural classification based on variograms and cross-variograms. The textural classification highlighted the land/sea contact better (more consistently) than does a pure spectral classification. Presence of sea and land in the training site resulted in significant spectral mixing, hindering a pure spectral classification. Classified pixels show as black in both maps

9.4 Summary

Texture is herein classified using variograms (and cross-variograms in the case of multispectral imagery). The more traditional approach to textural classification employs spatial co-occurrence matrices. In a related work (Carr and Miranda, 1998), the variogram method is shown to offer a substantial improvement in computational efficiency and provides comparable, or increased classification accuracy, when compared to an algorithm based on spatial co-occurrence matrices. The two examples presented, single waveband textural classification of topography adjacent to Lake Tahoe, Nevada (USA), and multiple waveband textural classification of fjords near Oslo, Norway, demonstrate the different classifications resulting when texture signatures are emphasized, in contrast to when spectral signatures are emphasized.

In both of these examples, texture that is visually evident as topography or land–sea contact, is associated with a larger spectral variability. The variogram is able to mathematically accommodate this spectral variability in the context of its relationship to spatial position of pixels; that is, classification of digital image texture using variograms is a neighbourhood operation, not a pixel-specific operation such as maximum-likelihood (Bayesian) classification. Spectral classification, however, because it is a pixel-specific operation and not a neighbourhood operation, is not as successful when used to classify textures such as those identified in this study. Larger statistical variances of image pixels selected for training sites inhibit the success of such pixel-specific algorithms.

A few final comments are offered regarding the algorithms that are presented for classification of texture. Only four spatial directions are accommodated. This is done for simplicity, as well as making use of the regular geometry of image pixels. Users must necessarily decide which spatial direction is most similar to directional characteristics exhibited by the texture of a particular class. Certainly, software can be modified to more precisely match a particular spatial direction. With respect to the multispectral classification of texture, for which cross-variograms are necessary, the notion of intercorrelation (inter-statistical correlation) between bands is discussed. What is important is how image pixel values vary spatially after two wavebands are summed; this is not necessarily related to the statistical correlation between image bands.

Acknowledgements

This research was supported by the NASA (National Aeronautics and Space Administration)-funded University of Nevada System Space Grant Consortium. Software used for this study is available free of charge from the author, or can be downloaded as described in the issue of *Computers and Geosciences* that contains Carr (1996). The Lake Tahoe image is a 1984 Landsat TM scene, part of a larger collection of Landsat TM scenes donated in 1984 to the Mackay School of Mines, University of Nevada (USA) by Chevron, USA, Inc. The SPOT image of Norwegian fjords was obtained from a CD-ROM: NPA SPOT Satellite Image Library for Schools, NPA Group Ltd, Edenbridge, Kent, TN8 6HS, UK.

References

Carr, J.R., 1995, *Numerical Analysis for the Geological Sciences* (Englewood Cliffs, NJ: Prentice-Hall).
Carr, J.R., 1996, Spectral and textural classification of single
and multiple band digital images, *Computers and Geosciences*, 22, 849–865.
Carr, J.R. and Miranda, F.P., 1997, The semi-variogram in comparison to the co-occurrence matrix for classification of image texture, *IEEE Transactions on Geoscience and Remote Sensing*, 36, 1945–1952.
Carr, J.R. and Myers, D.E., 1984, Application of the theory of regionalized variables to the spatial analysis of Landsat data, *Spatial Information Technologies for Remote Sensing Today and Tomorrow: Proceedings, October 2, 3, 4, 1984, Sioux Falls, South Dakota* (Silver Springs, MD: IEEE Computer Society), 55–61.
Curran, P., 1988, The semi-variogram in remote sensing: an introduction, *Remote Sensing of Environment*, 24, 493–507.
Haralick, R.M., 1979, Statistical and structural approaches to
texture, *Proceedings of the Fourth International Joint Conference on Pattern Recognition* (New York: Institute of Electrical and Electronics Engineers, IEEE), 45–60.
Haralick, R.M., Shanmugam, K. and Dinstein, I., 1973, Textural features for image classification, *IEEE Transactions on Systems, Man and Cybernetics*, 3, 610–621.
Matheron, G., 1963, Principles of geostatistics, *Economic Geology*, 58, 1246–1266.
Miranda, F.P. and Carr, J.R. 1994, Application of the semi-variogram textural classifier (STC) for vegetation discrimination using SIR-B data of the Guiana Shield, northwestern Brazil, *Remote Sensing Reviews*, 10, 155–168.

Miranda, F.P., MacDonald, J.A. and Carr, J.R., 1992, Semi-variogram textural classifier (STC) for vegetation discrimination using SIR-B data of Borneo, *International Journal of Remote Sensing*, **13**, 2349–2354.

Miranda, F.P., Fonseca, L., Carr, J.R. and Taranik, J.V., 1996, Analysis of JERS-1 (FUYO-1) SAR data for vegetation discrimination in northwestern Brazil using the semi-variogram textural classifier (STC), *International Journal of Remote Sensing*, **17**, 3523–3529.

Myers, D.E., 1982, Matrix formulation of co-kriging, *Journal of the International Association for Mathematical Geology*, **14**, 249–257.

Schowengerdt, R.A., 1983, *Techniques for Image Processing and Classification in Remote Sensing* (New York: Academic Press).

Serra, J., 1982, *Image Analysis and Mathematical Morphology* (London: Academic Press).

Woodcock, C., Strahler, A. and Jupp, D., 1988, The use of variograms in remote sensing: I. Scene models and simulated images; II. Real digital images, *Remote Sensing of Environment*, **25**, 323–348.

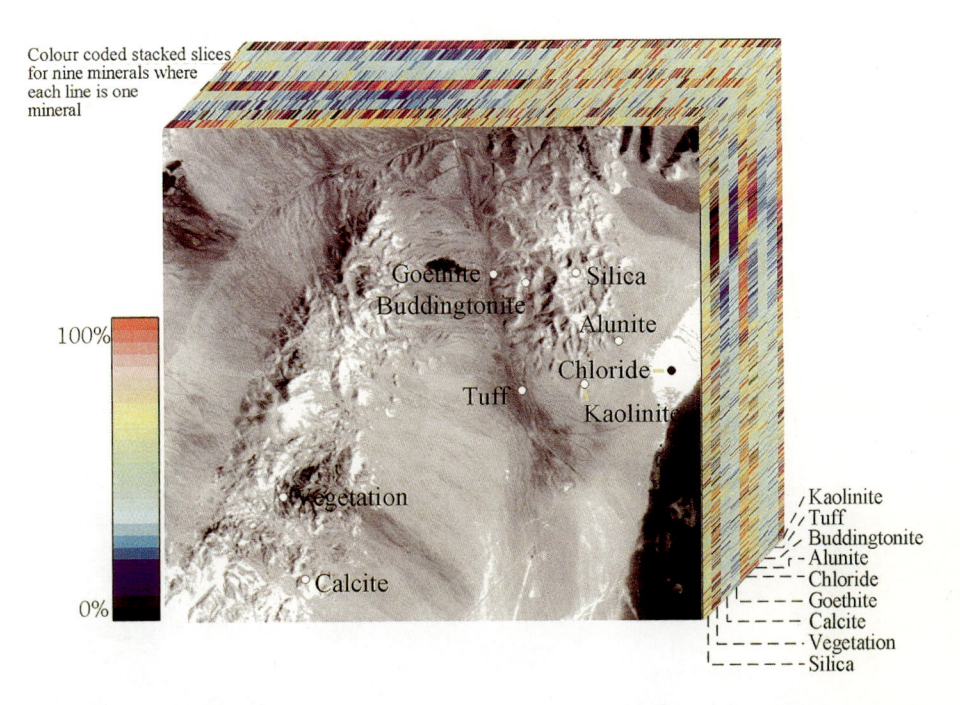

Plate I: Figure 10.5 *Three-dimensional data cube where the face of the cube is an AVIRIS image showing the pure mineral end-members for which probability images were calculated and the third dimensions are spectral slices (for the top and right-hand column of the image) in which the colour coding corresponds to the probability (interpreted as relative abundance) of the materials within a pixel*

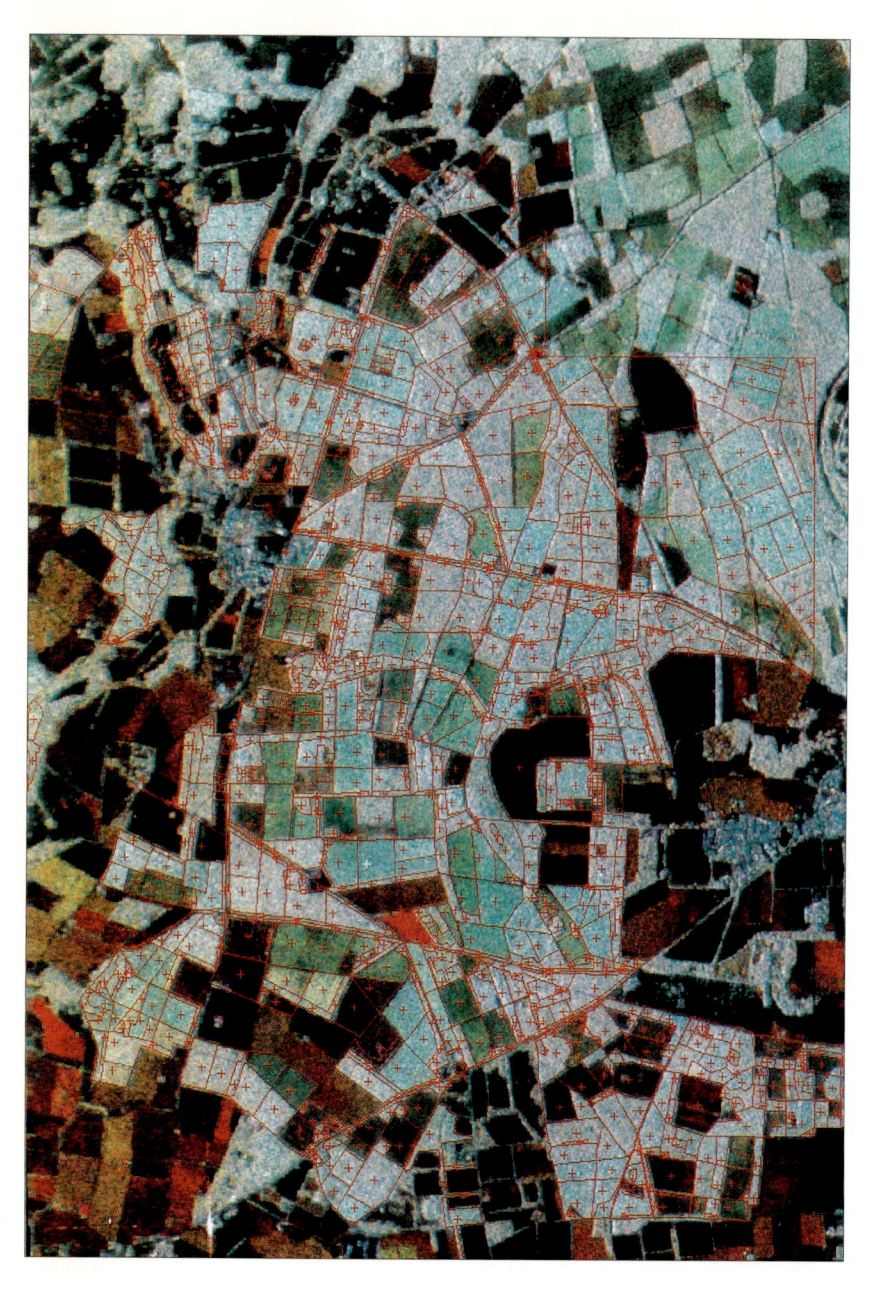

Plate II: Figure 13.2 *Three-band composite airborne synthetic aperture radar image of Thetford Forest, Norfolk, overlain with a vector data set of Forestry Commision stand boundaries. (Red – P band HV, Green – L band HV and Blue – C band HV)*

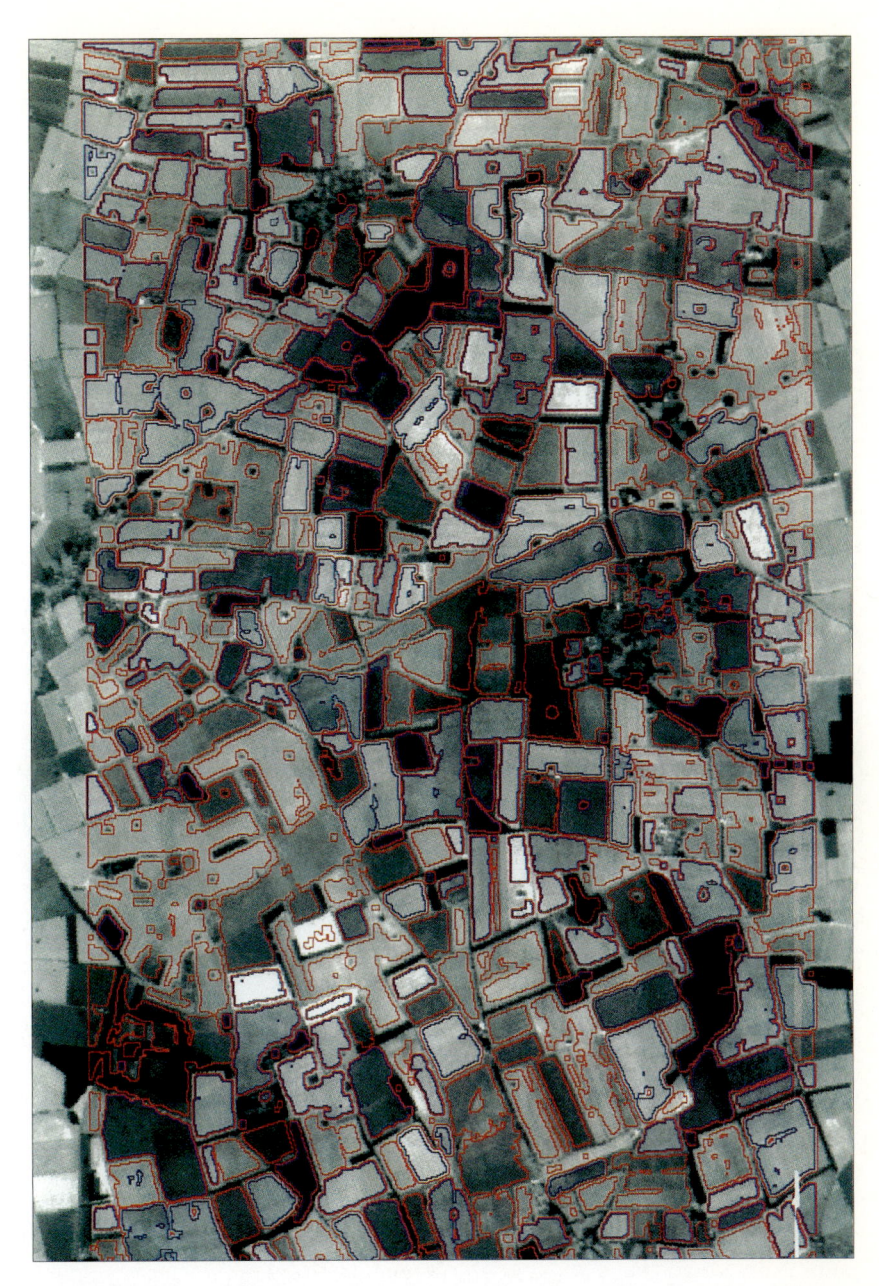

Plate III: Figure 13.5 *SPOT Panchromatic image of an agricultural area of Norfolk overlain with the results of an automated area extraction and vector editing procedure, described in Section 13.3.1. Blue vectors are originally derived vectors from the standard deviation filter. Red areas are the results of the vector editing procedure (in many fields where no editing was carried out the red areas overlie the blue)*

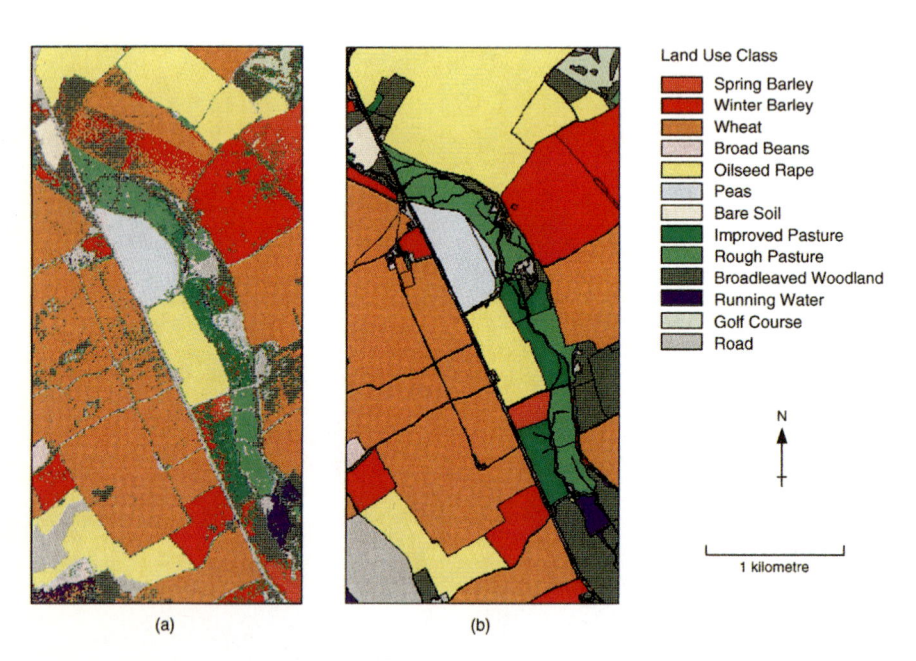

Plate IV: Figure 14.3 *(a) Per–pixel and (b) per–field classification of the study area (Land-Line vector data ©Ordnance Survey 1996)*

10
Geostatistical Approaches for Image Classification and Assessment of Uncertainty in Geologic Processing

Freek van der Meer

10.1 Introduction

Most terrestrial materials are characterized by spectral absorption features typically 20–40 nm in width. In optical remote sensing the spectral characteristics of Earth materials are studied by reviewing pixel reflectance in various wavelength wavebands either qualitatively in the form of image products (e.g., colour composites, ratio images, etc.) or quantitatively. Currently operational Earth observation satellites have waveband widths that do not allow the resolution of subtle absorption features that characterize geological materials of interest. Imaging spectrometry deals with the acquisition and analysis of image data in many, narrow contiguous spectral wavebands (Goetz, 1991) to enable the reconstruction of image spectra on a pixel-by-pixel basis. These image spectra can be compared directly either qualitatively or quantitatively with field or laboratory spectra (radiance or reflectance) to allow Earth surface mineralogy mapping based on the identification of unique spectral absorption features. This chapter will describe geostatistical techniques for the analysis of spectral information from imaging spectrometers.

Imaging spectrometry started in the early 1980s when the first airborne imaging spectrometer data were collected using a one-dimensional profiler. In 1983 NASA

Advances in Remote Sensing and GIS Analysis. Edited by Peter M. Atkinson and Nicholas J. Tate.
© 1999 John Wiley & Sons Ltd.

started airborne image acquisition with the Airborne Imaging Spectrometer (AIS) which was succeeded in 1987 by the Airborne Visible/Infrared Imaging Spectrometer (AVIRIS; Vane et al., 1993). While AIS was a proof-of-concept engineering test bed, AVIRIS was proposed as a facility that would supply well-calibrated data routinely for many different purposes. AVIRIS collects images in 224 contiguous wavebands resulting in a complete reflectance spectrum for each 20 × 20 m pixel in the 0.4 to 2.5 μm region with a sampling interval of 10 nm (Goetz et al., 1982; Vane and Goetz, 1988, 1993). The importance of imaging spectrometry for the future of remote sensing is evident from several spaceborne imaging spectrometer systems that are in the planning, design and commissioning phase. The European Space Agency (ESA) is designing MERIS (i.e., Medium Resolution Imaging Spectrometer; Rast, 1992) planned for Envisat which is to be launched in 1999, while NASA is developing the Advanced Spaceborne Thermal Emission and Reflectance Radiometer (ASTER; Kahle et al., 1991); a high spectral resolution imaging spectrometer planned for the first (A) platform of NASA's Earth Observing System (EOS-A). The first operational satellite imaging spectrometer to be expected was the hyperspectral scanning instrument LEWIS designed by the TRW company within the scope of NASA's Satellite Technology Initiative (NASA STI) which was planned to be launched in 1997 and should cover the 400–1000 nm range with 128 wavebands and the 900–2500 nm range with 256 wavebands of 5 nm and 6.25 nm width respectively. Unfortunately, LEWIS was lost in the atmosphere a few days after a successful launch. Several instruments are in the planning stage, such as the Chinese CIS, NASDA's GLI instrument and the Australian (e.g., CSIRO) ARIES-I.

Raw imaging spectrometer data, as provided to the user, are generally at-sensor radiance data that have been converted to physical units in the laboratory by calibration using reference target reflectors of known reflectance and a fully controlled environment. Conversion of the at-sensor radiance data to reflectance is done using atmosphere models (radiative transfer codes) or by standardizing reflectance relative to known targets in the scene. Techniques available for thematic analysis of reflectance data are:

- Binary encoding (Goetz et al., 1985; Jia and Richards, 1993)
- Waveform characterization (Crowley et al., 1989; Okada and Iwashita, 1992)
- Spectral feature fitting (Mackin et al., 1991)
- Spectral angle mapping (Kruse et al., 1993)
- Spectral unmixing (Boardman, 1991; Settle and Drake, 1993)
- Constrained energy minimization (Farrand and Harsanyi, 1997)
- Classification (Cetin et al., 1993; Lee and Landgrebe, 1993)

Imaging spectrometry has changed optical remote sensing from traditional interpretation based mainly on spatial image characteristics to identification based on the physical properties of reflectance. Another development in imaging spectrometry is the shift from qualitative identification by comparison of image and laboratory spectra to quantitative mapping of surface characteristics. A common feature of the techniques listed above is that they correlate the reflectance found in a pixel with the reflectance of known materials to enable a labelling of the unknown pixel responses. In this process, qualitative results are obtained (i.e., the pixel response looks similar to this mineral spectrum from the library) rather than quantitative results (i.e., with a probability of 90% that the two spectra are similar). Furthermore, the techniques use only spectral information without considering the location of the pixel in the scene and the relation to its neighbours. In general, existing techniques have the following shortcomings:

- Ground information is often needed which is sometimes unavailable or difficult to obtain.
- The uncertainty or the likelihood that the pixel actually represents a certain material is not quantified.
- The methods use only the *spectral* characteristics of an unknown pixel in comparison with that of a set of library samples without considering the *spatial* aspects or the relative location of an unknown pixel with respect to pixels from the training data set.
- The methods work on a pixel support (support is a geostatistical term that is synonymous to spatial resolution in remote sensing) without allowing estimation on larger or smaller volumes.

In this chapter, I shall present different approaches in which geostatistics helps in resolving such shortcomings. However, I first consider simple correlation to quantify the relation between pixel spectra of unknown origin and laboratory spectra of which the mineralogy is known. Next, I shall explore the spatial arrangement of the data through a non-parametric extension of kriging to remotely sensed data. Cokriging will be suggested as a technique for exploring the correlation between ground-based measurements and image data, thus allowing integration of field and image data. Finally, conditional simulation is used to assess the uncertainty in the estimates. Each section starts with the development of the theory followed by an example related to imaging spectrometer data analysis.

10.2 Matching Pixel and Laboratory Spectra; Correlation Theory

10.2.1 Theory

Spectral matching is done using the correlation between a pixel spectrum (test spectrum) and a laboratory spectrum (reference spectrum) of known mineralogy at different matching positions resulting in a correlogram (i.e., a function describing correlation at different spectral shifts). By convention, the reference spectrum is moved and negative match positions are shifts towards shorter wavelengths. Thus match position -1 means that we are calculating the cross-correlation between the test spectrum and the reference spectrum in which all channels have been shifted by one channel position number to the lower end of the spectrum. The cross-correlation, r_m, at each match position, m, is equivalent to the linear correlation coefficient and is defined as the ratio of the covariance to the product of the sum of the standard deviations. If we denote the test and reference spectrum as λ_t and λ_r respectively, and define n as the number of overlapping positions, the cross-correlation for match position m can be calculated as

$$r_m = \frac{n\Sigma\lambda_r\lambda_t - \Sigma\lambda_r\Sigma\lambda_t}{\sqrt{[n\Sigma\lambda_r^2 - (\Sigma\lambda_r)^2][n\Sigma\lambda_t^2 - (\Sigma\lambda_t)^2]}} \qquad (1)$$

This is similar to the classical definition of correlation. The statistical significance of the cross-correlation coefficient can be assessed by the following t-test

$$t = r_m\sqrt{\frac{n-2}{1-r_m^2}} \qquad (2)$$

which has $(n-2)$ degrees of freedom and tests the null hypothesis stating that the correlation between the two spectra at a specific match position is zero.

10.2.2 Example

In Figure 10.1 (after Van der Meer and Bakker, 1997) the results of calculating cross-correlograms are shown for spectra re-sampled to the AVIRIS waveband passes (using the full-width-half-maximum of the instrument's response function), demonstrating the potential of using the cross-correlogram for mineral mapping by spectral shape matching. As a reference spectrum, kaolinite is used and compared with two other minerals: alunite and buddingtonite.

10.3 Incorporating Spatial Information in the Matching: Non-parmetric Geostatistics

10.3.1 Theory

Non-parametric geostatistics comprises techniques that are probabilistic in nature but do not rely on a prior multivariate (normal) distribution hypothesis about the random function used to model regionalized variables. Indicator kriging is one such technique. A method for classification based on indicator kriging (Journel, 1983) was introduced in 1994 (Van der Meer, 1994a), which uses both spectral information and spatial information (in the classification process). The method is carried out through six basic steps (Figure 10.2).

Classification starts by defining the spectral wavebands that contain 'key information' on the spectral response of a certain ground class of interest, defined either from field or laboratory spectroscopic investigations (step I). The next step (step II) is to define the spectral range in each waveband by means of setting upper and lower limits for the DN values in those spectral wavebands. For all locations $\mathbf{s}$ in the selected wavebands, b (b being any number of wavebands out of the entire set of B spectral wavebands), the data are transformed into indicator variables (step III). Binary indicator values $i(\mathbf{s}_b; z_b)$ for cutoffs z_b at location $\mathbf{s}$ are defined as

$$i(\mathbf{s}_b; z_b) = \begin{cases} 1, & \text{if } z_{bl} < z_b(\mathbf{s}) \leq z_{bu} \\ 0, & \text{otherwise} \end{cases} \qquad (3)$$

where z_{bl} and z_{bu} are the lower and upper threshold values for cutoff z_b respectively, and

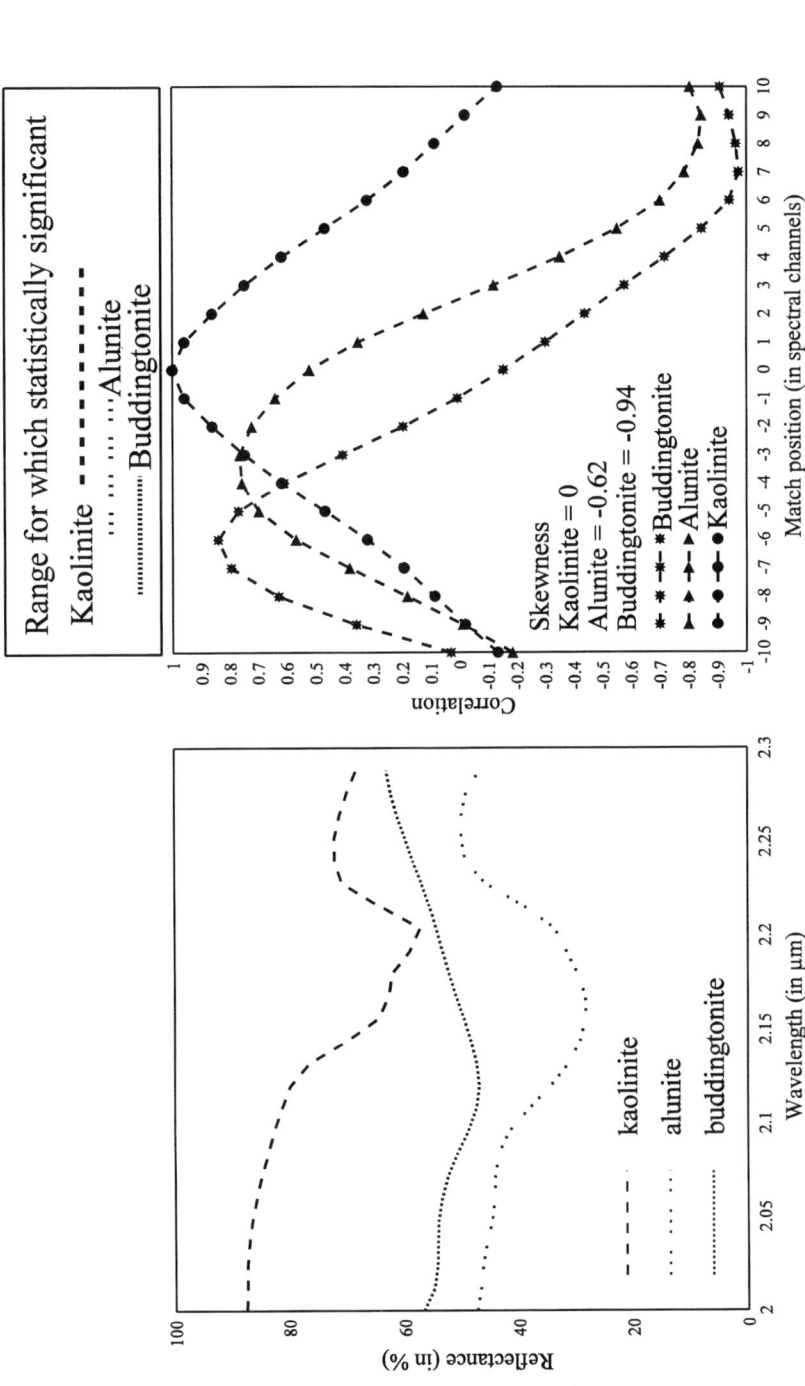

Figure 10.1 Library and test spectra (left diagram) re-sampled to AVIRIS waveband passes and corresponding cross-correlograms (right diagram) for kaolinite (as reference spectrum) versus kaolinite, alunite and buddingtonite (as test spectra). The vertical dashed and dotted lines above the right-hand diagram indicate the match positions for which the cross-correlation calculated was found significant (according to the t-test at a 95% confidence level). A perfect match is found for kaolinite while the negative skewness values for alunite and buddingtonite indicate that these minerals have absorption features in the wavelength range tested but at shorter wavelength relative to kaolinite (after Van der Meer and Bakker, 1997)

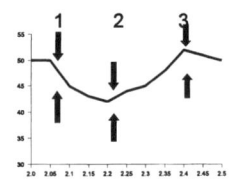

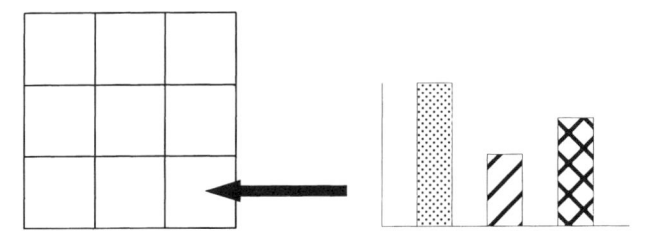

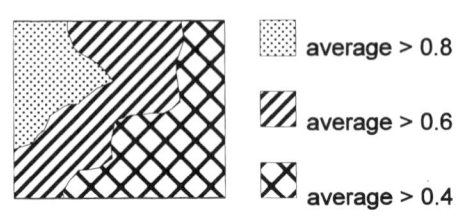

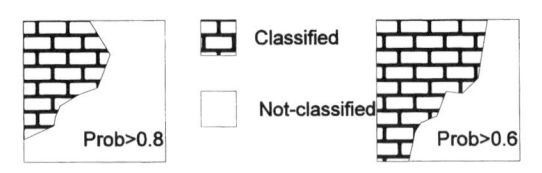

Figure 10.2 Outline of the indicator kriging based classification technique for extraction of absorption features and mineral mapping from image data (modified after Van der Meer, 1994a). See text for explanation

$z(\mathbf{s})$ is the data value at location $\mathbf{s}$ for waveband b. Indicator kriging (Journel, 1983) estimates the conditional expectation of the indicator transform of the random variable z_b given the realizations of n other neighbouring data values resulting in a cumulative distribution function for each cutoff z_b

$$[i(\mathbf{s}_b; z_b)]^* = E\{I(\mathbf{s}_b; z_b|(n)\}^* \qquad (4)$$
$$= Prob^*\{(Z_{bl} < Z_b(\mathbf{s}) \le Z_{bu})|(n)\}$$

where (n) is the conditioning information available in the neighbourhood of location $\mathbf{x}$. Subsequently (step IV), indicator estimation is performed for each of the class indicators in the different wavebands to obtain estimates of the local mean indicator giving an estimate of the quasi-point support values within a local area (i.e., the indicator kriging estimator predicts the probabilities assuming the support of the control points is equivalent to that of the output support). If the process is stationary and independent of $\mathbf{x}$, then the indicator kriging predictor is

$$[i(\mathbf{x}_b; z_b)]^* = \sum_{i=1}^{n} \lambda_i(\mathbf{x}_b; z_b) \, i(\mathbf{x}_{bi}, z_b) \qquad (5)$$

where

$$\sum_{i=1}^{n} \lambda_i(\mathbf{x}_b; z_b) = 1 \qquad (6)$$

A block indicator kriging algorithm was used, thus discretizing the block area into point samples yielding

$$i_v(\mathbf{x}_b; z_b) \ne \frac{1}{|V|} \int_{v(\mathbf{x})} i(\mathbf{x}_b'; z_b) \, d\mathbf{x}' \cong \frac{1}{N} \sum_{j=1}^{N} i(\mathbf{x}_{bj}'; z_b) \qquad (7)$$

where $v(\mathbf{x})$ is a block of size $|V|$ centred at $\mathbf{x}$, and the $i(\mathbf{x}_{bj}'; z_b)$ are the N-points discretizing (i.e., within) the volume $v(\mathbf{x})$. Block indicator kriging gives the average $1/N\Sigma[i(\mathbf{x}_{bj}; z_b)]_{ik}^*$ which is an estimate of the proportion of point values $z(\mathbf{x}_{bj}')$ within the block area $v(\mathbf{x})$ that are outside the class interval enclosed by the lower and upper thresholds (z_{bl} and z_{bu}) defined for waveband b. The ordinary block indicator kriging (OBIK) estimator using n control points takes the form

$$[i(\mathbf{x}_b; z_b)]_{obik}^* = [prob\{z_{bl} < z_b(\mathbf{x}) \le z_{bu}|(N)]_{obik}^*$$
$$= \sum_{j=1}^{n} \lambda_j(\mathbf{x}_b; z_b) \, \gamma([i(\mathbf{x}_{bi}; z_b)], [i(\mathbf{x}_{bj}; z_b)]) + \mu = \bar{\gamma}([i(\mathbf{x}_{bi}; z_b)], v(\mathbf{x})) \text{ for all } i = 1,...,n \quad (8)$$

and its estimation variance

$$\sigma_{obik}^2 = \sum_{i=1}^{n} \lambda_j(\mathbf{x}_b; z_b) \, \bar{\gamma}([i(\mathbf{x}_{bi}; z_b)], v(\mathbf{x})) + \mu - \bar{\gamma}(v(\mathbf{x}), v(\mathbf{x})) \qquad (9)$$

where the weights $\lambda_j(\mathbf{x}_b; z_b)$ are required to sum to 1. The terms $\gamma([i(\mathbf{x}_{bi}; z_b)], [i(\mathbf{x}_{bj}; z_b)])$ are the point-to-point (or sample-to-sample) semivariances, μ is the Lagrange multiplier, the terms $\bar{\gamma}([i(\mathbf{x}_{bi}; z_b)], v(\mathbf{x}))$ are the point-to-block semivariances, and the term $\bar{\gamma}(v(\mathbf{x}), v(\mathbf{x}))$ is the block-to-block semivariance. In the actual calculation, covariances were used for numerical stability. These values for all wavebands b used in the analysis are integrated by simple averaging (step V) and thus give the final estimated proportion or probability of a pixel belonging to a certain class

$$prob_{vj} = \sum_{j=1}^{b} i_{vB}(\mathbf{x}_b; z_b)/b \tag{10}$$

The final classification (step VI) follows from thresholding these averaged proportions or probabilities at a user-defined value. The level of accuracy of the classification is proportional to the average proportion of a block area chosen as a threshold level.

10.3.2 Example

Classification using the indicator kriging classifier was applied to 1994 AVIRIS data (corrected to reflectance) from the Cuprite mining area situated some 30 km south of the town of Goldfield in western Nevada (Figure 10.3). The area contains both hydrothermally altered rocks within a sequence of rhyolitic welded ash flow and air fall tuffs, and unaltered rocks (Albers and Stewart, 1972; Ashley, 1977). As a result of the alteration, several clay minerals (i.e., kaolinite, alunite, illite, buddingtonite) form a mineral zonation characteristic of a fossilized hot-spring deposit which often contains gold. In this example I will focus on mapping kaolinite and alunite which are the most abundant minerals in the Cuprite AVIRIS scene. Figure 10.3 shows AVIRIS single-pixel spectra from areas known to represent the minerals of interest and the upper and lower limits of the reflectance values in the spectral wavebands selected for the analysis. For all of these limits the data are transformed into 0 and 1 values; 1 if they are below the threshold and 0 if they are above the threshold. The result is two binary (0,1) maps for each spectral waveband used in the classification; one for the upper limit and one for the lower limit. Subsequently, the binary maps are interpolated to give a map of the probability that the value of a block lies in between the indicated limits for the selected channel. Repeating this procedure for all key wavebands results in a set of probability maps (one for each waveband) which can be integrated by calculating the joint probability which is a measure of the likelihood that a pixel belongs to a certain class. The joint probability is obtained as

$$\Pr\{joint\} = \Pr\{map_i\}*\Pr\{map_j\}*\Pr\{map_k\}*...\Pr\{map_n\} \tag{11}$$

where map_i to map_n are the probability maps for key wavebands i to n used. In Figure 10.4 the final integrated probability image for kaolinite is shown. This information is then used as input to classify kaolinite. By setting tolerances on the minimum probability acquired for each class pixels can be classified with a predefined accuracy (given by the error terms for all individual images). Comparing probability maps for different pure materials reveals that each pixel is made up of several materials whose

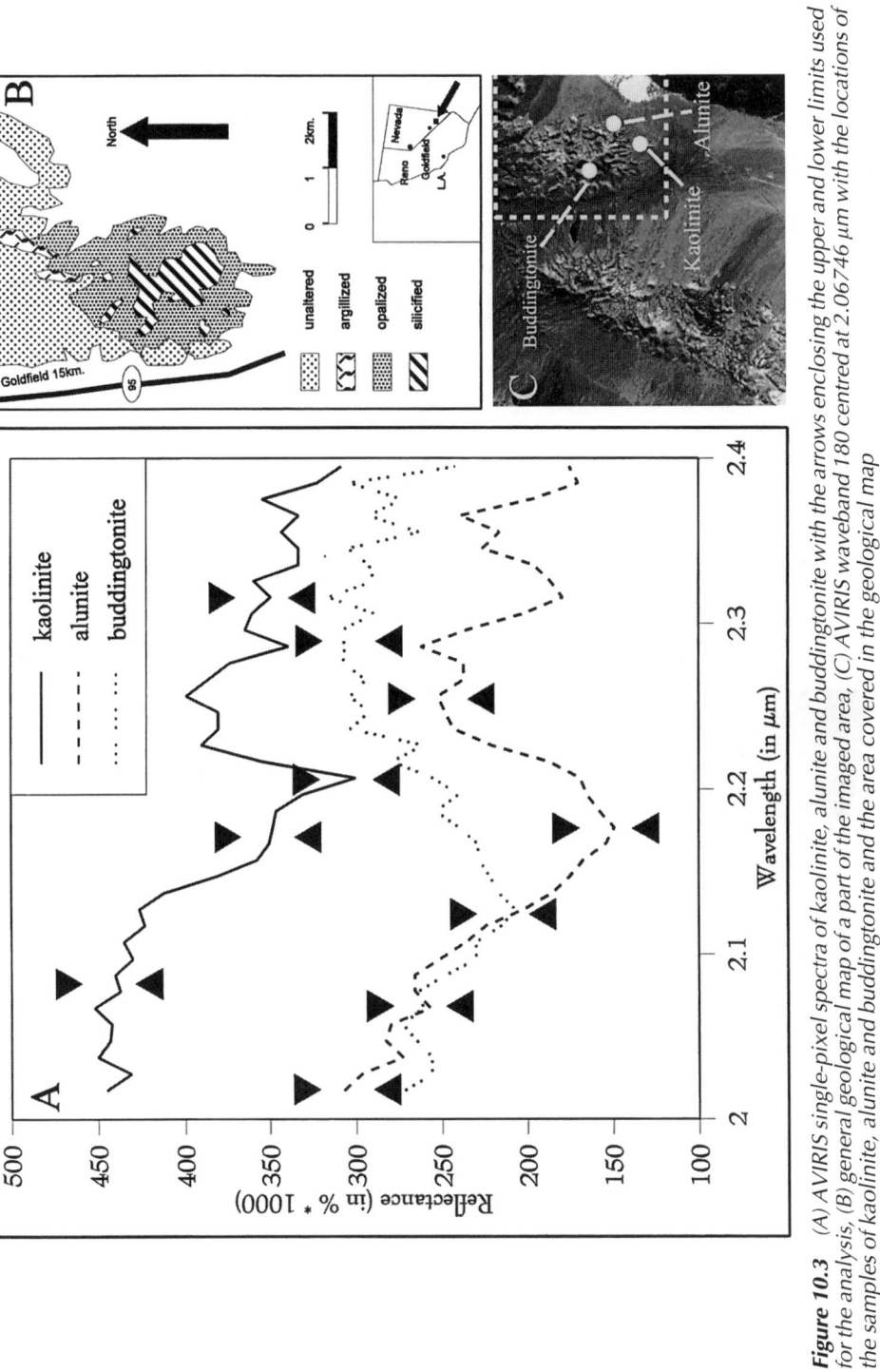

Figure 10.3 (A) AVIRIS single-pixel spectra of kaolinite, alunite and buddingtonite with the arrows enclosing the upper and lower limits used for the analysis, (B) general geological map of a part of the imaged area, (C) AVIRIS waveband 180 centred at 2.06746 µm with the locations of the samples of kaolinite, alunite and buddingtonite and the area covered in the geological map

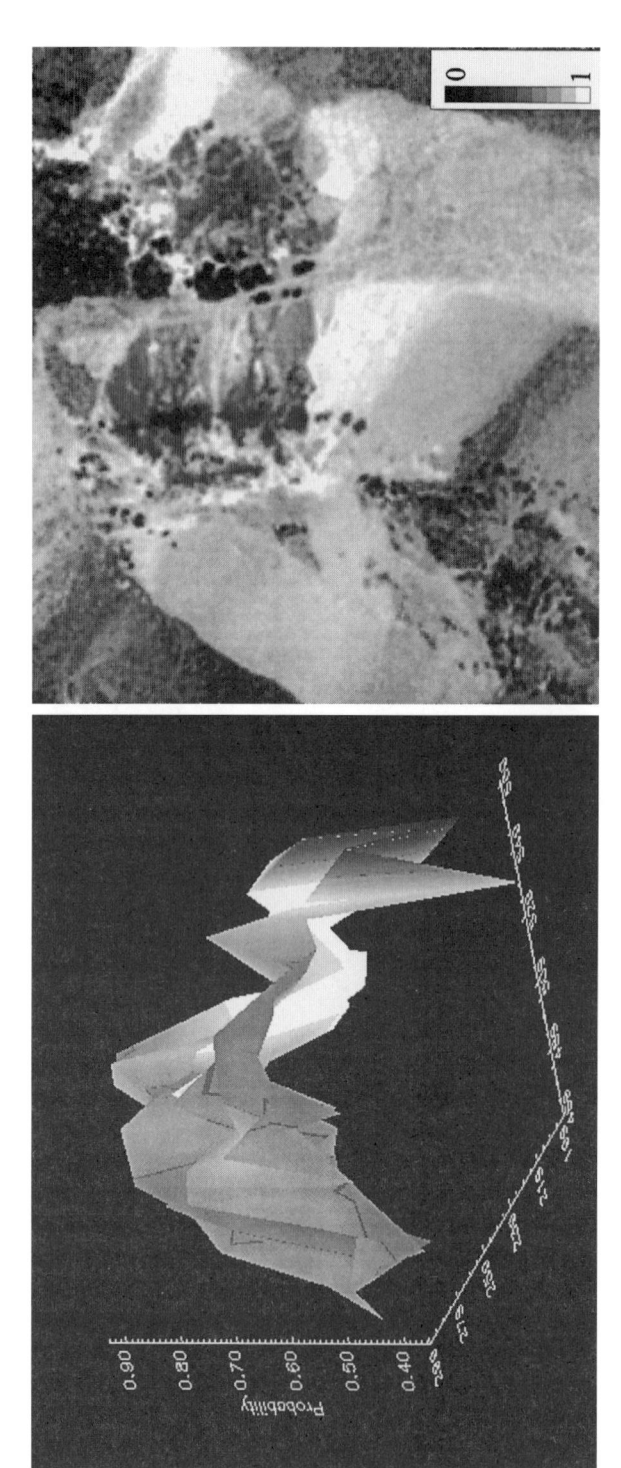

Figure 10.4 *Probability image for kaolinite and pixel classified as kaolinite (right) and kaolinite surface (left) where the probability of a pixel belonging to the class kaolinite determines the height of the surface and the grey shading determines the overall error in the assessment (dark = low, bright = high)*

probabilities sum up to 100%. Thus, by evaluating the individual probability fields, one may gain insight into the relative abundances of materials within a pixel. In classical remote sensing this is referred to as 'spectral unmixing' or 'sub-pixel analysis'. The idea is summarized in Figure 10.5 (Plate I) in the form of a three-dimensional data cube where the face of the cube is an AVIRIS image showing the pure mineral end-members for which probability images were calculated and the third dimension comprises spectral slices (for the top and right-hand column of the image) in which the colour coding corresponds to the probability (interpreted as relative abundance) of the materials.

Incorporating spatial as well as spectral information into image classification improves the classification accuracy. This is demonstrated in another study (Van der Meer, 1996), where simulated images are used with complete control over the spatial variation to compare the performance of conventional classification techniques and the extension to non-parametric geostatistics. A total of 2000 pixel spectra were characterized in three image wavebands (two wavebands on the absorption shoulder and one in the centre of the absorption feature) to represent a mineral with a fictive absorption feature. These points were used as input data for a geostatistical simulation to create three spectral image wavebands to be used for a classification. From the resulting image wavebands 1000 of the pixels known to exhibit the absorption feature were used to train the classifiers. The remaining 1000 image pixels from the conditioning data set were used to assess the accuracy of the classifications. The simulated image wavebands were created using various nugget variances (representing the noise in the data) and various correlation lengths (ranges; representing the spatial continuity of the data). The experiments clearly showed that only with very short range values or very high nugget variance did classification (i.e., without considering spatial arrangement) perform better than kriging; in all other cases, the contribution of spatial information was noticed in the accuracy.

10.4 Incorporating Spatial Information and Ground Control in the Matching: Coregionalization

10.4.1 Theory

The statistics introduced so far treat the matching of image spectra only, either with or without considering the spatial arrangement of the data. However, in many cases ground-based radiometer data are available and could be used. In most remote sensing studies field spectral measurements are used to validate the results of image processing or to aid in the interpretation of the data. In this section I suggest cokriging as a means of using the correlation that exists between ground radiometer data and spectra in imaging spectrometer data. Ordinary kriging uses linear combinations of values at sampled locations to estimate the value at an unsampled location. In cokriging, values at unsampled locations are estimated from the combination of values at sampled locations and other correlated variables. This is particularly useful where values of the variable of interest (generally referred to as the 'primary variable') are scarce and values from the spectrally correlated variable (generally referred to as the 'secondary variable') are relatively easily assessed. As with all kriging methods, the estimates can be derived for selected points or locations of interest or for whole images.

Coregionalization exploits the covariance between two or more regionalized variables to improve the prediction of a variable at unknown locations. It uses an extension of the variogram to two variables, u and v, as

$$\gamma_{uv}(\mathbf{h}) = \tfrac{1}{2} E\{z_u(\mathbf{x}) - z_u(\mathbf{x}+\mathbf{h})\}\{z_v(\mathbf{x}) - z_v(\mathbf{x}+\mathbf{h})\} \quad (12)$$

defining the cross-variogram between variables u and v. In the general case, the cross-variogram for two variables (u and v), which are both known for all locations $\mathbf{x}_i$ and $\mathbf{x}_i + \mathbf{h}$ becomes

$$\gamma_{uv}(\mathbf{h}) = \frac{1}{2n(\mathbf{h})} \sum_{i=1}^{n(\mathbf{h})} \{z_u(\mathbf{x}_i) - z_u(\mathbf{x}_i+\mathbf{h})\}\{z_v(\mathbf{x}_i) - z_v(\mathbf{x}_i+\mathbf{h})\} \quad (13)$$

Assume that we have sampled k variables, $m = 1,2,...,k$, and we wish to estimate one of these, variable u. The cokriging system of equations now becomes

$$z_u^*(\mathbf{x}) = \sum_{m=1}^{k} \sum_{i=1}^{n_m} \lambda_{im} z(\mathbf{x}_{im}) \quad (14)$$

where n_m is the number of sampling points for variable m. Minimizing the estimation variance leads to the following cokriging equations

$$\sum_{v=1}^{k} \sum_{j=1}^{n_m} \lambda_{im} \gamma_{mv}(\mathbf{x}_{im},\mathbf{x}_{jv}) + \mu_v = \lambda_{uv}(V,\mathbf{x}_{jv}) \quad (15)$$

for all $v = 1,2,...,k$, and $j = 1,2,...n_v$, where $\gamma_{mv}(\mathbf{x}_{im},\mathbf{x}_{jv})$ is the cross-semivariance between variables m and v at sites i and j, $\gamma_{uv}(V,\mathbf{x}_{jv})$ is the cross-semivariance between the variables at site j and the block volume, V, to be estimated, and μ_v is the Lagrange multiplier for the vth variable. The cokriging variance is given by

$$\sigma^2_u(V) = \sum_{v=1}^{k} \sum_{j=1}^{n_v} \lambda_{jv} \gamma_{uv}(V,\mathbf{x}_{jv}) - \mu_u - \gamma_{uu}(V,V) \quad (16)$$

where $\gamma_{uu}(V,V)$ is the within-block variance of variable u.

10.4.2 Example

In a series of articles (Webster et al., 1989; Atkinson et al., 1992, 1994) the theory of cokriging in remote sensing is derived. In the first article (Webster et al., 1989) the theory is described and implemented for estimating green leaf area index and percentage cover of clover using field spectral measurements of vegetation index as a secondary variable (Atkinson et al., 1992). Finally Atkinson et al. (1994) apply cokriging to the problem of estimating properties at the ground using airborne multispectral scanner data. Cokriging was found to be nine times more efficient than kriging for creating maps of percentage cover of clover. Furthermore these authors argue that the use of cokriging and kriging algorithms for estimating field data:

(a) incorporates the spatial arrangement of the data as opposed to simple regression methods;
(b) yields an assessment of error;
(c) can be used to lay out sampling schemes prior to aid field studies; and
(d) can provide data at places where none is available.

In the present example, cokriging is applied to the problem of dolomitization using a 100-by-100 pixel subset of airborne imaging spectrometer data from the GERIS 63 channel instrument flown over an area in southern Spain (see Van der Meer, 1994b, for details on the area). Laboratory spectral measurements of mixtures of pure minerals of calcite and dolomite have revealed a linear relationship between the position of the carbonate absorption feature in the 2.30–2.35 μm range and the percentage weight of calcite of a sample (Table 10.1). A similar relationship can be found between the depth of the carbonate absorption feature and the relative amount of calcite. The first relationship will be further exploited here. The primary variable of interest in this cokriging example is the *degree of dolomitization of limestones on the ground* expressed as the percentage of calcite to dolomite in a sample (assuming that these are the bulk constituents of pure limestones) derived from field spectral measurements. This variable is of economic importance because dolomitization increases the porosity of limestones thus making the rocks potential reservoir rocks for oil. Furthermore, skarn-type deposits are often associated with dolomites. Field spectral measurements were conducted with a GER Mark V spectroradiometer on rock surfaces of 1×1 m. The position of the carbonate absorption waveband was extracted and converted to relative calcite abundance using the linear equation derived from Table 10.1. There were 350 measurements on the ground, of which 250 were used as the primary variable and 100 were excluded from the analysis so that they could be used as a verification set. The secondary variable was the reflectance in GERIS waveband centred at 2.3536 μm (i.e., on the absorption minimum for calcite) for the 100-by-100 pixel window.

Table 10.1 Absorption-band depth and position of synthetic calcite/dolomite mineral mixtures (after Van der Meer, 1994b)

Weight % calcite	Carbonate-band position (μm)			Carbonate-band depth (% refl.)		
	Minimum	Average	Maximum	Minimum	Average	Maximum
0	2.3031	2.3039	2.3049	31.4	31.6	31.8
10	2.3049	2.3056	2.3066	30.4	30.7	30.9
20	2.3079	2.3082	2.3089	29.7	30.1	30.5
30	2.3101	2.3116	2.3125	29.2	29.6	30.0
40	2.3151	2.3165	2.3176	28.4	28.7	28.9
50	2.3219	2.3228	2.3239	27.9	28.0	28.4
60	2.3271	2.3284	2.3291	27.0	27.4	27.6
70	2.3340	2.3351	2.3360	26.2	26.4	26.8
80	2.3390	2.3403	2.3415	25.5	25.7	26.0
90	2.3436	2.3445	2.3453	25.0	25.2	25.5
100	2.3451	2.3465	2.3473	24.7	24.9	25.2

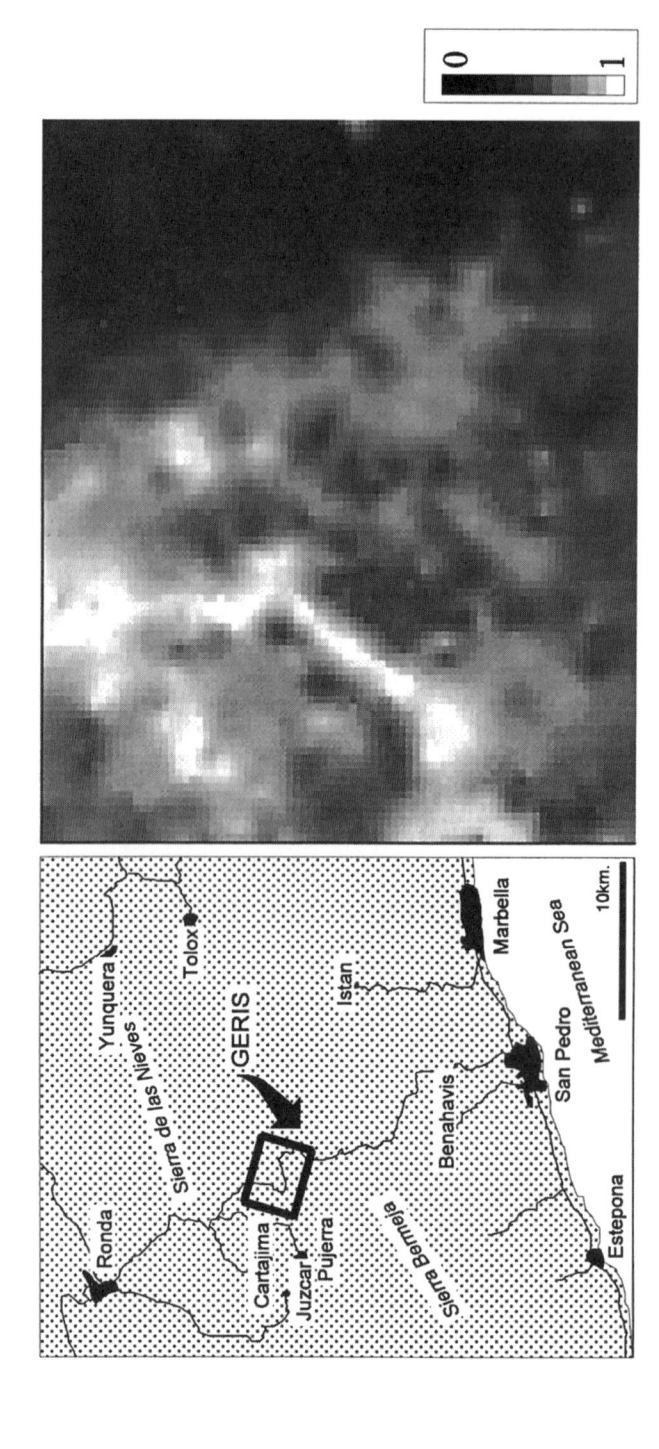

Figure 10.6 Cokriged estimates of calcite percentage from GERIS data acquired in southern Spain. Box outlines the imaged area

However, first an NDVI was used to mask those pixels that were moderately to highly vegetated. A good correlation exists between the 250 ground samples and the corresponding reflectances of the GERIS data. Three variograms were calculated, the variogram for the primary variable, the variogram for the secondary variable and the cross-variogram, and used for the cokriging (Figure 10.6). The comparison of cokriged calcite percentages and ground-based measurements for the 100 test samples (Table 10.2) demonstrates the potential of the technique. In a recent article, Dungan (1995) compared regression, cokriging and probability field simulation for the mapping of vegetation quantities using 300 ground measurements of vegetation quantity and different radiance imagery that had different levels of correlation with the true data. This study showed clearly that cokriging and probability field simulation performed much better than the regression method. Only at correlations (between the 300 ancillary data points and corresponding image radiance values) of 0.9 or higher was regression more accurate than the geostatistical approaches.

10.5 Assessing Uncertainty: Stochastic Simulation

10.5.1 Theory

Kriging produces the best possible estimate of a variable by minimizing the local error occurring in the estimation process; however, it also tends to smooth the degree of variability. In contrast, stochastic simulation produces many (as many as required) equally probable realizations (i.e., images) displaying the variation of a property of interest. The realizations are all different, but have in common that:

1. They reproduce the statistics of the original data (i.e., the histogram).
2. They reproduce the spatial dispersion of the original data (i.e., the variogram).
3. Conditional simulation honours the original data (i.e., returns the original values).

Since simulation reproduces the variogram it leaves the complex spatial variability intact.

Sequential simulation (see Deutsch and Journel, 1992, for an overview) builds on the principle of drawing values of a variable $Z(\mathbf{x})$ from its conditional distribution using all of the data within the neighbourhood of $\mathbf{x}$ including original (conditioning data) and previously simulated data. Consider the joint distribution of N random variables (RVs) Z_i that represent the same attribute at the N nodes of a dense grid discretizing field A. Next, consider the conditioning of these random variables by a set of n data of any type (symbolized by $|(n)$). The corresponding N-variate conditional cumulative distribution function (ccdf) is denoted

Table 10.2 Comparison of 100 field measurements of calcite percentage and predicted values through cokriging

	Mean	Standard deviation	RMSE
Test set ($n=100$)	46.8	10.6	
Cokriging	47.5	6.8	10.1

162   Advances in Remote Sensing and GIS Analysis

$$F_{(N)}(z_1,...,z_N|(n)) = Prob\{Z_i \leq z_i, i = 1,...,N|(n)\} \tag{17}$$

The conditioning is done as follows:

1. Draw a value $z_1^{(l)}$ from the univariate ccdf of Z_1 given (n) original data values where $z_1^{(l)}$ is considered a conditioning datum for all subsequent drawings. The information data set (n) is now updated to $(n + 1) = (n) \cup \{Z_1 = z_1^{(l)}\}$.
2. Draw a value $z_2^{(l)}$ from the univariate ccdf of Z_2 given the updated set $(n + 1)$, then update the information set to $(n + 2) = (n + 1) \cup \{Z_2 = z_2^{(l)}\}$.
3. Continue until all N RVs Z_i are considered.

The resulting set $\{z_i^{(l)}, i = 1, 2,..., N\}$ represents the joint realization of the N dependent RVs Z_i. Other realizations can be produced in another random process. The sequential simulation procedure requires the determination of N univariate ccdfs namely

$$\begin{aligned} & Prob\{Z_1 \leq z_1|(n)\} \\ & Prob\{Z_2 \leq z_2|(n + 1)\} \\ & Prob\{Z_3 \leq z_3|(n + 2)\} \\ & \quad ... \\ & Prob\{Z_N \leq z_N|(n + N - 1)\} \end{aligned} \tag{18}$$

10.5.2 Example

A 100-by-100 pixel window was selected from the calibrated GERIS data set of southern Spain consisting only of carbonate rocks covered partly with vegetation. Approximately 86% of the pixels that were moderate to heavily vegetated were removed. The position of the carbonate absorption-band was determined for the remaining pixels (Figure 10.7) using the procedure of continuum removal. The full range of pixels characterizing calcite/dolomite mineral mixtures is covered by spectral wavebands 53 to 56 of the GERIS data set (these wavebands are 0.017 μm wide). Using the semi-linear model for the calcite–dolomite weight percentages (Table 10.1) the absorption-band centres (λ_{Carb}) occurring in any GERIS spectral waveband (e.g., λ_{53}) could be converted to weight percentages calcite ($z(\mathbf{x})$) for a pixel at location $\mathbf{x}$ as:

$$\begin{aligned} & \lambda_{Carb} \in \lambda_{53} : 0\% < z(\mathbf{x}) \leq 26\% \\ & \lambda_{Carb} \in \lambda_{54} : 26\% < z(\mathbf{x}) \leq 57\% \\ & \lambda_{Carb} \in \lambda_{55} : 57\% < z(\mathbf{x}) \leq 86\% \\ & \lambda_{Carb} \in \lambda_{56} : 86\% < z(\mathbf{x}) \leq 100\% \end{aligned} \tag{19}$$

The definition of the variable $z(\mathbf{x})$ allows one to follow two approaches in determining the per cent calcite for the unknown pixels (e.g., the pixels previously removed as representing dense vegetation) treating them as a variable belonging to a discrete class (e.g., $z_k < z(\mathbf{x}) \leq z_j$ where z_k and z_j are two arbitrary cutoff values limiting the class values). If we regard variable $z(\mathbf{x})$ as a discrete or categorical variable, it will take on the value 1 when belonging to a class, else 0. In the case where we consider the variable $z(\mathbf{x})$ as a *threshold variable* (i.e., regarding the variable as continuous), it will take on the value 1 if the value at a location $z(\mathbf{x})$ is smaller than or equal to the threshold. Consider the spatial distribution of K mutually exclusive categories $s_k, k = 1, 2,..., K$. Let the

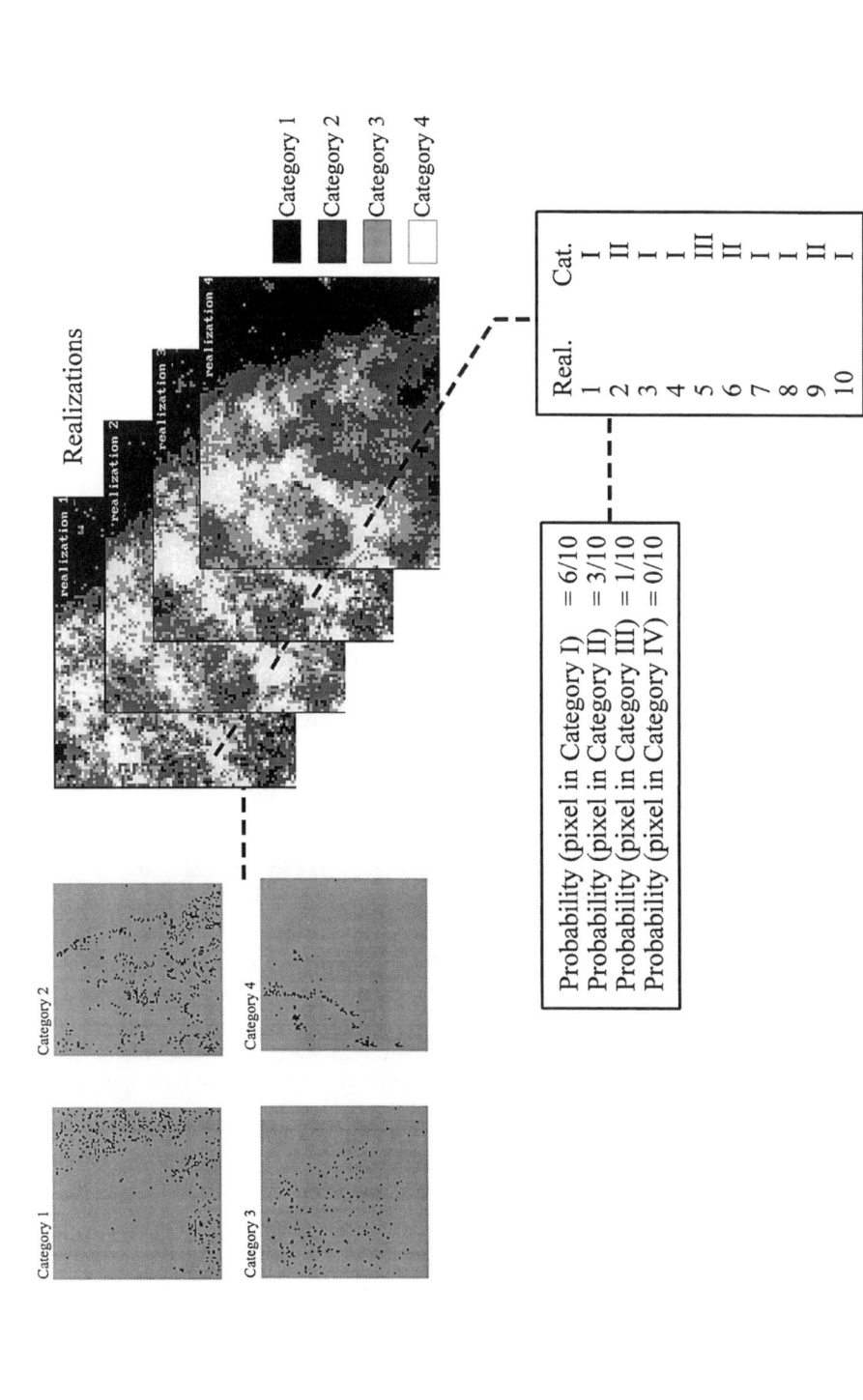

Figure 10.7 Variable z(**x**) data set used as input for SCIS (category variable; n = 1316) and four realizations of an SCIS (category variable) on an small 100–by–100 pixel area. It is shown is how simulated values from a set of realizations (in this example 10) can be used to derive the probability of a pixel belonging to a certain class. Probability values for different classes provide information on uncertainty related to class assignment

indicator of class s_k be 1 if $\mathbf{x} \in s_k$, else 0. Mutual exclusion yields $i_k(\mathbf{x})i_{k'}(\mathbf{x}) = 0, k \neq k'$ and

$$\sum_{k=1}^{K} i_k(\mathbf{x}) = 1 \tag{20}$$

Simple kriging (SK) or ordinary kriging (OK) of the indicator variable $i_k(\mathbf{x})$ provides an estimate for the probability that s_k prevails at location $\mathbf{x}$. Using SK thus gives

$$Prob^*\{I_k(\mathbf{x}) = 1|(n)\} = p_k + \sum_{k=1}^{n} \lambda_\alpha [I_k(\mathbf{x}_\alpha) - p_k] \tag{21}$$

where $p_k = E\{I_k(\mathbf{x})\} \in [0,1]$, and λ_α are the weights of the SK system of equations. Four realizations of an SCIS (category variable) for the four categories are shown in Figure 10.7. To enhance the differences between the realizations, cross tables were calculated from the frequency of pixel value combinations occurring in two realizations. Table 10.3 is a cross table of SCIS (category variable) realizations 2 and 4. From this table realizations 2 and 4 are similar: 68% of the pixels are of the same class, 29.1% differ by one class, 2.9% differ by two classes. No differences of three classes occurred between these realizations. In this examples four different realizations result in four estimates of the classes to which a pixel belongs. Simulation can provide an infinite number, n, of realizations and thus provide n predictions of the category to which each pixel belongs. These can be interpreted in two ways:

1. Extract for each pixel the modal value and assign the pixel to that category.
2. Use the n predictions to calculate a probability of a pixel belonging to a class.

The latter can be calculated as the number of times a pixel is assigned to a category divided by the number of realizations from which we derive the prediction. Thus for each category we derive a probability; the four probability values give insight into the uncertainty associated with the class assignment. Note that simulation of continuous variables and simulation exploiting the cross-correlation (covariance) of two or more variables, analogous to cokriging, can also be conducted. An example of the procedure is shown in Figure 10.7.

Table 10.3 Cross table for realizations 2 and 4 resulting from the SCIS (category variable). The rows consist of the categories simulated in realization 4, and the columns consist of the categories simulated in realization 2. The numbers are percentages

	1	2	3	4
1	23.8	3.0	0.2	0.0
2	2.6	22.3	7.8	1.0
3	0.1	6.6	12.0	3.6
4	0.0	1.6	5.5	9.9

10.6 Conclusions

This chapter gives examples of applying spatial statistics in spectral analysis for geology. While spectral matching algorithms and classification routines compare the *spectral* characteristics of unknown pixels with known materials, geostatistics allows one to include *spatial* aspects in this process also (e.g., the location of a sample in space and its relation to the neighbouring pixels). Examples of such analysis on imaging spectrometer data presented throughout this chapter have shown that this approach increases the prediction accuracy. Furthermore, through cokriging a link is created between field spectroscopic measurements and image data and stochastic simulation has been shown to provide insight into data uncertainty. Through the change of support corrections that geostatistical estimation techniques typically provide, the link to different spatial scales can be made allowing upscaling and downscaling. Many variables of interest to the geologist change in time. The examples shown so far are static because they provide a result that may be valid at a certain instance in time only. To describe and monitor dynamic processes (e.g., desertification, crop growth, fault movements, volcanic processes) the time factor needs to be considered and predictions have to be linked with their timeliness. With the advent of orbital spaceborne imaging spectrometers that provide spectral information at regular time intervals, future research work should concentrate on space-time kriging (and cokriging) to allow the monitoring of changes and accurate prediction into the future.

References

Albers, J.P. and Stewart, J.H., 1972, Geology and mineral deposits of Esmeralda County, Nevada, *Nevada Bureau of Mines and Geology Bulletin*, **78**, 1–80.

Ashley, R.P., 1977, Geologic map and alteration map, Cuprite mining district, Nevada (unpublished).

Atkinson, P.M., Webster, R. and Curran, P.J., 1992, Cokriging with ground-based radiometry, *Remote Sensing of Environment*, **41**, 45–60.

Atkinson, P.M., Webster, R. and Curran, P.J., 1994, Cokriging with airborne MSS imagery, *Remote Sensing of Environment*, **50**, 335–345.

Boardman, J.W., 1991, Sedimentary facies analysis using imaging spectrometry, *Proceedings of the 9th Thematic Conference on Geologic Remote Sensing*, ERIM, Denver, USA, 29 April–2 May 1991 (Ann Arbor: Environmental Research Institute of Michigan), 1189–1199.

Cetin, H., Warner, T.A. and Levandowski, D.W., 1993, Data classification, visualization, and enhancement using n-Dimensional Probability Density Functions (nPDF): AVIRIS, TIMS, TM, and geophysical applications, *Photogrammetric Engineering and Remote Sensing*, **59**, 1755–1764.

Crowley, J.K., Brickey, D.W. and Rowan, L.C., 1989, Airborne imaging spectrometer data of the Ruby Mountains, Montana: mineral discrimination using relative absorption band-depth images, *Remote Sensing of Environment*, **29**, 121–134.

Deutsch, C.V. and Journel, A.G., 1992, *GSLIB: Geostatistical Software Library and User's Guide* (New York: Oxford University Press).

Dungan, J.L., 1995, Geostatistical approaches for spatial estimation of vegetation quantities using ground and image data, in P.J. Curran and Y.C. Robertson (eds), *Proceedings of the 21st Annual Conference of the Remote Sensing Society*, 11–14 September 1995, Southampton, UK, 947–954.

Farrand, W.H. and Harsanyi, J.C., 1997, Mapping the distribution of mine tailings in the Coeur d'Alene River valley, Idaho, through the use of a Constrained Energy Minimization Technique, *Remote Sensing of Environment*, **59**, 64–76.

Goetz, A.F.H., 1991, Imaging spectrometry for studying earth, air, fire and water. *EARSel Advances in Remote Sensing*, **1**, 3–15.

Goetz, A.F.H. and Rowan, L.C., 1981, Geologic remote sensing, *Science*, **211**, 781–791.

Goetz, A.F.H., Rowan, L.C. and Kingston, M.J., 1982, Mineral identification from orbit: initial results from the Shuttle Multispectral Infrared Radiometer, *Science*, **218**, 1020–1031.

Goetz, A.F.H., Vane, G., Solomon, J.E. and Rock, B.N., 1985, Imaging spectrometry for earth remote sensing, *Science*, **228**, 1147–1153.

Jia, X. and Richards, J.A., 1993, Binary coding of imaging spectrometer data for fast spectral matching and classification, *Remote Sensing of Environment*, **43**, 47–53.

Journel, A.G., 1983, Nonparametric estimation of spatial distributions, *Mathematical Geology*, **17**, 445–468.

Kahle, A.B., Palluconi, F.D., Hook, S.J., Realmuto, V.J. and Bothwell, G., 1991, The Advanced Spaceborne Thermal Emission and Reflectance Radiometer (ASTER), *International Journal of Imaging Systems and Technology*, **3**, 144–156.

Kruse, F.A., Lefkoff, A.B., Boardman, J.W., Heidebrecht, K.B., Shapiro, A.T., Barloon, P.J. and Goetz, A.F.H., 1993, The Spectral Image Processing System (SIPS) – interactive visualization and analysis of imaging spectrometer data, *Remote Sensing of Environment*, **44**, 145–163.

Lee, C. and Landgrebe, D.A., 1993, Analysing high-dimensional multispectral data, *IEEE Transactions on Geoscience and Remote Sensing*, **31**, 792–800.

Mackin, S., Drake, N., Settle, J. and Briggs, S., 1991, Curve shape matching, end-member selection and mixture modelling of AVIRIS and GER data for mapping surface mineralogy and vegetation communities, *Proceedings of the 3rd AVIRIS Workshop*, Pasadena, USA, 20–21 May 1991, JPL Publication 91-28, 158–162.

Okada, K. and Iwashita, A., 1992, Hyper-multispectral image analysis based on waveform characteristics of spectral curve, *Advanced Space Research*, **12**, 433–442.

Rast, M. 1992. ESA's activities in the field of imaging spectroscopy, in F. Toselli and J. Bodechtel (eds), *Imaging Spectroscopy: Fundamentals and Prospective Applications* (Dordrecht: Kluwer), 167–191.

Settle, J.J. and Drake, N.A., 1993, Linear mixing and the estimation of ground cover proportions, *International Journal of Remote Sensing*, **14**, 1159–1177.

Van der Meer, F., 1994a, Extraction of mineral absorption features from high-spectral resolution data using non-parametric geostatistical techniques, *International Journal of Remote Sensing*, **15**, 2193–2214.

Van der Meer, F., 1994b, Sequential indicator conditional simulation and indicator kriging applied to discrimination of dolomitization in GER 63-channel Imaging Spectrometer data, *Nonrenewable Resources*, **3**, 146–164.

Van der Meer, F., 1996, Performance characteristics of the Indicator Classifier tested on simulated image data, *International Journal of Remote Sensing*, **17**, 621–627.

Van der Meer, F. and Bakker, W., 1997, CCSM: Cross Correlogram Spectral Matching, *International Journal of Remote Sensing*, **18**, 1197–1201.

Vane, G. and Goetz, A.F.H., 1988, Terrestrial imaging spectroscopy, *Remote Sensing of Environment*, **24**, 1–29.

Vane, G. and Goetz, A.F.H., 1993, Terrestrial imaging spectrometry: current status, future trends, *Remote Sensing of Environment*, **44**, 117–126.

Vane, G., Green, R.O., Chrien, T.G., Enmark, H.T., Hansen, E.G. and Porter, W.M., 1993, The Airborne Visible/Infrared Imaging Spectrometer (AVIRIS), *Remote Sensing of Environment*, **44**, 127–143.

Webster, R., Curran, P.J. and Munden, J.W., 1989, Spatial correlation in reflected radiation from the ground and its implications for sampling and mapping ground-based radiometry, *Remote Sensing of Environment*, **29**, 67–78.

11

A Syntactic Pattern-Recognition Paradigm for the Derivation of Second-Order Thematic Information from Remotely Sensed Images

Stuart L. Barr and Michael J. Barnsley

11.1 Introduction

It has been suggested that, while *land cover* mapping from digital multispectral images is a comparatively straightforward task, obtaining accurate and reliable information on *land use* from remotely sensed data is much more problematic (Barnsley *et al.*, 1993). This is because land use is an abstract concept – an amalgam of social, economic and environmental factors – one that is defined in terms of *function* rather than physical *form*. By comparison, the data contained in a multispectral image represent sampled (both spatially and spectrally) and quantized measures of the radiation emanating (whether reflected, emitted, or scattered) from a collection of three-dimensional objects on the Earth's surface. Consequently, while the spectral patterns recorded in a multispectral image may be related in a reasonably *direct* way to the physical and chemical properties of the principal objects present within the scene (i.e., to the land cover), the relationship with the land use is, in most instances, both complex and *indirect*. In that sense, we might refer to land cover as being 'first-order' information about the scene, while land use is 'second-order' (implying that it is further removed from the basic

---

Advances in Remote Sensing and GIS Analysis. Edited by Peter M. Atkinson and Nicholas J. Tate.
© 1999 John Wiley & Sons Ltd.

data). Despite this, certain types of land use have a characteristic spatial expression that can often be identified in fine spatial resolution images. Thus, for example, residential districts in many Western European cities are frequently characterized by a complex spatial assemblage of buildings with concrete, tile or slate roofs, as well as tarmac or concrete-surfaced roads, interspersed with areas of open space comprising lawns, trees, bare soil and water. Identification of these features and an analysis of their spatial pattern is, of course, the basis of human photo-interpretation. If this process could be formalized, and the features and patterns measured in digital multispectral images, it might be possible to develop an automated or semi-automated system with which to infer land use from remotely sensed data, or, more generally, to derive second-order information from first-order data. Syntactic (or structural) pattern recognition techniques appear to offer considerable potential in this context (Ballard and Brown, 1982). The remainder of this chapter, therefore, discusses the potential of these techniques, particularly those based on graph-theoretic methods, to provide a practical set of tools for deriving land use (second-order) information from an initial land cover (first-order) image. Much of the discussion is based around a syntactic pattern-recognition data model, known as XRAG (eXtended Relational Attribute Graph), developed by Barr and Barnsley (1997); however, a number of alternative approaches are also raised. The reader is referred to the paper by Barr and Barnsley (1997) for a detailed description of the XRAG data model and some of its associated data processing algorithms, while Barnsley and Barr (1997) illustrate the application of these to the analysis of land use in large-scale digital map data.

11.2 Relating Image Structure to Land Use

In this section we present some of the arguments outlined above in a more formal manner. The example of inferring land use from the spatial pattern (or structure) of land cover is used as a special case of the more general problem of deriving 'second-order' thematic information from a 'first-order' spatial data set.

Let us start by assuming that the original digital, multispectral image has been divided into a set of discrete regions – labelled and assigned to one of a set of mutually exclusive land cover classes – such as might typically be performed using a per-pixel, multispectral classification algorithm (Aplin *et al.*, Chapter 14, this volume). This process can be expressed by the general mapping

$$f : I \mapsto F \tag{1}$$

where *I* is the set of multispectral responses of the pixels in the image and *F* is the set of land cover classes (first-order themes). To derive information on land use (second-order themes), a further mapping must be defined. This has the general form

$$g : F \mapsto S \tag{2}$$

where *S* is the set of land use categories. The second mapping, Equation (2), might be informed through the use of data which are either external (i.e., ancillary) or internal (i.e., intrinsic) to the image, or some combination of the two. The external information might include various socioeconomic indicators, such as digital map data on population density, areas of open space, and administrative boundaries (Sadler *et al.*, 1991; Barnsley *et al.*, 1993). The internal information might be derived from an analysis of the morphological properties (e.g., size and shape) of the land cover regions, or from the spatial and semantic relations between them (e.g., adjacency, containment, distance and direction). Taken together, these may convey information on the structural composition of the underlying scene from which land use may be inferred (Laurini and Thompson, 1992; Barr and Barnsley, 1997). In this chapter, we focus on the use of internal information only. The relationship between land use (second-order thematic information) and the multispectral image from which it is derived can therefore be expressed formally by the composition

$$I \circ S = \{(i \mapsto s) : i \in I, s \in S, \exists f \in F([i \mapsto f] \wedge [f \mapsto s])\} \qquad (3)$$

which states that there is a mapping from the image domain to the second-order thematic domain that is itself a composition of two mappings: (i) from the image domain to the first-order thematic domain and (ii) from the first-order to the second-order domains.

Clearly, there are several assumptions implicit in the discussion above. First, the regions identified in the land cover map must accurately represent and delimit the key spatial entities in the corresponding scene, such as fields, forests, lakes, roads and buildings: thus, the derivation of the members of the set $S(s \in S)$ will be sensitive to (i.e., a function of) the accuracy with which $i \mapsto f$ (in other words, to the initial land cover classification). Second, these must be spectrally separable in the original multispectral image. Third, there must be discernible differences between the spatial structure of the land cover parcels in each of the land use categories of interest, so that they can be identified, delineated and labelled in an automated or semi-automated manner (Barnsley and Barr, 1997). The general validity and applicability of these assumptions has yet to be examined rigorously, although numerous studies have accepted them implicitly (Wharton, 1982; Gurney and Townshend, 1983; Moller-Jensen, 1990; Johnsson, 1994; Barnsley and Barr, 1997).

It is important to define more precisely the sets F and S given in Equations (1)–(3). In doing so, it should be evident that their definition must incorporate both thematic and spatial elements. To explore this point, consider the possible outcomes if the set S was defined solely in thematic terms: these comprise a one-to-one, an exclusive many-to-one, a one-to-many, or a many-to-many relation between the land cover classes $f \in F$, and the land use categories, $s \in S$. In the first two outcomes, each and every $f \in F$ is related to one, and only one, $s \in S$. In the case of urban areas this implies that, for example, all regions of the land cover type building *must* form part of, say, the residential land use category, and no other land use category. This is clearly untenable, since commercial and industrial districts also contain buildings. Although the second two relations appear to overcome this problem, in reality they produce unresolvable ambiguities – e.g., a single building region cannot simultaneously form part of the mutually exclusive land use categories residential, commercial and industrial. The solution is to retain some reference to the geometric elements of the regions when defining the sets F and S. This can be achieved by defining three further sets, R, C and U; where $R = r_1, r_2, r_3, ..., r_n$, the members of which refer to the geometric elements (i.e., centroid and boundary) of the land cover

170   Advances in Remote Sensing and GIS Analysis

and land use regions, and the sets C and U comprise the potential land cover and land use themes. The sets F and S can now be defined as

$$F = R \times C$$
$$S = R \times U$$

and consist of a series of ordered pairs defining the geometric elements of each region and, for each member of this series, the entire set of themes to which it may be assigned (Figure 11.1).

Defining the sets F and S in this way allows the derivation of land use information

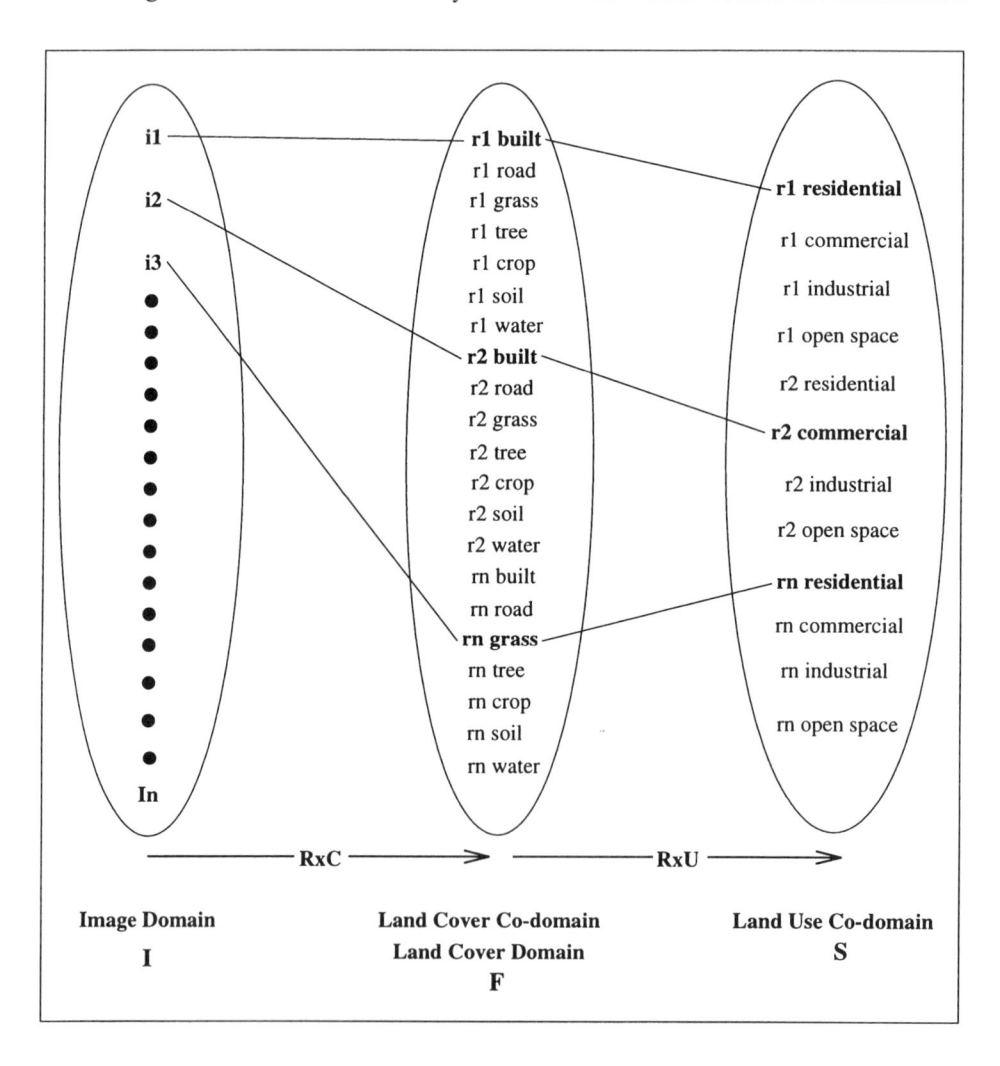

Figure 11.1 *Definition of the sets F and S which allows the composition* I ∘ S = {(i ↦ s) : i ∈ I, s ∈ S, ∃f ∈ F([i ↦ f] ∧ [f ↦ s])} *to be used to derive second-order thematic information (land use). The text in bold (e.g.* **r1 residential***) indicates the member of the subset in any given co-domain to which a mapping applies*

from an analysis of the spatial structure of land cover regions to be expressed as a decision-theoretic problem. In the multi-class case, the decision rule is

$$\forall r \in R\{f_r \in F \mapsto s_r \in S \; iff \; g_i(s_r) > g_j(s_r), \forall j \in U \wedge i \neq j\} \quad (4)$$

where $g(s_r)$ is the functional agreement between the spatial structure of the land cover regions in the area concerned and the expected spatial composition for a given land use category. The latter may be based on predefined examples sampled from within the image – loosely equivalent to 'training areas' in per-pixel, multispectral classification – or on more abstract models of expected scene structure. For the special case of a binary decision rule, Equation (4) can be modified to give

$$f_r \in F \mapsto s_r \in S \stackrel{def}{=} \begin{cases} 1 & iff \; g(s_r) > \mathcal{T}_s \\ 0 & \text{otherwise} \end{cases} \quad (5)$$

where $g(s_r)$ is as above and $\mathcal{T}_s$ is some predefined threshold or similarity measure (defined in spatial structural terms) to be achieved with respect to s.

11.3 A Syntactic Pattern Recognition Paradigm

The previous section outlined a formal statement of a method for deriving land use information from the spatial structure of discrete, land cover regions identified in a satellite sensor image. In this section, we review the applicability of syntactic pattern-recognition techniques as a practical framework within which this might be achieved.

Broadly speaking, syntactic pattern recognition (hereafter abbreviated to 'SyntPR', after Schalkoff, 1992) is concerned with the analysis of the spatial and semantic structure of digital images. Although it has received relatively little attention to date in the context of analysing and processing satellite sensor images, SyntPR has been widely used in the interpretation of digitized aerial photography (Matsuyama, 1987; Nicolin and Gabler, 1987; McKeown, 1988; Mehldau and Schowengerdt, 1990). In this latter context, SyntPR techniques have been used to recognize the basic spatial entities (e.g., tarmac runways) present within simple scenes, to analyse these as spatial agglomerations, and hence to identify the principal *functional* groupings (e.g., airports; McKeown et al., 1989). The comparative lack of interest in these techniques in the broader field of Earth observation owes much to the relatively coarse spatial resolution of past and, to a lesser extent, present satellite sensors. This places severe limits on the accuracy with which the primary spatial entities (e.g., fields, roads and buildings) in the observed scene can be identified and delineated, with concomitant impacts on the measurement of their morphological properties and structural relations. This situation is, however, set to change dramatically with the imminent launch of a number of commercial satellite sensors which will produce image data with a spatial resolution of between 1 and 4 m in panchromatic mode and between 4 and 15 m in multispectral mode (McDonald, 1995; Fritz, 1996). Interest in syntactic pattern recognition techniques is therefore expected to grow considerably over the next few years as data from these sensors become increasingly widely available (Guindon, 1997).

One feature of the few previous attempts to use SyntPR techniques to analyse satellite sensor images is that they have typically employed a rather limited subset of structural properties and relations (Gurney and Townshend, 1983; Moller-Jensen, 1990; Johnsson, 1994). This, in turn, restricted the range of spatial analysis techniques available to them. If SyntPR techniques are to be used to their full potential to analyse satellite sensor images, then a much wider range of properties and relations (and hence data processing techniques) must be exploited. This raises several important issues that will be dealt with in the following sections, namely:

- what is the most suitable geometric representation scheme to derive and encode the structural information contained within first-order (i.e., land cover) thematic maps produced from satellite sensor images;
- what structural information can be derived from this geometric representation and how appropriate is this to the task of deriving the second-order thematic (e.g., land use) information about the corresponding scene;
- how should this structural information be organized and represented in a computational form to enable flexible access and analysis in subsequent data processing stages; and,
- what data processing techniques are most appropriate to derive second-order thematic information from these structural data?

11.3.1 Geometric Representation

Perhaps the most important stage in deriving structural information from an image is the recognition, geometric representation and encoding of the basic spatial entities present in the corresponding scene (e.g., fields, woods, lakes, roads and buildings). This is because all other structural information will be derived from these data. The fundamental spatial unit in remote sensing is, of course, the pixel, in as much as it defines the area of ground from which the sensor samples the exitant radiation at a given instant in time. Partly as a result of this, and partly because of their computational simplicity, the overwhelming majority of measures of spatial pattern and structure used in remote sensing are based on the 'moving-window' approach – focusing on variability that can be determined within a multi-pixel (typically rectangular) window passed across the image. These range from estimates of image texture based on grey-level co-occurrence matrices (Haralick, 1979) to measures of the spatial pattern of land cover in first-order thematic maps (Barnsley and Barr, 1996). The pixel is, however, an arbitrary spatial unit – an artefact of the spatial resolution of the sensor – which bears little resemblance to the geometric form of the principal spatial entities in the corresponding scene. By the same token, the moving-window approach artificially constrains the geographical focus within which spatial pattern and structure are explored (Dillworth et al., 1994). Moreover, the optimum size of the window must normally be determined empirically and often varies across the scene (Barnsley and Barr, 1996).

The alternative to a pixel-based geometric representation scheme is a region-based approach. In the context of this chapter, a region is defined as the maximal set of spatially contiguous pixels which have been assigned the same label in an initial (land cover) classification of the image (Barr and Barnsley, 1995a). Regions defined in this way should correspond more closely (in geometric terms) to the principal spatial entities in the scene, allowing for the effects of the spatial and spectral sampling of the sensor and errors or deficiencies in the classification. A further advantage of the region-based approach is that it is possible to derive a much wider range of structural information, compared to that available via the pixel-based approach, although there is a small computational overhead in deriving the regions from the land cover map.

There are many ways to derive and represent a set of discrete, non-overlapping regions from a land cover map (Gonzales and Wintz, 1987; Schalkoff, 1989; Sonka et al., 1993). These include quadtrees, polygonal approximations, the use of one-dimensional signatures, B-spline representation, convex hull representations and techniques based on computational morphology. The approach preferred here is to use a boundary-tracing algorithm and to encode the elements of the boundary by means of Freeman chain-codes (Freeman, 1975; Gonzales and Wintz, 1987; Barr, 1992; Barr and Barnsley, 1993). In addition to its simplicity, this approach has the advantage that the region boundaries are not generalized in any way; although, by the same token, they retain the 'saw-tooth' effect resulting from the regular tessellation (pixelization) of the scene. Figure 11.2 illustrates the process of deriving and encoding the boundary of a single region using Freeman chain-codes. The resultant information – namely, (i) the unique, ordinal reference (ID) for the region, (ii) the location of the first (start) element of the region boundary in image space, (iii) the label assigned to the region in the land cover classification, and (iv) the Freeman chain-code description of the region boundary – is stored in a Region Search Map (RSM) (Barr and Barnsley, 1997).

11.3.2 Structural Information

Having derived a geometric description of the regions in the land cover map, the next task is to extract information on their structural characteristics. Regions embedded in a two-dimensional plane can convey a variety of structural information. Broadly speaking, we can divide this information into two main types, namely *properties* of a region and *relations* between two or more regions.

Table 11.1 lists several of the *properties* that can be derived for a discrete, labelled region. Most of these are *morphological* descriptors which can be derived from an analysis of either the region's boundary or its interior elements. The other properties are derived from further analysis of the land cover map or the original multispectral image, in conjunction with information on the spatial extent of each region. It is important to note that the properties listed in Table 11.1 are expressed in a variety of forms, ranging from a character string to an $N \times N$ matrix of real numbers. The issue of how such diverse information should be represented and stored in computational form is discussed further in Section 11.3.3.

The spatial relations between two or more regions are similarly diverse in form and nature (Table 11.2). The spatial topological relations listed in Table 11.2 are obtained using a relational algebra based on the intersection of the boundaries and interiors of two *objects* (Egenhofer and Franzosa, 1991; Egenofer et al., 1994; Sharma et al., 1994; Worboys, 1995). A set of quantitative, non-topological spatial relations also exist,

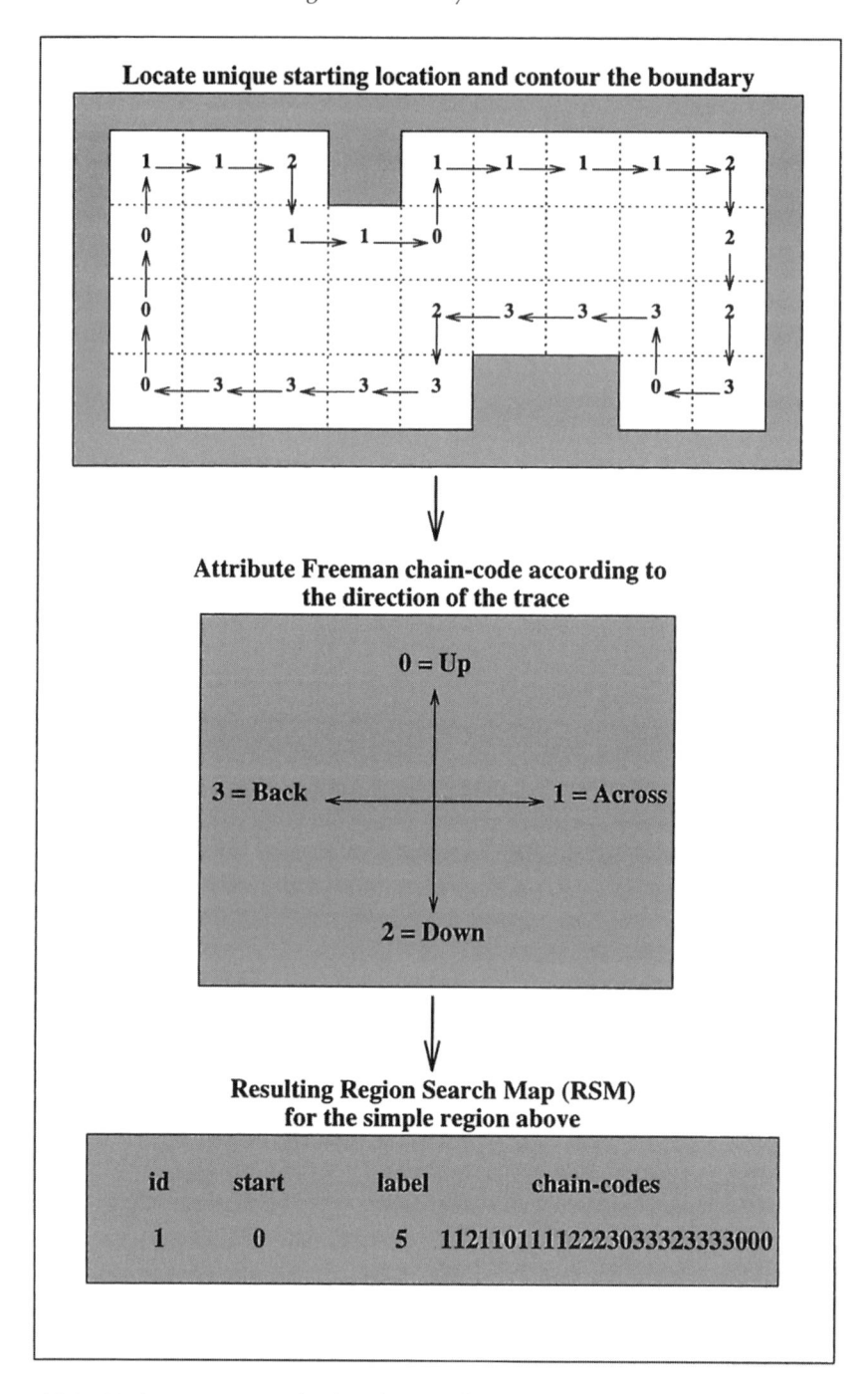

174 *Advances in Remote Sensing and GIS Analysis*

Figure 11.2 *Various stages involved in deriving the Freeman chain-code representation of a first-order (land cover) region*

Table 11.1 Selected region properties. Adapted from Gonzalez and Wintz (1987), Sonka et al. (1993) and Heijden (1994)

Group	Property	Form	Units
Interior	area	single real	pixel or geographic
	compactness	single real	unitless
	aspect ratio	single real	ratio
	centroid (1st moment)	2-element real vector	pixel or geographic
	Nth-order moments	N-element real vector	pixel or geographic
	eccentricity	single real	pixel or geographic
	subjective characterization	string (e.g., huge, small)	symbolic
Boundary	perimeter	single real	pixel or geographic
	curvature	single real	unitless
	chain-code shape	chain of reals	unitless
	width	single real	pixel or geographic
	height	single real	pixel or geographic
	diameter	single real	pixel or geographic
	elongatedness	single real	ratio
	rectangularity	single real	ratio
	direction	single real	angular
	bounding rectangle	4-element real vector	pixel or geographic
	subjective characterization	string (e.g., narrow, long)	symbolic
Grey-level	mean grey-level	N-element real vector	digital number
	median grey-level	N-element real vector	digital number
	covariance	$N \times N$ real matrix	digital number
	texture	single real	unitless
Thematic	assigned class	string	symbolic
	confidence of assignment	single real	probability/fuzzy

including the distance between two regions and the angular orientation (or direction) between them (Peuquet and Ci-Xiang, 1987). Similarly, a set of qualitative, non-topological spatial relations may also be derived as a result of assessing the relative projected positions of the regions with respect to one or more coordinate axes; these are expressed as a set of extended symbolic projections (Holmes and Jungert, 1992). Further spatial relations, which may be either topological or non-topological, can be produced as combinations of the spatial relations given above on the basis of some subjective criteria. For example, if we describe the spatial relationship between two regions in terms of being '*near*' or '*far*', the relationship depends on the subjective characterization of distance.

11.3.3 Structural Representation

The ability to exploit the structural information contained within an image divided into a set of discrete, non-overlapping regions is reliant on the availability of a flexible means of representing and computationally encoding this information. One way to achieve this is to use the concepts and algorithms arising from graph theory – a

Table 11.2 *Selected spatial relations between regions. Adapted from Egenhofer and Franzosa (1991), Frank and Mark (1991), Holmes and Jungert (1992) and Worboys (1995)*

Group	Relation
Spatial topological	A and B are disjoint
	A and B touch (adjacent)
	A equals B
	A is inside B (B contains A)
	A is covered by B (B covers A)
	A contains B (B is inside A)
	A covers B (B is covered by A)
	A and B overlap with disjoint boundaries
	A and B overlap with intersecting boundaries
Spatial metric	boundary distance between A and B
	centroid distance between A and B
	orientation and direction between A and B
Spatial ordering	A is above B (B is below A)
	A is below B (B is above A)
	A is left of B (B is right of A)
	A is right of B (B is left of A)
	A is in front of B (B is behind A)
	A is behind B (B is in front of A)
	A is between B and C
Subjective spatial relations	A is near B (B is near A)
	A is far from B (B is far from A)
	other symmetric, subjective relations of A and B
	other non-symmetric, subjective relations of A and B

well-established branch of set-theoretic, discrete mathematics (Schalkoff, 1992; Sonka et al., 1993; Heijden, 1994). Thus, it is possible to represent image structure in the form of a graph

$$G = \{N, R\} \qquad (6)$$

where N is the (non-empty) set of nodes (or vertices), for which each $n \in N$ corresponds to a single, discrete region identified in the land cover map, and R is the set of edges, $R \subset N \times N$, for which each $r \in R$ indicates the existence of a relationship between a pair of nodes (regions) in the set N; so that, $(a, b) \in R$ signifies that a relationship exists between regions a and b. Figure 11.3 illustrates the graph representation of the spatial relation `adjacency` for a simple scene comprising six regions.

Simple graphs, such as the one defined in Equation (6), can be extended to represent more than one spatial relation at a time and to represent selected properties (e.g., some morphological descriptor) of the regions. This is commonly referred to as an '*attribute graph*'

$$G_a = \{N, R, P\} \qquad (7)$$

where R is the set of relations that exist between nodes in the set N, and P is the set of properties of these nodes (Schalkoff, 1989, 1992). Barr and Barnsley (1997) have further

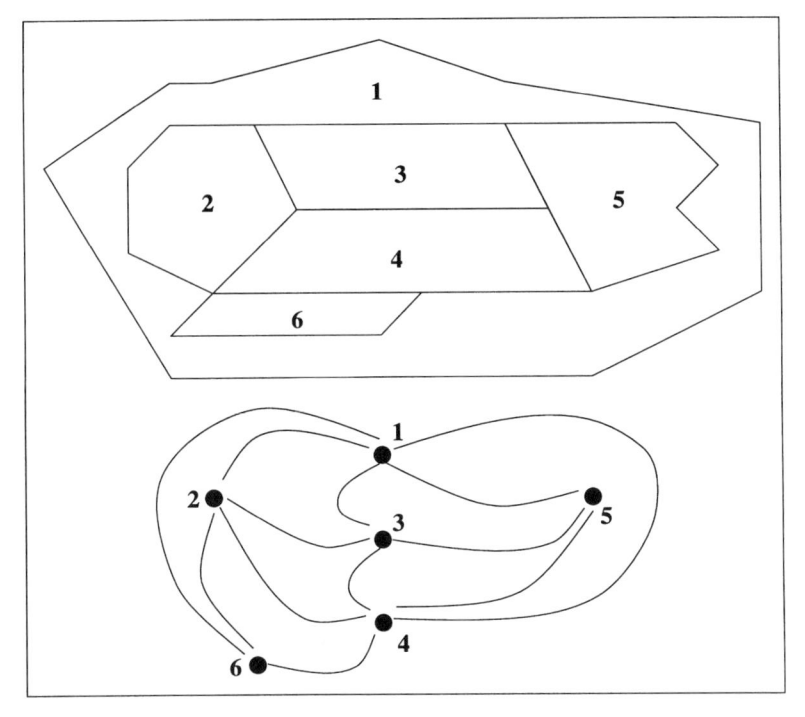

Figure 11.3 *Graph representation of the spatial relation* adjacency *for a simple scene comprising six discrete regions*

extended the basic attribute graph to derive a data model, referred to as XRAG (eXtended Relational Attribute Graph), which is defined as the heptuple

$$XRAG = \{N, E, EP, I, L, G, C\} \quad (8)$$

where

- N is the set of nodes, such that $N \neq \{\}$;
  E is the set of (*extrinsic*) spatial relations between $n \in N$ (e.g., adjacency and containment);
- EP is the set of (*extrinsic*) spatial properties associated with the relations in E (e.g., distance and direction);
- I is the set of (*intrinsic*) properties relating to $n \in N$ (e.g., area and perimeter);
- L is the set of labels (*interpretations*) assigned to $n \in N$ (e.g., grass, tree, urban, non-urban);
- G is the set of groups '*binding*' each $l \in L$ to an interpretational context (e.g., the label tarmac may be 'bound' to the group land cover, while the label residential is bound to the group land use); and,
- C is the set stating the confidence to which $l \in L \rightarrow n \in N$ (e.g., 'there is a 0.9 probability that region X belongs to land cover class Y').

This represents a much more complete and explicit model of image structure than either the simple graph or the attribute graph. It is divided into seven, logical subsets (groupings), each representing a different form of structural information (Barr and Barnsley, 1995a, 1995b). XRAG has been fully implemented as a computational data structure in a series of abstracted data types (Barr and Barnsley, 1997; Figure 11.4). In addition, a series of associated data processing modules has been developed that can be used to derive information on the properties and relations listed in Tables 11.1 and 11.2 from a Freeman chain-code representation of the regions identified in an initial land cover classification. These are used to populate the XRAG data structure.

Semantic networks represent a closely related alternative to XRAG (Ballard and Brown, 1982; Schalkoff, 1989, 1992). They also utilize a graph-based representation of image structure, but do so in a less rigorous and formal manner than either basic attribute graphs or the XRAG model. In a semantic network, all relations and properties – whether numeric or symbolic – are represented as directed edges between nodes. While this provides a similar degree of flexibility to XRAG, in terms of representing image structure, problems can be encountered when attempting to use the information represented within them. In particular, the information contained within a standard semantic network is not ordered or structured in any way. The result of this, even for simple scenes, is often a large, highly interconnected network in which it may be difficult to examine the subtle variations in image structure that are likely to exist between different land use categories or other second-order themes (Schalkoff, 1992).

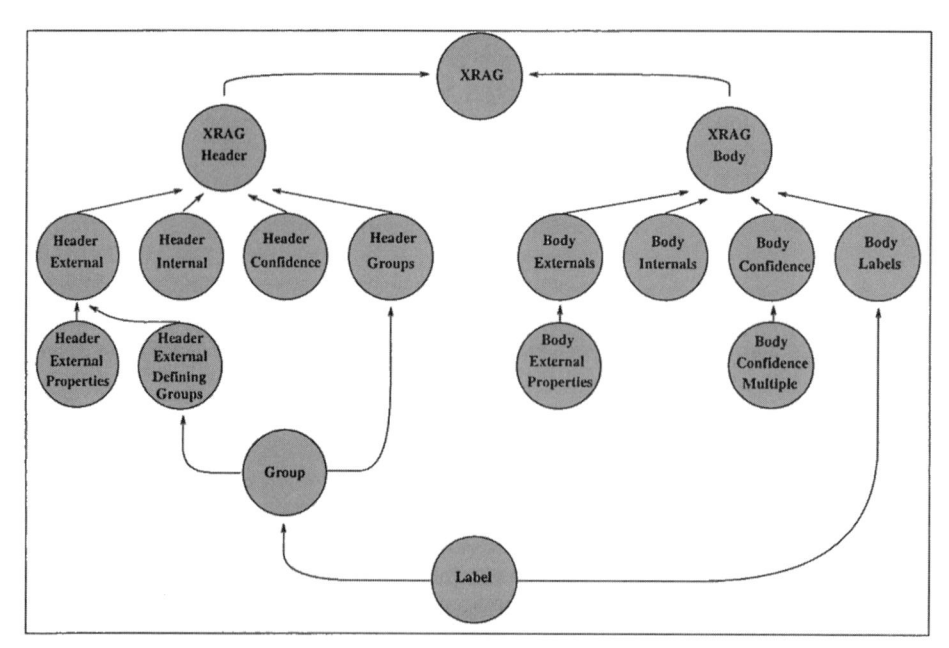

Figure 11.4 *XRAG data structure, showing the hierarchy and inheritance structure of the various data types that it contains*

11.3.4 Structural Inference

Ultimately, the purpose of deriving and representing the structural information contained in a land cover map derived from a remotely sensed image is to infer additional information about the corresponding scene; for example, that pertaining to land use (Barnsley and Barr, 1997). Relatively little attention has, however, been given to the development of the data processing techniques needed to perform this type of inference in the context of digital, multispectral images obtained by Earth-orbiting satellite sensors. Indeed, there is a general paucity of formal models on the structural operators, their semantics and expected results required to underpin the development of such techniques. While a full discussion of the necessary models and techniques is beyond the scope of this chapter, a number of general SyntPR approaches (that might act as a starting point for the development of structural inference tools) are reviewed.

We start by returning to Equation (4), which presented a decision rule that might be used to determine the land use category associated with a single region (or set of regions) identified in an initial land cover map derived from a remotely sensed image. The decision rule involves an assessment of the *match* or *similarity* between the spatial structure of the land cover regions in any given part of the image and that of a set of predefined (candidate) land use categories. This was simplified in Equation (5) to produce a binary decision rule, measured against some structural threshold, $\mathcal{T}_s$, for a specific land use category. The comparison between $g(s_r)$ and $\mathcal{T}_s$ might be based on searching the structural information to satisfy either a single goal or a series of related goals that characterize the particular land use in question. Techniques such as this are often termed *condition searching/goal satisfaction* approaches.

Several techniques have been developed to assess the similarity between graph models and graphically represented spatial data (Schalkoff, 1992). In general, these examine whether the model and the data are *isomorphic* (i.e., have the same graph structure) or, if they are not, the degree to which they are non-isomorphic (e.g., by determining the number of structural differences that exist) (Ballard and Brown, 1982; Schalkoff, 1992; Sonka *et al.*, 1993; Heijden, 1994). Graph matching and graph similarity measures have been widely used in a number of computer vision/machine vision tasks, though these often involve a relatively small number of basic structural descriptions of the spatial entities in the observed scene (Ballard and Brown, 1982; Sonka *et al.*, 1993; Heijden, 1994). Their applicability to the analysis and interpretation of images acquired by Earth observation sensors is likely to be more problematic, largely because of the greater structural complexity of the corresponding scenes. This results in much greater variability – and, hence, uncertainty – not only in terms of the geometric properties of scene primitives (regions), but also in their structural properties and relations. Consequently, several important questions need to be addressed before graph matching or graph similarity measures can be used routinely to derive second-order information from Earth observation images, namely:

- do second-order themes, such as different categories of land use, differ discernibly in terms of their structural composition at the spatial resolution of current and planned future satellite sensors;
- how variable are the characteristic structural properties and relations of specific

180 *Advances in Remote Sensing and GIS Analysis*

second-order themes at these spatial resolutions; and,
- is a consistent subset of the structural properties and relations of the (first-order) land cover regions appropriate across a range of different sensor spatial resolutions?

Several studies have investigated, either implicitly or explicitly, the potential of using condition-based graph searching as the basis for a goal-satisfaction approach to derive thematic information from Earth observation images (Gurney and Townshend, 1983; Nichol, 1990; Barr, 1992; Barr and Barnsley, 1995b). For example, Gurney and Townshend (1983) identified areas of urban and non-urban land by searching for regions with a predefined size and spatial arrangement (adjacency and containment), and within given geographical distance criteria, in a land cover map derived from a Landsat MSS image (Gurney and Townshend, 1983). Nichol (1990) used a similar approach to generalize land cover data produced from a Landsat MSS image. This was achieved by searching the data for regions below a predetermined size and merging them (i.e., re-labelling and dissolving the shared boundary) with the adjacent region with which they shared the longest common boundary (Nichol, 1990). More recently, Barr (1992) showed that accurate maps of the urban–rural boundary can be derived from SPOT HRV multispectral images from an analysis of the structural characteristics of the constituent land cover regions. The extent of the urban areas was inferred by searching the resulting structural graphs for adjacent 'built' regions (and areas of open space wholly contained within these) according to the minimum area and distance criteria used by the Department of the Environment for England and Wales in their definition of urban land (Barr, 1992). This process is illustrated in Figures 11.5 and 11.6. This work has been extended by Barr and Barnsley (1995a, 1995b), who attempt to build a query-language interface to the graph searching and criteria assessment modules, and by Barnsley and Barr (1997), who use XRAG to examine differences in the spatial structure of various urban land uses based on an analysis of Ordnance Survey digital map data.

An alternative to condition-based graph searching is to use techniques drawn from the field of artificial intelligence, in particular knowledge-based techniques. In these, domain-specific knowledge is used to build models of the composition and structure of

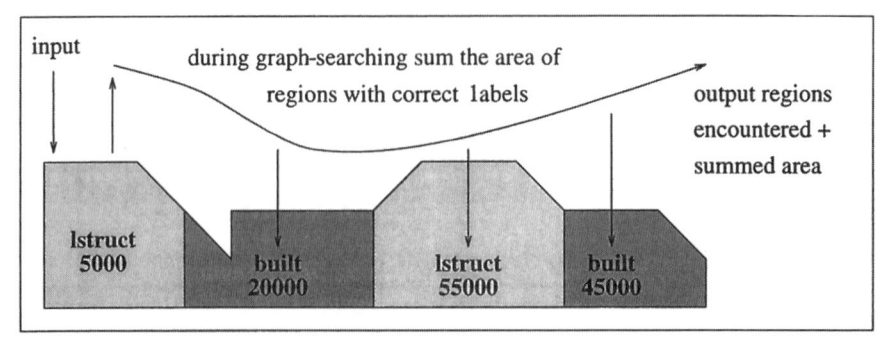

Figure 11.5 *Dynamic graph-searching for* adjacent built (built *and* lstruct) *land cover regions to derive 'urban' land*

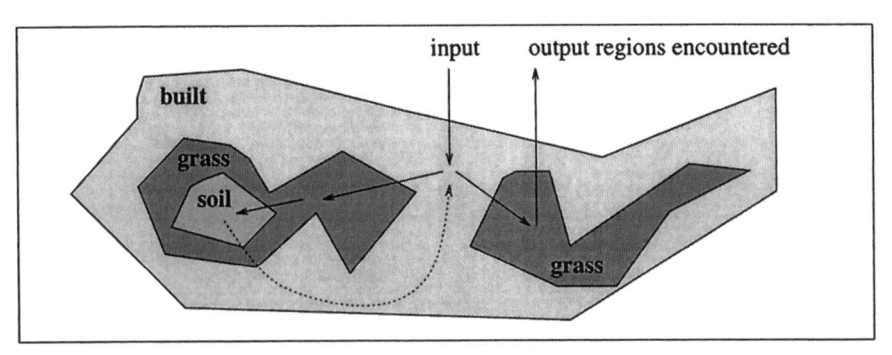

Figure 11.6 Graph-searching multiply contained regions of vegetation in built-up areas to derive 'urban' land

the scene and, hence, the image (Argialas and Harlow, 1990). The models include descriptions and definitions of the constituent objects in the scene (spatial entities in the image) and the relationships between them, as well as the concepts and criteria to be used during data processing. While knowledge-based techniques have been used widely in the interpretation of digitized aerial photography (Matsuyama, 1987; Nicolin and Gabler, 1987; McKeown, 1988; Mehldau and Schowengerdt, 1990), their application to data from satellite sensors is much less common. In those studies which have used a knowledge-based approach, Prolog (Moller-Jensen, 1990) or a commercial expert-system shell (Johnsson, 1994) is used to build the production rules for deriving second-order information from the expected structural composition of the first-order (land cover) regions. Alternatives to production rule-based systems include *frames* and *blackboard models*. These have formed the basis of knowledge representation and system control in several image understanding systems applied to digitized aerial photography (Matsuyama, 1987; Nicolin and Gabler, 1987; Argialas and Harlow, 1990). One advantage that they have over production rule systems is that knowledge and the interpretation process (i.e., the flow of control within the system) can be structured according to the task at hand, allowing knowledge to be embedded within hierarchically organized frameworks (Argialas and Harlow, 1990).

11.4 Conclusions

This chapter has presented the case for deriving 'second-order' thematic information (principally that relating to land use) about a scene from an analysis of the composition and spatial structure (or pattern) of 'first-order' (land cover) regions identified in fine spatial resolution remotely sensed images. The discussion has been placed in the context of a syntactic pattern-recognition paradigm, since this provides an appropriate data processing framework and set of computational techniques. More specifically, a region-based, graph-theoretic data model – known as XRAG (eXtended Relational Attribute Graph) – and its associated data processing algorithms that are appropriate to the task of inferring second-order thematic information have been reviewed. A fuller discussion of the XRAG data model is given in Barr and Barnsley (1997), while the application of XRAG and related data processing algorithms to the analysis of land use in Ordnance Survey digital map data is described in Barnsley and Barr (1997). Finally, this chapter has highlighted: (i) the potential of syntactic pattern-recognition approaches, and, (ii) the need to develop new or revised methods for structural inference – such as graph similarity measures, graph searching procedures and condition searching/goal satisfaction techniques – that can handle both the complex scenes and the thematic and positional uncertainties associated with satellite sensor images.

References

Argialas, D.P. and Harlow, C.A., 1990, Computational image interpretation models: an overview and perspective, *Photogrammetric Engineering and Remote Sensing*, **56**(6), 871–886.

Ballard, D.H. and Brown, C.M., 1982, *Computer Vision*. (Englewood Cliffs, NJ: Prentice-Hall).

Barnsley, M.J. and Barr, S.L., 1996, Inferring urban land use from satellite sensor images using kernel-based spatial reclassification, *Photogrammeric Engineering and Remote Sensing*, **62**(8), 949–958.

Barnsley, M.J. and Barr, S.L., 1997, A graph-based, structural pattern recognition system to infer urban land use from fine spatial resolution land cover data, *Computers, Environment and Urban Systems*, **21**(3/4), 209–225.

Barnsley, M.J., Barr, S.L., Hamid, A., Muller, J-P., Sadler, G.J. and Shepherd, J.W., 1993, Spatial analytical tools to monitor the urban environment, in P. Mather (ed.), *Geographical Information Handling – Research and Applications* (Chichester: Wiley), 147–184.

Barr, S.L., 1992, Object-based re-classification of high resolution digital imagery for urban land-use monitoring, *Proceedings of XXIX Conference of the International Society for Photogrammetry and Remote Sensing, International Archives of Photogrammetry and Remote Sensing: Commission 7*, Washington, DC, 1–14 August 1992, 969–976.

Barr, S.L. and Barnsley, M.J., 1993, Object-based spatial analytical tools for urban land-use monitoring in a raster processing environment, *Proceedings of the Fourth European GIS Conference (EGIS'93)* (Utrecht: EGIS Foundation), 810–822.

Barr, S. L. and Barnsley, M.J., 1995a, A region-based spatial analysis and modeling system for urban land use mapping, *Proceedings of the Annual Conference of the Remote Sensing Society*, Southampton, September 1995, 1179–1186.

Barr, S.L. and Barnsley, M.J., 1995b, A spatial modeling system to process, analyse and interpret multi-class thematic maps derived from satellite images, in P. Fisher (ed.), *Innovations in GIS 2* (London: Taylor & Francis), 53–65.

Barr, S.L. and Barnsley, M.J., 1997, A region-based, graph-theoretic data model for the inference of second-order thematic information from remotely-sensed images, *International Journal of Geographical Information Science*, **11**(6), 555–576.

Blamire, P. and Barnsley, M.J., 1995, Information extraction from very high spatial resolution images of urban areas, *Proceedings of the Annual Conference of the Remote Sensing Society*, Southampton, September 1995, 1212–1220.

Dillworth, M.E., Whister, J.L. and Merchant, J.W., 1994, Measuring landscape structure using geographic and geometric windows, *Photogrammetric Engineering and Remote Sensing*, **60**(10), 1215–1224.

Egenhofer, M.J. and Franzosa, R.D., 1991, Point-set topological spatial relations, *International Journal of Geographical Information Systems*, **5**(2), 161–174.

Egenhofer, M.J., Clementini, E. and Di Felice, P., 1994, Evaluating inconsistencies among multiple representations, in T.C. Waugh and R.G. Healey (eds), *Advances in GIS Research, Proceedings of the Sixth International Symposium on Spatial Data Handling*, Volume 2, University of Edinburgh, 901–920.

Frank, A.V. and Mark, D.M., 1991, Language issues for GIS, in D.G. Magmire, M.F. Goodchild and D.W. Rhind (eds), *Geographical Information Systems: Principles and Applications* (London: Longman), 147–163.

Freeman, J., 1975, The modeling of spatial relations, *Computer Graphics and Image Processing*, **4**, 156–171.

Fritz, L.W., 1996, The era of commercial earth observation satellites, *Photogrammetric Engineering and Remote Sensing*, **62**, 39–45.

Gonzalez, R.C. and Wintz, P., 1987, *Digital Image Processing* (New York: Addison-Wesley).

Guindon, B., 1997, Computer-based aerial image understanding: a review and assessment of its application to planimetric information extraction from very high resolution satellite images, *Canadian Journal of Remote Sensing*, **23**, 38–47.

Gurney, C.M. and Townshend, J.R.G., 1983, The use of contextual information in the classification of remotely sensed data, *Photogrammetric Engineering and Remote Sensing*, **49**, 55–64.

Haralick, R.M., 1979, Statistical and structural approaches to texture, *Proceedings of the IEEE*, **67**, 786–804.

Heijden, F., van der, 1994, *Image Based Measurement Systems: Object Recognition and Parameter Estimation* (Chichester: Wiley).

Holmes, P.D. and Jungert, E.R.A., 1992, Symbolic and geometric connectivity graph methods for route planning in digitised maps, *IEEE Transactions on Pattern Analysis and Machine Intelligence*, **14**(5), 549–565.

Johnsson, K., 1994, Segment-based land-use classification from SPOT satellite data, *Photogrammetric Engineering and Remote Sensing*, **60**(1), 47–53.

Laurini, R. and Thompson, D., 1992, *Fundamentals of Spatial Information Systems* (London: Academic Press).

McDonald, R.A., 1995, Opening the Cold War sky to the public: declassifying satellite reconnaissance imagery, *Photogrammetric Engineering and Remote Sensing*, **61**, 385–390.

McKeown, D.M., 1988, Building knowledge-based systems for detecting man-made structures from remotely-sensed imagery, *Philosophical Transactions of the Royal Society London, Series A*, **324**, 423–435.

McKeown, D.M., Harvey, W.A. and Wixson, L.E., 1989, Automated knowledge acquisition for aerial image interpretation, *Computer Vision, Graphics and Image Processing*, **46**, 37–81.

Matsuyama, T., 1987, Knowledge-based aerial image understanding systems and expert systems for image processing, *IEEE Transactions on Geoscience and Remote Sensing*, **25**(3), 305–316.

Mehldau, G. and Schowengerdt, R.A., 1990, A C-extension for rule-based image classification systems, *Photogrammetric Engineering and Remote Sensing*, **56**(6), 887–892.

Moller-Jensen, L., 1990, Knowledge-based classification of an urban area using texture and context information in Landsat-TM imagery, *Photogrammetric Engineering and Remote Sensing*, **56**(6), 899–904.

Nichol, D.G., 1990, Region adjacency analysis of remotely-sensed imagery, *International Journal of Remote Sensing*, **11**(11), 2089–2101.

Nicolin, B. and Gabler, R., 1987, A knowledge-based system for the analysis of aerial images. *IEEE Transactions on Geoscience and Remote Sensing*, **25**(3), 317–329.

Peuquet, D. and Ci-Xiang, Z., 1987, An algorithm to determine the directional relationship between arbitrarily shaped polygons in the plane, *Pattern Recognition*, **20**, 65–74.

Sadler, G.J., Barnsley, M.J. and Barr, S.L., 1991, Information extraction from remotely-sensed images for urban land analysis, *Proceedings of the Second European GIS Conference (EGIS'91)* (Utrecht: EGIS Foundation), 955–964.

Schalkoff, R.J., 1989, *Digital Image Processing and Computer Vision* (New York: Wiley).

Schalkoff, R.J., 1992, *Pattern Recognition: Statistical, Structural and Neural Approaches* (New York: Wiley).

Sharma, J., Flewelling, D.M. and Egenhofer, M.J., 1994, A qualitative spatial reasoner, in T.C. Waugh and R.G. Healey (eds), *Advances in GIS Research, Proceedings of the Sixth International Symposium on Spatial Data Handling*, Volume 2, University of Edinburgh, 665–681.

Sonka, M., Hlavac, V. and Boyle, R., 1993, *Image Processing, Analysis and Machine Vision* (London: Chapman & Hall).
Wharton, S.W., 1982, A contextual classification method for recognising land use patterns in high resolution remotely-sensed data, *Pattern Recognition*, **15**, 317–324.
Worboys, M.F., 1995, *GIS: A Computing Perspective* (London: Taylor & Francis).

12
The Rôle of Classified Imagery in Urban Spatial Analysis

Victor Mesev and Paul A. Longley

12.1 Introduction

Over the last 30 years, a wide range of models and techniques have been developed to quantify generalizations which vary significantly across space. These models and techniques have been given the collective name of 'spatial analysis'. GIS, on the other hand, is a more recent innovation, which has emerged in part out of remote sensing and which is becoming increasingly associated with spatial analysis methods. In this chapter we will discuss two of the principal ways in which the association between spatial analysis and GIS is developing: first, through the development of so-called 'data models' (Martin, 1996), whereby model-based assumptions are used to represent and depict spatial distributions of geographical phenomena (e.g., Langford and Unwin, 1994); and, second, through the development and application of spatial analytic measures to summarize the distribution of geographical phenomena (e.g., Batty and Longley, 1994; Bailey and Gattrell, 1995).

We will examine how the spatial information from classified remotely sensed images can be analysed to reveal changes in the urban morphological structure. We will particularly examine how density profiles, calculated from the urban core to the urban periphery, can infer urban processes and prescribe planning scenarios. As a check on the usefulness of classified imagery, comparisons will be made between density profiles generated from more conventional zonal-based urban sources (i.e., population census, and postal). Following Batty and Longley (1994), we will also look at how the complex scale-dependent form of classified urban pixels may be further characterized using

---

Advances in Remote Sensing and GIS Analysis. Edited by Peter M. Atkinson and Nicholas J. Tate.
© 1999 John Wiley & Sons Ltd.

fractal analysis, which is in itself the study of complicated phenomena manifesting self-similarity at many scales (Goodchild and Mark, 1987; De Cola, 1989).

12.2 Spatial Analysis of Urban Areas

The context to this work is the urgent need to develop appropriately detailed urban morphologies suitable for a wide range of policy problems and scenarios. Many of the issues and problems that face cities are consequences of broad secular economic and social changes, and as such affect all settlements within any given settlement system. With regard to the UK such issues include, or have included: decentralization and deconcentration of population throughout the 1960s, 1970s and 1980s (e.g., Shepherd and Congdon, 1990); three decades of retail decentralization (Wrigley, 1996); changes in the pattern and scale of residential development; the implications of increasing car ownership and changed public transport pricing for congestion; the effects of greenbelt policies upon urban development patterns and densities (Longley *et al.*, 1992); and the energy efficiency and broader 'sustainability' of different urban forms (Breheny and Archer, 1996). Following recent developments in GIS and remote sensing, it is now possible to address these and other problems through system-wide, controlled comparisons of detailed urban morphologies.

Given the resurgence of interest in cities and the problems of city systems that is presently under way, it might at first sight appear strange that generalizable comparative analysis of urban morphologies has not become more established. Changing academic fashions aside, there are two core reasons for this which, we will argue, may each be tackled through 'RS–GIS'-based spatial analysis: (i) data quality and their fitness for purpose, and (ii) poverty of spatial measurement methods.

12.2.1 Data Quality and Fitness for Purpose

Most of the potential applications that we have identified above are substantially socio-economic in focus, and as a consequence the quantitative information that forms the basis to analysis is fraught with difficulties of definition, scale and aggregation. The definitional issue focuses upon fitness for purpose with respect to particular 'urban' applications – as, for example, with the definition of 'irreversibly urban' land adopted by the British Office of Population Censuses and Surveys (OPCS, 1984) for land inventory analysis. Spatial analysis of 'urban' land use is founded upon definitions which are inherently more subjective and functionalist than those that are the outcome of satellite imagery classifications but, as we shall see, the difference is one of degree rather than kind. The most appropriate definition of 'urbanity' will vary widely between applications, suggesting the need to tailor data models to specific applications – in other words, a 'horses for courses' approach towards definition and measurement. With regard to both satellite and socio-economic sources, it is only since the mid-1990s that a wide range of appropriately detailed digital sources suited to such a multifaceted task have become available.

A second important aspect of the definitional and measurement issues concerns the problems of scale and aggregation of socio-economic data. The restrictions imposed by confidentiality and other constraints dictate that socio-economic data are made available only for areal aggregations, yet it is a fact of life that, in general, such aggregations are not 'natural' units for socio-economic analysis. Empirical understanding of 'modifiable areal unit' effects upon spatial analysis is developing with ongoing improvements in computing power, the associated developments in GIS software and the proliferation of digital data sets (Openshaw and Alvanides, 1999), while Mitchell *et al.* (1996) have described the ways in which analogous approaches to satellite image classification might be used to 'unmix' areally aggregate data and hence reveal sub-unit variation.

The issues of scale and aggregation effects are present, albeit in different guises, in the classification of the pixels that make up satellite images. As we will see in Section 12.3, the development of satellite technology has resulted in improvement in image resolutions, particularly spatial resolutions – from Landsat Thematic Mapper (TM) of the mid-1980s (30 m) through SPOT HRV (20 m/10 m) to the EyeGlass, EarthWatch and Space Imaging instruments (1–5 m) of the late 1990s (Corbley, 1996). Even though the spatial resolutions are much higher than most socio-economic sources, statistical classifications nevertheless generate error and fuzziness from urban images (Fisher, 1998).

12.2.2 Measuring and Scaling Urban Morphology

Attempts to measure real-world settlements have been based upon idealized shape and size indices (see Haggett *et al.*, 1977), but are evidently inappropriate for representing the irregularity that characterizes developing city structures. As a consequence, they have not been pursued with sustained vigour.

More recently, Batty and Longley (1994) have argued that fractal geometry (Mandelbrot, 1983), a geometry of the irregular, presents a better way of summarizing the spatial structure of real-world forms, and have developed a range of associated morphometric measures of shape, form, scale and dimension. These are based upon a range of scaling relations between urban area, perimeter and radial distance measures, which are used in the calculation of fractal dimensions. This range of scaling relations has been used to assemble fractal measures of urban growth and development patterns, transport networks and utility networks (see, *inter alia*, Frankhauser, 1990; Batty and Longley, 1994). However, these studies have also been shackled to different data definitions, of the kind identified in Section 12.2.1 above.

Moreover, despite Goodchild and Mark's (1987) assertion that empirical evaluation of the fractal dimension 'may be the most important parameter of an irregular ... feature, just as the arithmetic mean and other measures of central tendency are often used as the most characteristic parameters of a sample', progress towards routine calculation of such measures has been rather slow, in part because of the slow pace of quantitative morphological analysis of urban areas in general. Wider use of these measures in a descriptive sense is likely to be a prerequisite to their wider user in inferential analysis of form–process relations.

The unfortunate consequence of these two considerations – that available data have not been fit for particular and specialized purposes and that urban system properties have not routinely been detected through analysis of their forms – has been the retreat from the quest of morphological generalization. In the next section we describe how recent developments in the supply and analysis of spatial data have now fundamentally changed the context to urban morphology research.

12.3 Satellite Data and Urban Analysis

Remote sensing by satellite data has not been applied to the urban landscape quite to the same degree as in the natural environments of the biosphere, hydrosphere or atmosphere (Gurney *et al.*, 1993). Where work has resulted in reliable means for monitoring crop blight, ocean sedimentation and water vapour accumulation, work on cities has been constrained to more large-scale approximations of peripheral, population density (Langford and Unwin, 1994), spatial arrangement of urban land use (Barnsley and Barr, 1996) and, to a lesser extent, road network detection (Wang *et al.*, 1992). Although estimations of land cover and population distributions such as these may be directly inferred from raw satellite data, the bulk of research into urban mapping is primarily based on the recognition of consistent thematic patterns observable in the image. This is known as the classification of like pixels into categorical urban inventories. Pixels can also be grouped by proximity in order to compute the contextual or textural properties of an urban image. For example, the spectral values of roads may be very similar to flat-topped roofs, but if roads are recognized as linear arrangements of pixels, this confusion can be lessened by a contextual decision. Similarly, in textural classifiers, groups of contiguous pixels may be used to characterize spectral variations, or the roughness of an image. Examples of textural classifiers applied to urban areas have documented improvements in the detection of changes in residential development over time, and the detection of housing. More recently, neural networks have been trained to infer land cover classification, including an optimized neural net technique processing both spectral signatures and textural features based on grey-level histograms. However, if a neural net is trained with limited and incomplete examples of the classification, results may tend to be noisy and even unreliable. Examples and further reading on classification techniques can be found in Campbell (1996). Research has also recognized the inherent limitations of within-image classification and instead examined the introduction of information from external sources in order to improve classifications. This external, or ancillary information is most efficiently handled by GIS, and so links between remote sensing and GIS (Ehlers *et al.*, 1989) have seen breakthroughs in both the quality and quantity of classified urban categories. Numerous examples have documented improvements in not only the spatial accuracy of broad urban categories (Harris and Ventura, 1995; Barnsley and Barr, 1996) but also in the number and level of detail of sub-urban categories: non-residential, residential, and levels of residential density (Mesev, 1998), resulting in planning prescriptions.

However, no matter how the urban image is interpreted, the inherently intricate spatial variability of the urban surface along with the complex range of human activities, continues to limit progress in the mapping of urban areas from remotely sensed data. The problem has always been two-fold: to disentangle the mixed nature of urban landscapes, and at the same time to reconcile the physical layout of buildings and structures with definitions of human activity (i.e., land cover into land use). Instead of pursuing this double-edged problem, which in any case is asymptotic (successive improvements will be small but complete classification accuracy will never be achieved), we will concentrate in this chapter on the spatial, and not spectral, properties of classified imagery.

Compared to the spectral domain, very little research has been based on the wealth of information from the spatial form of classified land use categories. This oversight may be partly due to the inherent spatial complexities and overall noisiness associated with classified urban images. Nevertheless, the classified image of a settlement still possesses immense information on the spatial structure of the urban morphology and can also act as a backcloth to many spatial analysis techniques. In the rest of Section 12.3 we will build a case for the use of spatial information from classified imagery in the calculation of conventional density attenuation profiles.

12.3.1 Other Digital Sources of Urban Data

Before we begin to evaluate the potential contribution of classified imagery to urban spatial analysis, it is important to first examine the alternative sources of digital urban data presently available. Our discussion will centre upon data presently available in the UK, though similar digital information is available in most developed countries (see Mesev *et al.*, 1996, for the case of the USA). For the analysis of two- and three-dimensional urban geometry, the simplest measurement unit is the aggregation of population and housing characteristics into administrative areal units. The most popular types of zonal data include the Census of Population, which holds individual household characteristics as aggregated enumeration districts (EDs), and the Central Postcode Directory (CPD) (using the Postzon digital file), holding individual delivery addresses as units, sectors and districts. In both cases, ordinal, interval and ratio data are directly related to areal or volumetric units and are conventionally represented as choropleth generalizations.

In the present context, the most serious disadvantage of zonal units is that they frequently conceal intra-zonal variations in land use types and land use density. As a consequence, zonal data provide only a small number of widely spaced observations. Such observations do not provide sufficient detail to identify density functions, with the consequence that density gradients can be represented, at best, as crude generalizations devoid of fine-scale detail. The problems associated with choropleth population representations can be alleviated, to a certain degree, by interpolating a surface from either the geographically weighted or population weighted zonal centroids (Martin and Bracken, 1991; Martin, 1996). However, such representations are not strongly constrained by the detailed intra-urban geography of built structures.

An early digital data set concerned with the explicit representation of built form along with population distributions was created by vector digitizing urban boundaries from Ordnance Survey (GB) maps and matching these with geographically referenced population data from the UK Census. This urban boundary database was created by the UK Department of the Environment (DoE) (OPCS, 1984) for both 1981 (1:50 000 scale) and 1991 (1:10 000 scale), and delineates simple urban/non-urban boundaries based on DoE/OPCS definitions of what constitutes 'irreversibly urban' land. Land deemed as urban within four or more EDs constituted urban areas, with population totals based on figures from EDs which have 50% or more of their population within the designated urban area. This highly subjective definition of urban land, tied with the problems associated with the digital capture of what is essentially secondary data, means that the DoE database is at best a rather crude approximation of the extent of urban land use. Furthermore, urban land use is far more complicated than a simple urban/non-urban dichotomy, and what exactly constitutes land that is 'irreversibly urban' is heavily dependent upon recorded population totals – missing, perhaps, the peripheral non-residential development of Garreau's (1991) late twentieth century 'edge cities'. A clearer understanding of urban growth and density must therefore be based not only on a general definition of 'urban' but also on the many different land uses which together make up the urban mosaic.

12.3.2 Urban Classified Imagery

Given the limitations of the digital urban data sets discussed thus far, there is clearly scope for information that pertains explicitly to the physical morphology of urban areas.

The success of urban models inevitably depends on the generation of source data. In this chapter, remote sensing is seen as the most appropriate means for producing coverages of urban areas at consistent and readily updatable intervals. Given its rapid retrieval and global availability, satellite remote sensing is an ideal means for producing measurements from which to monitor various aspects of urban dynamics (e.g., Lo, 1986; Donnay and Barnsley, 1999). The advantages of remotely sensed data may be summarized in terms of the type of data representation, data accuracy, temporal flexibility, spatial coverage, and appropriateness in terms of modelling expediency.

Accuracy refers as much to the amount of correct detail available in the data as to spatial precision. In both cases, satellite data have traditionally been considered to lag behind data from topographic maps or aerial photographs. This is because data extraction methods have not fully harnessed the large amounts of information on the land surface that are held by multivariate satellite data sets. However, improved information extraction techniques hold the prospect of enhancing satellite data classifications so that they are at least as accurate as conventional survey-based maps. The recent improvements in the spatial resolutions of satellite data, moreover, suggest that spatial precision is now approaching that of survey-based methods. Finer spatial resolutions invariably mean that more spatial variation in land use can be inferred from remotely sensed data than those based on zonal representations. Many urban density models measure densities from concentric rings successively radiating out from the urban core. Data based on census tracts unfortunately produce rather coarse concentric rings, frequently between 1 km and 50 km in radius (Parr, 1985). Commonly used satellite data from Landsat and SPOT are now capable of 30 m and 10 m spatial resolutions respectively. As a result, concentric zones can now be defined at similar spatial resolutions, and hence pick up greater spatial variability in land use.

The ability to measure finer spatial variabilities is also a function of data representation. Here, methods that produce *net density* are understandably more appropriate than those that produce *gross density* measures from aggregated zonal representations. It is worth noting that studies as early as 1955 pointed out that the basic model for population densities, the negative exponential function, fits data represented by net density better than it fits gross density patterns (McDonald, 1989). Despite this, gross density measurements still became the dominant type of source data used in urban functions.

Another important aspect of data representation concerns data type and data capture. Data represented digitally are seen as more convenient than analogue maps, particularly when subjected to numerical analysis. In contrast with other digital data, satellite sensors are primary data acquisition devices, and as such are relatively free from those potential errors associated with the digitization process.

With respect to temporal flexibility, remotely sensed data are the only source that can constantly, with relatively consistent accuracy, and allowing for atmospheric conditions, produce data over the same area as frequently as every 16 days from Landsat, and every 26 days from SPOT. Data based on the UK Census are unfortunately constrained to the usual 10-year cycle. In addition, the wide areal coverages associated with satellite data allow both large cities and urban regions to be modelled from a convenient single data set, thus keeping computer processing time within acceptable ranges, and eliminating possible errors from data merging.

The last advantage of using classified satellite data in density function testing, that of appropriateness, is directly related to the manner in which functions are formulated here. First, evidence throughout this chapter indicates that the spatial form of urban areas closely resembles the dendritic, self-similarity patterns associated with fractal simulations. It is argued here and by the work of De Cola (1989) that the spatial patterns exhibited by classified satellite imagery capture this fractal nature, and in this way are the most appropriate data set for testing urban density functions based on fractal concepts. Taking into account all of these circumstances, it is clear that satellite data are the best available source of accurate, detailed, consistent, digital, small-scale, and appropriate data for urban modelling.

As we have already suggested in our introduction, most research treats the classified image as the end product (the spectral result) and neglects the wealth of information available on the spatial form of classified image data. We will now turn towards an assessment of the abilities of fractal geometry to measure and summarize the highly irregular spatial patterns of urban land cover/use produced by image classification. In a similar vein to De Cola (1989), fractal geometry will be used to characterize the spatial properties of classified multi-dimensional feature space. However, unlike De Cola, the derived fractal dimensions will further be used for comparative analyses which are designed to evaluate how form and density of urban land use vary within settlements. Urban profiles will further provide a means with which incremental urban development is precisely monitored and hence help identify which urban processes may be in operation. It is hoped that remotely sensed data might rejuvenate the perception that urban density functions provide useful and appropriate measures of urban development (cf. Zielinski, 1980). In this light classified image data are seen as more accurate and more consistent than traditional analogue data sources, as well as being more frequently updatable than most digitized data. Much of the earliest work on urban functions was based on deriving density from the ratio of land use, obtained from analogue topographic maps, and population, extracted from census tracts (summarized in Mills and Tan, 1980). More recently, digital representations of urban areas have provided more accurate and detailed delineations of urban boundaries and as such have produced more reliable measurements of form and density (Longley *et al.*, 1991; Longley and Mesev, 1997). In the United States, digital representations of residential streets are held as TIGER files, and represent a convenient surrogate for population distributions (Batty and Xie, 1994). This chapter will examine the contribution that classified remotely sensed data can make as the most important source data to the analysis of urban density gradients.

12.4 An Empirical Investigation of Density Profiles, Space Filling and Urban Morphology

In this section, we will demonstrate how differences introduced by the type of data representation affect the recorded measurements of urban morphology. Our case study is Norwich, UK, and its immediate environs. The city has a population of just over 200 000, covering approximately 185 km^2, and in our density profile analysis we have taken its cathedral as the historic centre and initial point of development. The comparison involves 1991 Census data at the ED level, 1991 postcode data at the sector level, population surface based on 1991 Census data, and a near-contemporary satellite image. The remotely sensed image is derived from a subset of a Landsat-5 TM scene (203-024) taken on 15 July 1989. Standard re-sampling and registration procedures are administered using the ERDAS (*Imagine* 8.2) image processing system. In Section 12.5, the same image will be compared against 1991 vector boundary data from the DoE series. This time the comparison will demonstrate how fractal geometry can be used to reveal important indicators of urban form and density. In order to make sure comparisons, in both cases, are consistent across disparate data sets, both the type of urban information and the scale of analysis have been standardized as much as practicable.

There are two main objectives of the first empirical analysis. The first is to examine whether and how classifications built around remotely sensed data differ from digital socio-economic sources, and the second is to ascertain whether and how any such differences affect the outcome and interpretation of urban morphological analysis.

All the data sets used in our comparative analysis are illustrated in Figure 12.1, their descriptive statistics in Table 12.1, and their derivation in Figure 12.2.

12.4.1 1991 Census (Zonal Form)

This layer comprised enumeration district (ED) digital boundaries, extracted as an ARC interchange file from the UKBORDERS database held at the University of

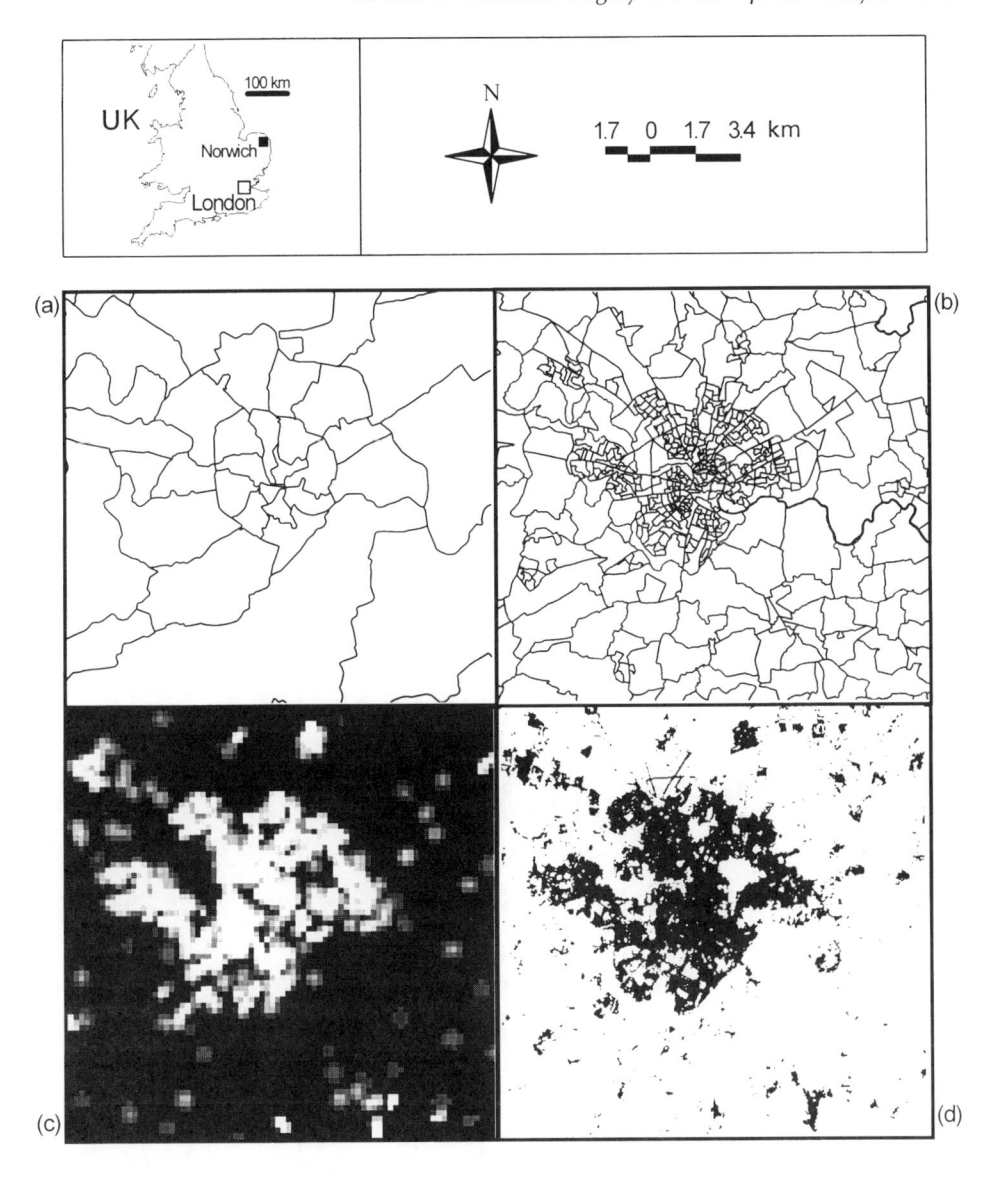

Figure 12.1 Four representations of the Norwich settlement: (a) postal sectors; (b) enumeration districts; (c) surface model; (d) classified satellite imagery

Edinburgh. These polygons are then linked to attribute data pertaining to the 'number of households' from the UK Census of Population database. A series of concentric zones was then imposed at regular intervals of 707 m (the diagonal calculation of 500 m). Although this distance is to some extent arbitrary, it does crudely approximate to the average spanning distance of an 'average' ED, and is similar to the scale of analysis in earlier work on density gradients (Mills and Tan, 1980). Density values for

Table 12.1 Spatial parameters for each of the four urban data representations

Parameter	Postal	Census	Surface	Image
Data type	Zonal (variable)	Zonal (variable)	Raster (202 m)	Raster (30 m)
Number (N)	27	437	5776	262 144
Zone width (R_w)	500 m	500 m	202 m	30 m
Diagonal zone width (R)	707.1 m	707.1 m	285.8 m	42.4 m
Number of zones (R_n)	11	11	27	182
Maximum radius (R_m)	7778.1 m	7778.1 m	7716.6 m	7716.8 m
No. of residential units (N_I)	66 382 delivery pts	81 332 households	83 292 households	80 925 households
Total area (A_F)	190.1 km^2	190.1 km^2	187.1 km^2	187.1 km^2
Area occupied (A_I)	190.1 km^2	190.1 km^2	34.6 km^2	31.7 km^2
Average density (N_I/A_I)	349.2 del.pts/km^2	427.8 hholds/km^2	2407.3 hholds/km^2	2552.8 hholds/km^2

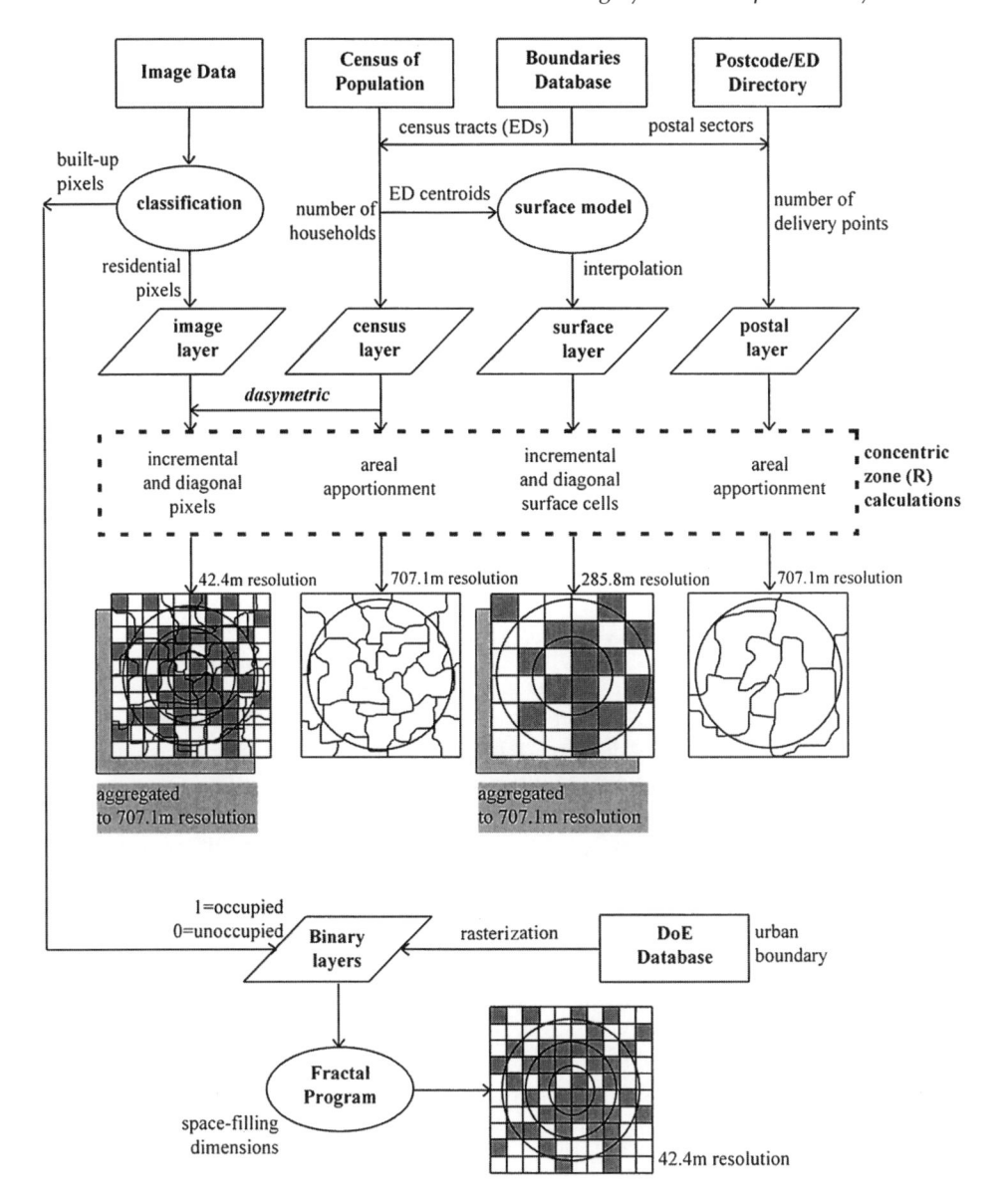

Figure 12.2 The flow of spatial analysis: data preparation and generation of concentric zones

each zone were then calculated by areal apportionment, whereby the populations of EDs falling entirely within any given zone were attributed to that zone. The populations of EDs that traversed two or more zones were allocated between zones in direct proportion to area.

12.4.2 Postal Geography (Zonal Form)

This was created in a similar manner to the census layer, again using a series of 707 m concentric zones, although at the much coarser resolution of the postcode sector. Digital postal sector boundaries were extracted as an ARC coverage from the Geoplan postcode database. The number of household (commercial omitted) delivery points for each sector was then determined using the 1991 Postcode/ED directory files held by MIDAS at the University of Manchester. Because postcodes and EDs are spatially incompatible, these files are used to calculate how many postcodes fall within each ED. Although the spatial unit used here is the postcode sector, the directory contains details of household numbers at the finest level, the unit postcode. Note that from Table 12.1 the total number of delivery points is lower than the total number of households. This is largely because of the presence of multiple households at single delivery points (i.e., flats).

Again, areal apportionment (Figure 12.3) was used to determine the density of delivery points for each 707 m concentric zone.

12.4.3 1991 Census (Population Surface Model)

As outlined above, the population surface model is a point-based areal interpolation technique which essentially constructs a raster grid from ED centroids and intercentroidal distances based on a distance–decay function (Martin and Bracken, 1991; Martin, 1996). ED centroids are digitized during compilation of the Census data products, and are subjectively defined representations of the probable centre of settlement in the ED – as such, they constitute a density measure. The surface model then distributes the 'number of households' attribute for each ED into grid cells using a distance–decay from the centroids: density values are thus high for centroids which are closely spaced and low where centroids are far apart. The improvement over conventional 'uniform area' ED representations is that the interpolated cells are derived from population-weighted centroids and so will more likely mirror the population distribution of a settlement. Thus although the location of specific built forms is not a component of the model, the model is less likely to misrepresent non-residential geography. It has been found that the interpolation assumptions operate most efficiently with square grid cells of around 200 m (Martin and Bracken, 1991). In this study, surface cells had a 202 m dimension, calculated to 285.8 m in the diagonal radius.

12.4.4 Image Layer

This layer is somewhat different from the other three. The image is not a consequence of administrative spatial divisions but a statistical classification of primary data pertaining to the Earth's physical surface. The classification of the image is able to determine the built-up land cover of the urban landscape and is completely independent of any functional characteristics on urban land use. Although a limited amount of

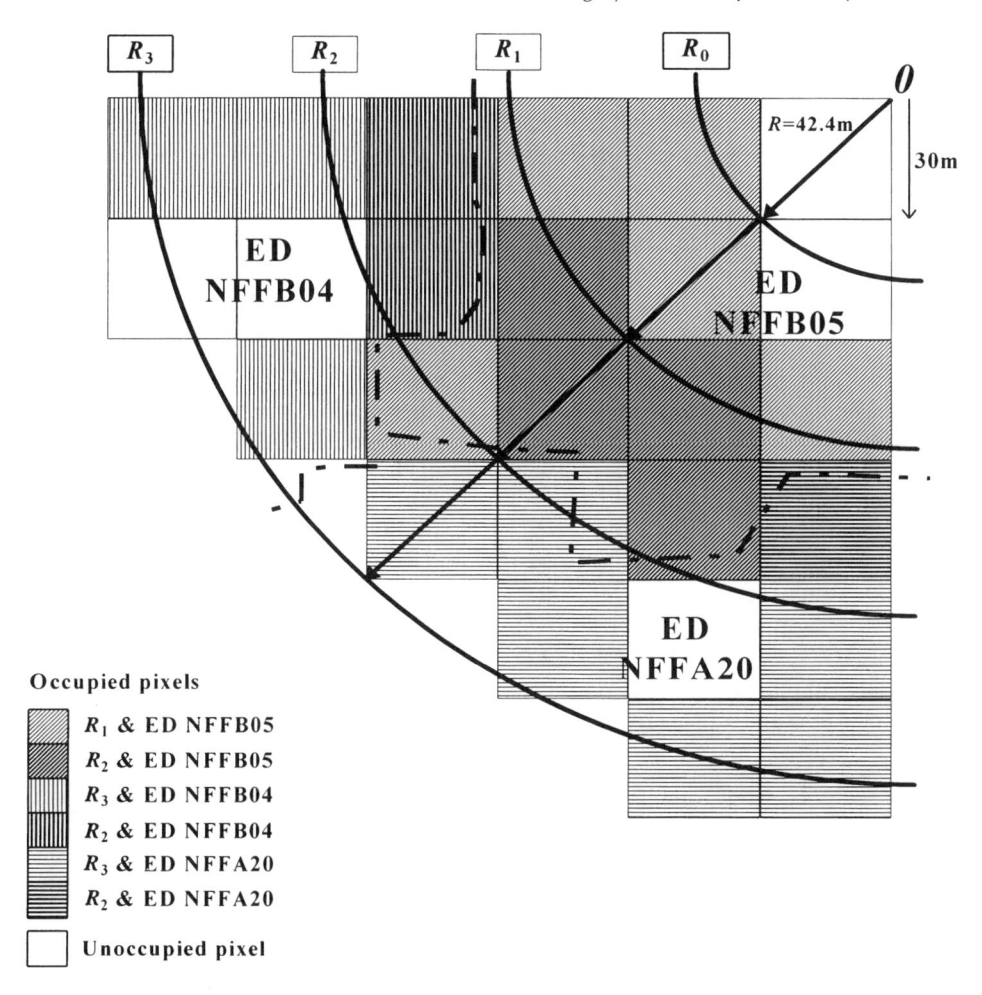

The Rôle of Classified Imagery in Urban Spatial Analysis 197

Figure 12.3 Allocation of enumeration district (ED) data to occupied pixels in concentric zones: for example, concentric zone R_2 contains seven occupied pixels composed of four from ED NFFB05, two from ED NFFB04, and one from ED NFFA20

functional urban land use can be imputed directly from the image, the more reliable means is by supplying the classification with additional socio-economic information. One such modification is to use data on the number of residential units from the Census in order to vary the *a priori* probabilities of the residential spectral classes during a maximum likelihood classification. The resulting classifications have accuracy levels of over 90%, and provide a good measure of the distribution of (mainly aggregations of) residential buildings. Each pixel is classified deterministically as either 'residential' or 'non-residential'. The technique is fully documented in Mesev (1998), and essentially involves the use of GIS data to first stratify urban images according to some spatial and contextual rules, and then to determine the area estimates of urban classes within each stratum. Area estimates are then normalized and inserted directly into the ML classifier as prior probabilities $\Pr(\mathbf{x}|w,z)$, that is the probability of pixel vector $\mathbf{x}$ belonging to class w is weighted by census variable z. All in all, ML classifications with *a priori* census weightings produce accuracy levels above those based simply on the standard equal prior probability assumption. In work elsewhere, favourable results have also been generated from area estimates which have been used as part of an iterative process for updating ML *a posteriori* probabilities (Mesev *et al.*, 1999). The notion of density for each pixel is here defined simply as the maximum likelihood land use allocation per pixel size (400 m² for SPOT, and 900 m² for Landsat).

Generating density profiles from the image layer required the assistance of dasymetric mapping (Langford and Unwin, 1994). Concentric zones of pixels were again partitioned by diagonal length (42.4 m for 30 m Landsat TM pixels). The number of occupied pixels (i.e., pixels that were classified as 'residential') in each zone were then overlaid with EDs, containing household numbers. Using dasymetric principles, occupied pixels were then allocated a proportion of an ED's count based on the spatial correspondence of the pixel and the ED. To simplify matters, over half the area of an occupied pixel had to lie within a corresponding ED for that pixel to be deemed part of the ED. An illustration of the principle is given in Figure 12.3. For example, concentric zone R_2 contains seven occupied pixels composed of four in ED NFFB05, two in ED NFFB04, and one in ED NFFA20. For each zone, the total number of households is calculated in this way, and then expressed as a density when the ratio against total area (occupied and unoccupied) of each zone is computed.

12.4.5 Density Profiles

The absolute and logged residential density profiles at the scales of resolution at which they were each created are shown in Figure 12.4. The finer scale image profile exhibits considerably greater small-scale variability than the surface and conventional Census geographies: the postal geography shows quite wide variability, although this is possibly an artefact of the coarse zoning system. The conventional choropleth representation of density from the Census differs from that of the surface representation of the same information, since the surface model does not presume that all space is populated. The logged profiles show considerable variability in the image at the extremities of the study area, representing the identification of outlying villages which are likely in the process of becoming assimilated into the residential fabric of the City. However, only the image profile suggests that the transition towards such discontinuous developments marks any kind of a boundary between the 'complete' and 'incomplete' parts of the settlement (Fotheringham *et al.*, 1989). This is not picked up in any of the other profiles.

All four profiles are standardized to a common scale of 500 m (707 m in the diagonal) in Figure 12.5. This figure clarifies that the densities recorded in the image are very much higher than those generated by the other data models – because the population of each zonal band is allocated only to those pixels in which the satellite classification records built structures. The population surface model implies densities that lie between those produced by the image and the ED-based uniform area model, sugges-

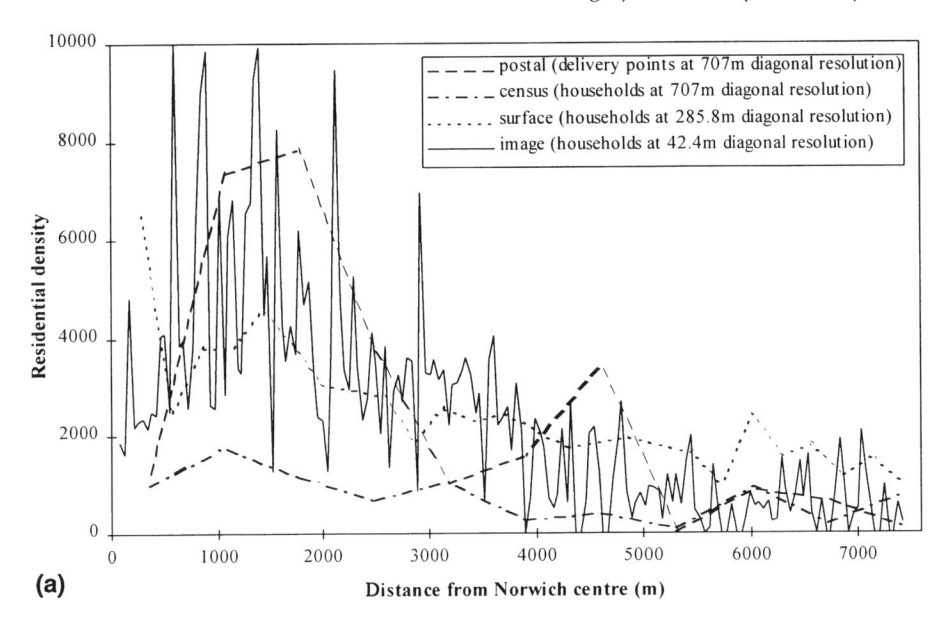

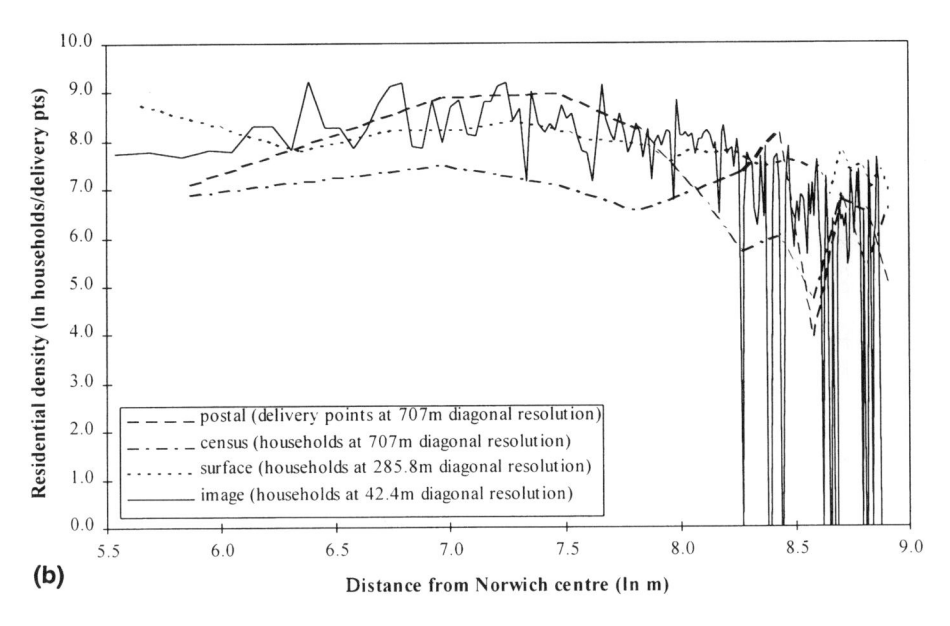

The Rôle of Classified Imagery in Urban Spatial Analysis 199

Figure 12.4 *Residential density profiles at finest available scales: (a) absolute, and (b) logged values*

ting that while it represents an improvement on the choropleth model through creation of unpopulated spaces, the distance–decay function produces a far less precise representation of the geographical extent of residential structures. The magnitude of difference between the two Census geographies is not constant through the profile,

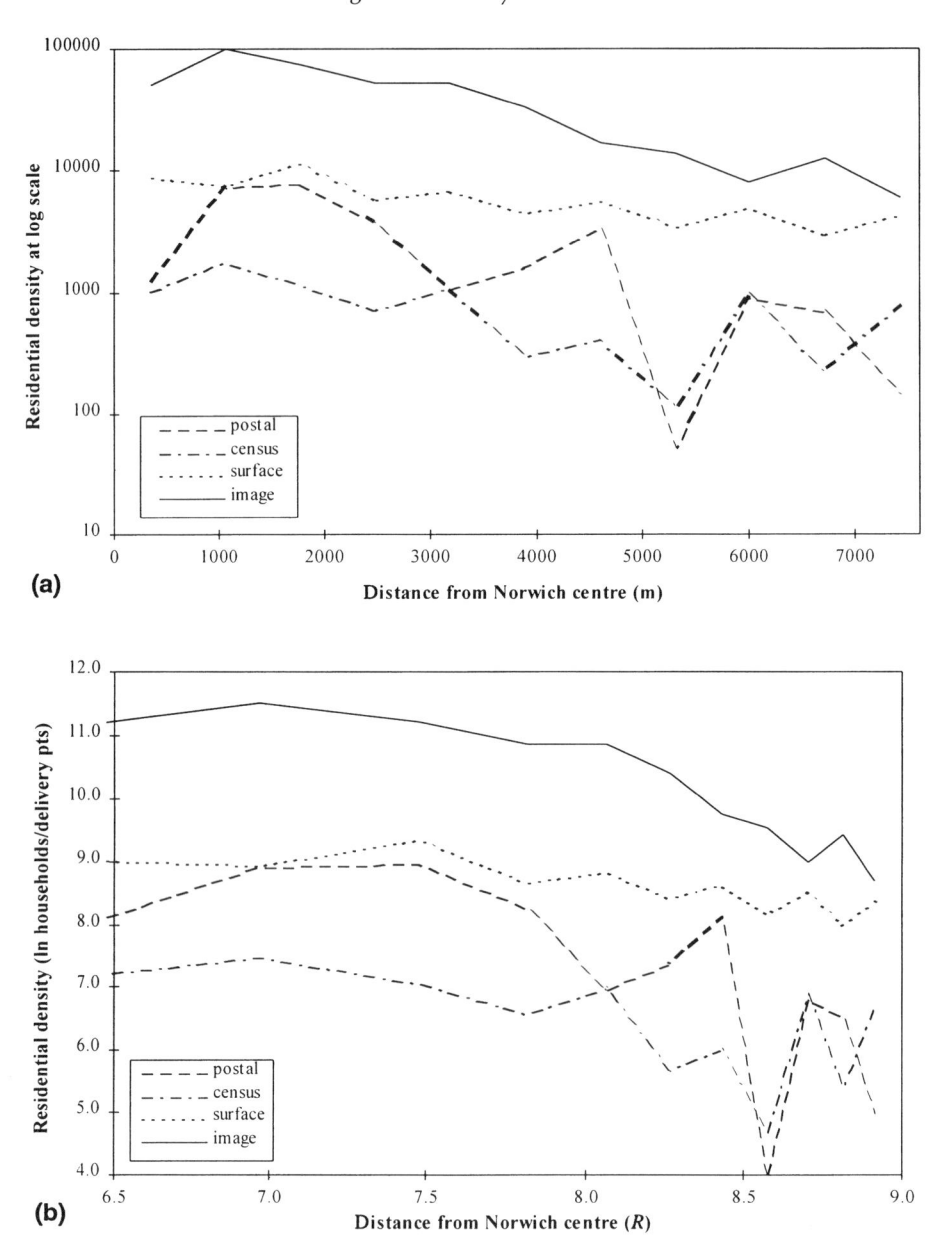

Figure 12.5 Residential density profiles at standardized 500 m resolution: (a) absolute, and (b) logged values

suggesting that the distortions that are inherent in one or both of these models is geographically variable. Lastly, postal geography has the steepest gradient. This is because of the small areas of postcode sectors around the urban core and the much larger areas towards the periphery; as well as the predominance of densely packed properties containing small households around the core.

In summary, examination of Figures 12.4 and 12.5 reveals how classified satellite imagery has created the most detailed morphological model of the residential structure of the settlement. Higher precision, and more objective measurement of physical urban land cover, over more conventional data sources has allowed the satellite images to reveal distinct irregularities (sharper peaks and troughs in the curve) in the density of urban development, and this may be used to inform theory in a data-led way.

12.5 Some Fractal Comparisons between Remotely Sensed Imagery and Digitized Urban Boundary Data

Finally, we will now examine how urban fractal models may also be used to generate density gradients. As noted earlier, the irregular shape of urban areas may be effectively summarized and modelled by fractal geometry.

The development of urban fractal models has been discussed by, *inter alia*, Frankhauser (1990) and Batty and Longley (1994). In this chapter, density and fractal dimension estimation is based purely on the idea of measuring the occupancy, or space-filling properties, of urban development. According to fractal theory, dimension, D, will fall between the established range of 1 and 2, where each land use (or occupancy) fills more than a line across space ($D = 1$) but less than the complete plane ($D = 2$). One way of calculating D refers to the incremental proportion of zone occupation and is expressed in terms of

$$D_{\text{DENSITY}} = 2 + \frac{\ln p(R')}{\ln R'} \qquad (1)$$

where $p(R')$ is the proportion of occupied cells at mean distance R'. However, this type of measure does not account for the variation in each land use. As such, it is impossible to speculate upon the shapes of these patterns with respect to density gradients or profiles. To circumvent the problem, regression lines will be fitted to the profiles generated from each surface in terms of counting land use cells, i, in each concentric zone from the urban centre, normalized, and expressed as densities p_i, producing

$$p_i = \ln \xi - \alpha \ln R_i \qquad (2)$$

where ξ is a constant of proportionality (which is not defined when the radius $R = 0$), and where α is the parameter of distance capable of accommodating scale independence observed in urban systems through the notions of fractal geometry (Batty and Kim, 1992). The fractal dimension is generated by the intercept parameter, which is, in turn, affected by the slope parameter, in Equation (2).

The main fractal program is a FORTRAN algorithm developed by Batty and Xie (1994) with minor modifications to the coding to allow successful implementation using UNIX-based software. Graphic output of urban density profiles is produced by a combination of subroutines by statistics and plotting software. Finally, a series of regression programs written in C have been applied to the natural logs of density profiles to determine the degree of linearity.

We will now begin to compare classified remotely sensed imagery with a secondary data set from the DoE, measuring urban boundaries. We will first examine some basic statistical indicators and then move on to fractal dimensions and density profiles.

12.5.1 Statistical Indicators

The basic statistics between the DoE urban boundary data and the classified image are shown in Table 12.2. Both the classified image and the DoE data measure the degree of urban built-up land associated with the Norwich settlement. Although it is very difficult to standardize 'built-up' between data sets that are fundamentally different, a compromise was achieved that proved to bridge most disparities. The DoE definition of 'urban' is rather lengthy but is essentially '... permanent structures; and transport corridors... which have built-up sites on one or both sides or which link built-up sites which are less than 50 m apart...', and is 'continuous area of land extending for 20 ha or more. Also separate areas of land are linked if less than 50 m apart.' (OPCS, 1984, p. 10). In the classified image, this type of urban land is simply all artificial built structures, and can be easily generated through simple spectral segmentations.

The preliminary analysis of Table 12.2 shows that just under 75% of the 'irreversibly urban' land delineated by the urban boundary envelope is classified as 'urban'. Moreover, the measured length of the perimeter, as well as the area–perimeter ratio, reveals the boundary delineated by the satellite image to be much more tortuous and intricate than its vector boundary counterpart. An obvious conclusion is that, while the urban boundary data provide a ballpark estimate of the extent of an urban area, a more precise estimate of the extent of urban land (as opposed, for example, to public open space and other non-artificial land covers) is obtained from the classified satellite image. Given that the resolutions of the two series are equal at 30 m (although the DoE data were rasterized to a slightly coarser resolution than its accuracy level of 25 m), this suggests that the satellite image discerns a more intricate boundary, and that the classification might be better suited to the calculation of indices of accessibility to non-urban land as a consequence.

Using the urban fractal model outlined earlier in this section, the fractal dimension of the classified image is 1.766, compared to 1.803 for the DoE data (Table 12.2). As the

Table 12.2 Comparison between image and boundary spatial statistics

Parameter	Classified image	DoE urban boundary
Data type	Raster (30 m)	Vector (rasterized to 30 m)
Total number of cells (N_t)	262 144	262 144
Number of occupied cells (N_1)	48 673	65 421
Total area of occupied cells (A_1)	43 805 700 m²	58 696 273 m²
Total perimeter length (P)	419 940 m	197 575 m
Area/perimeter ratio (A/P)	104.3 m²/m	297.1 m²/m
Maximum radius (R_m)	10 861.2 m	10 861.2 m
Tortuosity index ($P/2\pi R_m$)	6.154	2.895
Fractal dimension (D)	1.766	1.803

model is based on space occupancy, this lower dimension is a further indication of the ability of the image to generate more restrictive definitions of urban land, which, unlike the boundary data, take into account pockets or holes (such as open space and difficult relief) in the urban fabric.

12.5.2 Density Profiles

When the linear derivation of the fractal model is plotted, we obtain density profiles for the classified image and the DoE urban boundary data (Figure 12.6). The most striking aspect is that there is a close similarity between the two and that the image profile is on the whole consistently lower than the boundary data. The similarity is a check on the ability of both data sets as reasonable representations of the Norwich settlement, whereas the consistently lower profile for the image is a reflection of lower space occupancy (recognition of urban pockets). This is particularly apparent at the urban core, since the profiles begin to converge towards the periphery.

12.6 Conclusions

Progress in urban remote sensing has always been hampered by the quality of spectral information that can be reliably extracted from remotely sensed data. This chapter sought to demonstrate that spatial, not spectral, data may also be derived from classified imagery. However, unlike spectral information, spatial data may further be used in routine geographical analysis techniques, including the example of generating density gradients to reveal how the structure of urban morphologies changes from core to periphery. When compared to other digital urban data, spatial information from remotely sensed data is at a much finer resolution with a more objective definition of what constitutes urban land cover.

Work on urban density over the last 50 years has also been hampered by access to good data at the micro level, the inability to relate density to built form, and the inefficiency of manipulating the geographic system to find the correct scale at which density should be interpreted. Developments in satellite remote sensing offer a route beyond this impasse, and our preliminary assessment and analysis have shown that satellite imagery exhibits a number of strengths in the depiction of urban form. We are now in a position to reappraise the relevance of density, and we believe that the powerful techniques at our disposal (Batty and Longley, 1994) are able to help us understand the mosaic of densities that comprise contemporary cities. Urban morphology research has failed in the past because of inherently poor data quality and inappropriate representation of geographical objects; it is our view that it is still failing because it is not using the richness of available digital data to create fine-scale application-specific databases. Patterns are simply obscured in analysis which is scientifically loose in its control of the spatial system and its attributes. Examining densities using administrative geography is simply not good enough, and the preliminary analysis that we have set out here begins to demonstrate the need for better scientific standards in researching such questions.

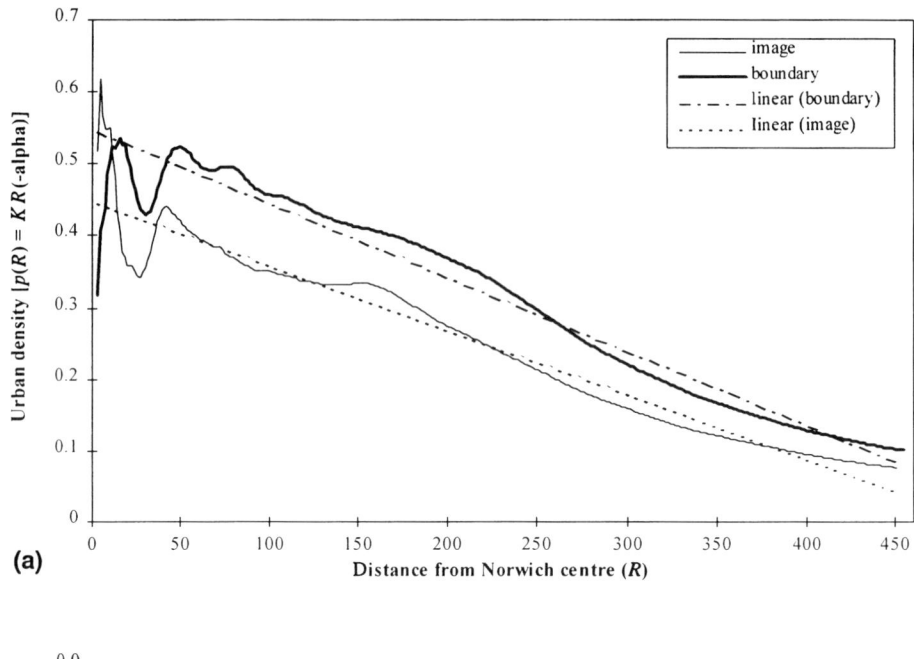

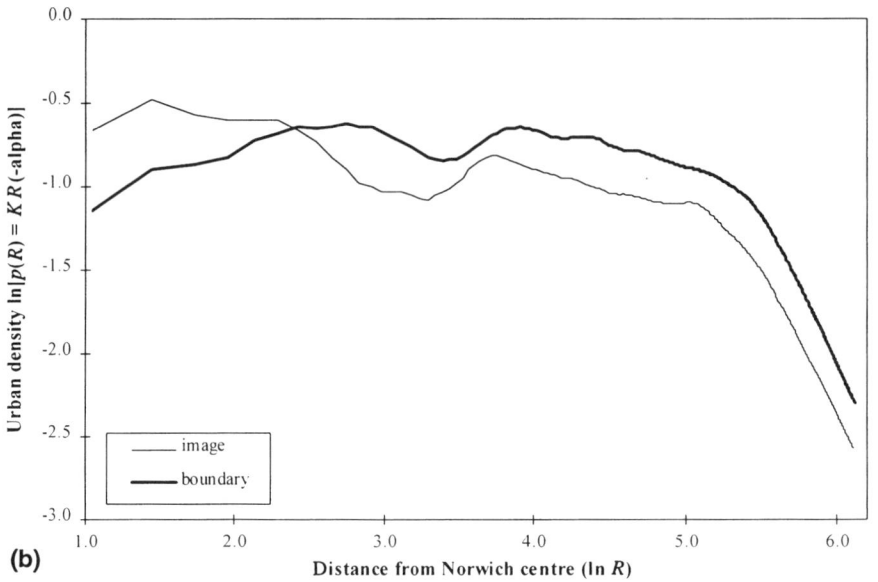

Figure 12.6 Classified image and boundary density profiles: (a) absolute values with linear approximation, and (b) logged values

Acknowledgements

This work was part funded by NERC grant GST/02/2241 under the URGENT Programme. Victor Mesev's work for this chapter was partly funded by Research Fellowship number H53627501295 from the UK Economic and Social Research Council (ESRC).

References

Bailey, T.C. and Gatrell, A.C., 1995, *Interactive Spatial Data Analysis* (Harlow: Longman).
Barnsley, M.J. and Barr, S.L., 1996, Inferring urban land use from satellite sensor images using kernel-based spatial reclassification, *Photogrammetric Engineering and Remote Sensing*, **62**, 949–958.
Batty, M. and Kim, K.S., 1992, Form follows function: reformulating urban population density functions, *Urban Studies*, **29**, 1043–1070.
Batty, M. and Longley, P., 1994, *Fractal Cities: A Geometry of Form and Function* (London and San Diego: Academic Press).
Batty, M. and Xie, Y., 1994, Modelling inside GIS: part 1. Model structures, exploratory spatial data analysis and aggregation, *International Journal of Geographical Information Systems*, **8**, 291–307.
Breheny, M. and Archer, S., 1996, Urban densities, local policies and sustainable development, Paper presented at the 36th European Congress of the European Regional Science Association, Zurich, Switzerland, 26–30 August.
Campbell, J.B., 1996, *Introduction to Remote Sensing* (London: Taylor & Francis).
Corbley, K.P., 1996, One-meter satellites, *Geo Info Systems*, July 1996, 28–42.
De Cola, L., 1989, Fractal analysis of a classified Landsat scene, *Photogrammetric Engineering and Remote Sensing*, **55**, 601–610.
Donnay, J.P. and Barnsley, M.J., 1999, *Remote Sensing and Urban Analysis* (London: Taylor and Francis) (in press).
Ehlers, M., Edwards, G. and Bédard, Y., 1989, Integration of remote sensing with geographic information systems: a necessary evolution, *Photogrammetric Engineering and Remote Sensing*, **55**, 1619–1627.
Fisher, P. 1998, Models of uncertainty in GIS data, in P.A. Longley, M.F. Goodchild, D.J. Maguire and D.W. Rhind (eds), *Geographical Information Systems: Principles, Techniques, Management, and Applications* (New York: Wiley), **1**, 191–205.
Fotheringham, A.S., Batty, M. and Longley, P., 1989, Diffusion-limited aggregation and the fractal nature of urban growth, *Papers of the Regional Science Association*, **67**, 55–69.
Frankhauser, P., 1990, Aspects fractals des structures urbaines, *L'Espace Géographique*, **19**, 45–69.
Garreau, J., 1991, *Edge City: Life on the New Frontier* (New York: Doubleday).
Goodchild, M.F. and Mark, D.M., 1987, The fractal nature of geographic phenomena, *Annals of the Association of American Geographers*, **77**, 265–278.
Gurney, R.J., Foster, J.L. and Parkinson, C.L., 1993, *Atlas of Satellite Observations Related to Global Change* (Cambridge: Cambridge University Press).
Haggett, P., Cliff, A.D. and Frey, A., 1977, *Locational Analysis in Human Geography* (New York: Wiley, and London: Edward Arnold).
Harris, P.M. and Ventura, S.J., 1995, The integration of geographic data with remotely sensed imagery to improve classification in an urban area, *Photogrammetric Engineering and Remote Sensing*, **61**, 993–998.
Langford, M. and Unwin, D.J., 1994, Generating and mapping population density surfaces within a geographical information system, *The Cartographic Journal*, **31**, 21–26.

Lo, C.P., 1986, *Applied Remote Sensing* (Harlow: Longman).
Longley, P. and Mesev, T.V., 1997, Beyond analogue models: space filling and density measurement of an urban settlement, *Papers in Regional Science*, **76**, 409–427.
Longley, P., Batty, M. and Shepherd, J., 1991, The size, shape and dimension of urban settlements, *Transactions, Institute of British Geographers*, N.S. **16**, 75–94.
Longley, P., Batty, M., Shepherd, J. and Sadler, G., 1992, Do green belts change the shape of urban areas? A preliminary analysis of the settlement geography of South East England, *Regional Studies*, **26**(5), 437–452.
McDonald, J.F., 1989, Econometrics of urban population density: a survey, *Journal of Urban Economics*, **26**, 361–385.
Mandelbrot, B.B., 1983, *The Fractal Geometry of Nature* (San Francisco: W.H. Freeman).
Martin, D., 1996, *Geographic Information Systems: Socioeconomic Applications* (London: Routledge).
Martin, D. and Bracken, I., 1991, Techniques for modelling population-related raster databases, *Environment and Planning A*, **23**, 1065–1079.
Mesev, V., 1998, The use of census data in urban image classification, *Photogrammetric Engineering and Remote Sensing*, **64**, 431–438.
Mesev, T.V., Longley, P. and Batty, M., 1996, RS-GIS: spatial distributions from remote imagery, in P. Longley and M. Batty (eds), *Spatial Analysis: Modelling in a GIS Environment*, (Cambridge: GeoInformation International), 123–148.
Mesev, T.V., Gorte, B. and Longley, P., 1999, Modified maximum-likelihood classifications and their application to urban remote sensing, in J.P. Donnay and M.J. Barnsley, (eds), *Remote Sensing and Urban Analysis* (London: Taylor & Francis) (in press).
Mills, E.S. and Tan, J.P., 1980, A comparison of urban population density functions in developed and developing countries, *Urban Studies*, **17**, 313–321.
Mitchell, R., Martin, D. and Foody, G., 1996, Unmixing aggregate data: estimating the social composition of enumeration districts, Paper presented in *GeoComputation '96*, University of Leeds.
Nordbeck, S. 1971, Urban allometric growth, *Geografiska Annaler*, **53B**, 54–67.
OPCS, 1984, *Key Statistics for Urban Areas, Office of Population Census and Surveys* (London: HMSO).
Openshaw, S., 1996, Developing GIS-relevant zone based spatial analysis methods, in P. Longley and M. Batty (eds), *Spatial Analysis: Modelling in a GIS Environment* (Cambridge: GeoInformation International), 55–73.
Openshaw, S. and Alvanides, S., 1999, Applying geocomputation to the analysis of spatial distributions, in P.A. Longley, M.F. Goodchild, D.J. Maguire and D.W. Rhind (eds), *Geographical Information Systems: Principles, Techniques, Management and Applications* (New York: Wiley), **1**, 267–282.
Parr, J.B., 1985, A population-density approach to regional spatial structure, *Urban Studies*, **22**, 289–303.
Shepherd, J. and Congdon, P., 1990, *Small Town England: An Investigation into Population Change among Small and Medium-Sized Urban Areas*, Progress in Planning Series (Oxford: Pergamon).
Wang, F., Treitz, P.M. and Howarth, P.J., 1992, Road network detection from SPOT imagery for updating geographical information systems in the rural–urban fringe, *International Journal of Geographical Information Systems*, **6**, 141–157.
Whitehand, J.W.R., 1992, Recent advances in urban morphology, *Urban Studies*, **29**, 619–636.
Wrigley, N., 1996, Sunk costs and corporate restructuring: British food retailing and the property crisis, in N. Wrigley and M. Lowe (eds), *Retailing, Consumption and Capital: Towards the New Retail Geography* (Harlow: Longman).
Zielinski, K., 1980, The modelling of urban population density: a survey, *Environment and Planning A*, **12**, 135–154.

13

Image Classification and Analysis Using Integrated GIS

Jackie C. Hinton

13.1 Introduction

The integration of remote sensing and geographic information systems (GIS) in environmental applications has become increasingly common in recent years. Remotely sensed images are an important data source for environmental GIS applications and conversely GIS capabilities are being used to improve image analysis procedures. When image processing and GIS facilities are combined in an integrated system vector data can be used to assist in image classification and raster image statistics within vectors can be used as criteria for vector query and analysis.

This chapter describes applications of an integrated raster/vector processing environment for the analysis of Earth observation (EO) data. In particular the benefits of an integrated object-oriented GIS to the analysis of radar data and to the derivation of a vector data set from images is discussed.

Increasingly EO-derived information is finding a market in real applications. Generally the use of remote sensing achieves greatest success when the data are combined with other environmental data. Such spatially referenced environmental data are stored in a variety of formats, and integrating data sets which are in different formats has, until recently, been a stumbling block in most analysis systems. In the late 1980s 'integration' was still seen as a research area, but gradually increased levels of integration were achieved by data conversion or simply at the level of raster or vector backdrop. Increasingly the term 'integrated' is attached to research papers describing the use of remote sensing and GIS. Most such 'integrated' approaches have involved separate image processing systems with classified image results being either (a) converted to

---

Advances in Remote Sensing and GIS Analysis. Edited by Peter M. Atkinson and Nicholas J. Tate.
© 1999 John Wiley & Sons Ltd.

vector format and imported into a vector GIS (e.g., Johansen *et al.*, 1994; Mattikali, 1994) or (b) transferred in the form of a classified coded image to a GIS capable of handling raster data, for integration with vector map data (e.g., Pathan *et al.*, 1993), because the required more fully integrated approach was not possible using the systems available (Janssen *et al.*, 1990). As the need for closer integration of raster and vector processing became apparent, commercial constraints meant it has become more usual for processing packages to be adapted to be able to receive data from other systems and convert them, rather than to rethink the design of software. This has meant that the format used for data in the past has been dictated by the system to be used, rather than the format most suited to the data. Considering that data transfer between systems and formats is a potential source of error (van der Knaap, 1992), this is clearly a less than optimal situation. All users of environmental data should expect to make routine use of remotely sensed data and to integrate them seamlessly with their other data sets. A fully integrated system should involve a single software unit with combined processing (Ehlers *et al.*, 1989). No format conversions should be necessary and it should be possible to transfer information seamlessly between the images and the vector database in both directions (Hinton, 1996). In other words each object (point, line, polygon) in the vector data set must 'know' which pixels in an image fall within/beneath it. Conversely each pixel should 'know' which vector object it falls within.

The analyses described in this chapter were carried out using the Laserscan IGIS software, a fully integrated GIS incorporating object-based database technology. GIS analyses deal with real-world objects and therefore we need a data model best suited to represent that real world and the way objects in it relate to their environment and the objects around them. For this, an object-oriented database design is conceptually the most natural. The advantages of an object-oriented design to GIS is explained fully in excellent articles elsewhere (Worboys *et al.*, 1990; Milne *et al.*, 1993; Worboys, 1994; Raper and Livingstone, 1995). Any digital information can be incorporated into the object database. The logical data model is defined by the user as object characteristics related to geometry, attributes and 'methods' which operate on data associated with the objects. These user-defined methods exhibit 'behaviours', which define how the object relates to information, to the other objects in the database and to the value of the image pixels within the object. The examples described in this chapter will show how this type of data model is of benefit for integrated raster/vector analysis.

13.2 Image Classification in an Integrated GIS

There is a commonly held conception that the main processing tasks in remote sensing are concerned with the labelling of each pixel, but this is not necessarily so. It is more meaningful to look at areal characteristics of radar data for example, because the speckle in such data make them difficult to analyse on a per-pixel basis. Remotely sensed images are also being used to extract landscape characteristics such as road locations or field boundary information to correct vector line maps. There is no longer a need to work just in pixel mode with remotely sensed data.

One form of remotely sensed data that is of increasing interest is the use of radar sensors for environmental and agricultural monitoring. Different crops are thought to have distinct backscatter behaviour, with features in the temporal backscatter curve corresponding to stages in crop development (Wooding et al., 1994). The ability of radar sensors to penetrate cloud make them suitable for tropical regions with frequent cloud cover and they are increasingly being used for forest biomass studies in many parts of the world. However, radar data are particularly difficult to analyse because of the speckle caused by the coherent radar imaging process. Many previous attempts to classify radar data have required both pre-classification image smoothing to reduce speckle and post-classification noise removal through a majority filter (e.g., Ranson and Sun, 1994a). However in areas where there are small parcels of different radiometric response (e.g., different land use) filters will blur the edges between parcels. A more appropriate classification method for radar data is to classify a table of the mean image values within each land parcel. This requires the calculation of statistics from pixels within each parcel (Hinton, 1996). Using a vector data set defining the boundaries around homogeneous regions the calculation of statistics can be carried out only over pixels known to be from the same region.

A schematic representation of the functionality of a per-parcel classification process is shown in Figure 13.1. For the classification processes to be carried out in a GIS the system must have the facilities to display and enhance multispectral image data and calculate statistics of image pixels within selected vector polygons. Given the usefulness of such a capability it is perhaps surprising that very few systems include this functionality without requiring any format conversions and overlay operations. The system must then be able to apply a classification algorithm to a table of image values and return the resulting class to the vector polygon attribute table.

13.2.1 Example

An airborne SAR image from Thetford Forest, Norfolk, and a 1985 Forestry Commission stock map are shown in Figure 13.2 (Plate II). Normal per-pixel classification is not feasible because of the speckle that is visible in the image. These data have been used to determine whether radar images could be classified using the per-parcel method described above to separate forest stands of different species. Vector polygons had associated attributes of compartment number (for reference to the original map), tree species (or alternative land use) and planting year. Over 950 attributed vector areas were used in the analyses, covering an area of 10 × 10 km.

For this example, polygons representing forest stands of single species were selected using basic GIS select-by-attribute functionality and assigned to training areas in the following classes: Corsican pine stands, Scots pine stands and oak woodland. Polygons were randomly split into two sets, one of which was used for training and one for verification of the classification results. There were over 200 stands of each of Scots pine and Corsican pine and 10 of oak woodland.

In the example described here the mean of the image data was calculated for each of three available SAR bands (polarizations) for pixels in each polygon (forest parcel). It would equally have been possible to calculate additional measures, for example, standard deviation of the image data or a texture measure, and use these as input to a classification (Figure 13.1). The statistics were saved as attributes of each polygon and

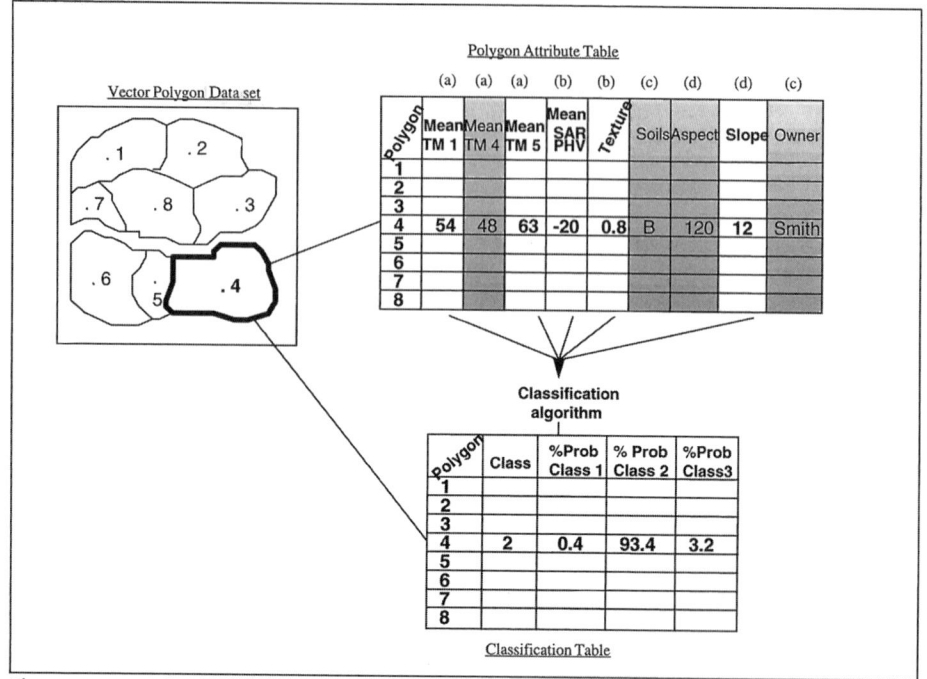

Figure 13.1 *Schematic representation of polygon classification functionality. Attributes stored in a database table are used in the classification, rather than values of pixels in a series of image layers. Attributes could be derived from images, for example (a) mean value of TM bands within each polygon, (b) mean SAR backscatter signal or image texture over the area defined by the polygon, (c) typical attributes attached to the vector data sets, or (d) values obtained from DEM within each polygon. The output of the classification is in the form of an attribute table, through the unique polygon identifier. (Reproduced from Hinton, 1996 with permission from Taylor & Francis)*

were used to examine the relationships between mean radar backscatter coefficient and other attributes of the forest stand such as species or age. For some polygons, examination of the histogram of SAR pixels within the polygon indicated noise from edge pixels arising from changes in land use across the boundaries, roads, tracks and rides along the edge of the stand and possibly georeferencing errors. In these areas a user-defined buffer from the edge could be used to exclude edge pixels in the calculation of image statistics. With the data model used for this analysis, it was possible to perform the new calculation in a selected subset of polygons without actually editing the boundaries. A 'shrunken' version of an individual polygon geometry could be calculated in system memory and used to constrain the query and calculation of image statistics. It has been suggested that edge pixels should be excluded from the calculation of statistics for classification as a matter of course (Archibald, 1987) but this seems wasteful both of processing time and of potentially useful information. It would be more useful if the only polygons to be shrunk were those where analysis of the image data within the area shows such edge-exclusion to be necessary. Within IGIS, a method was developed for each selected polygon to follow the procedure shown in Figure 13.3. While it could be argued that this would be possible in a system with the

Image Classification and Analysis Using Integrated GIS 211

more traditional GIS thematic data model it would be at least cumbersome, and certainly computationally expensive.

Plots of mean backscatter within the polygon against stand age for Corsican pine and Scots pine stands in the study area for SAR P band HV polarization are shown in Figure 13.4. A similar relationship between biomass (which is directly related to age) and backscatter has been shown previously in this area (Baker *et al.*, 1994) and elsewhere (Ranson and Sun, 1994b) and this agreement shows that the procedures developed to calculate image statistics within polygons worked correctly. There is a strong relationship between stand age and backscatter for this band for young stands of Corsican pine (Figure 13.4a). A similar relationship for Scots pine (Figure 13.4b) is less obvious because there are very few young stands of Scots pine in the forest.

A maximum likelihood classification was performed with the mean image value of three bands within the polygons as input. Classification accuracy, given in Table 13.1,

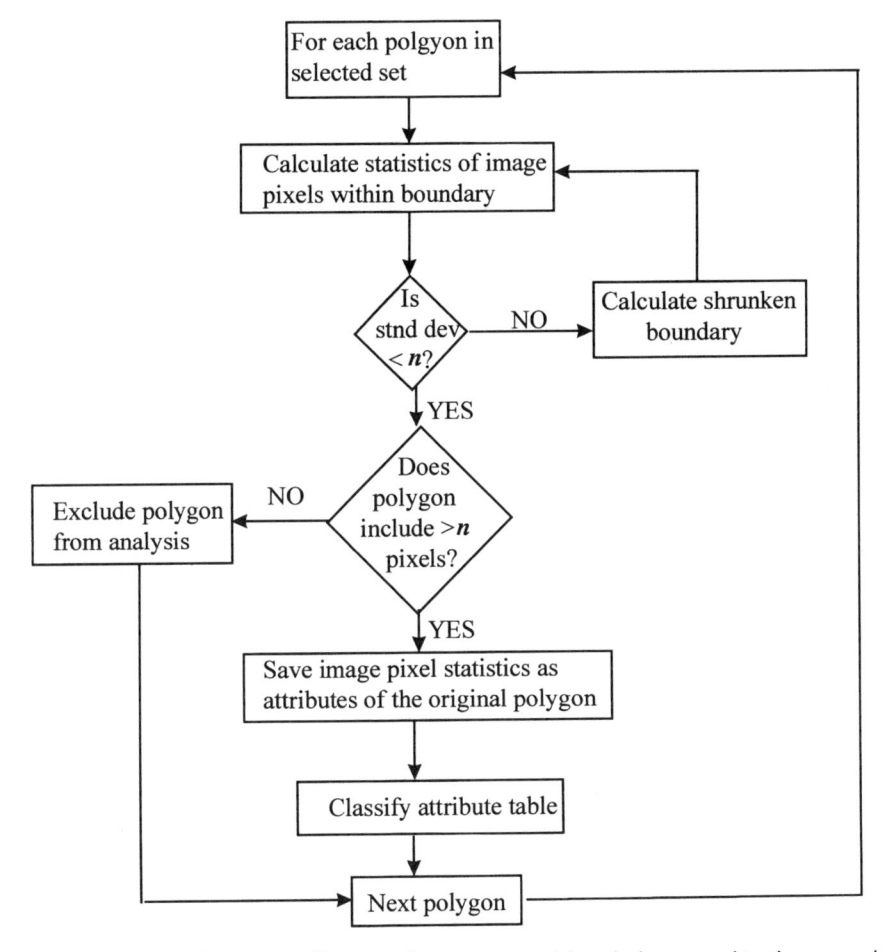

Figure 13.3 *Flow diagram to illustrate the programmable rule-base used in the per-polygon classification*

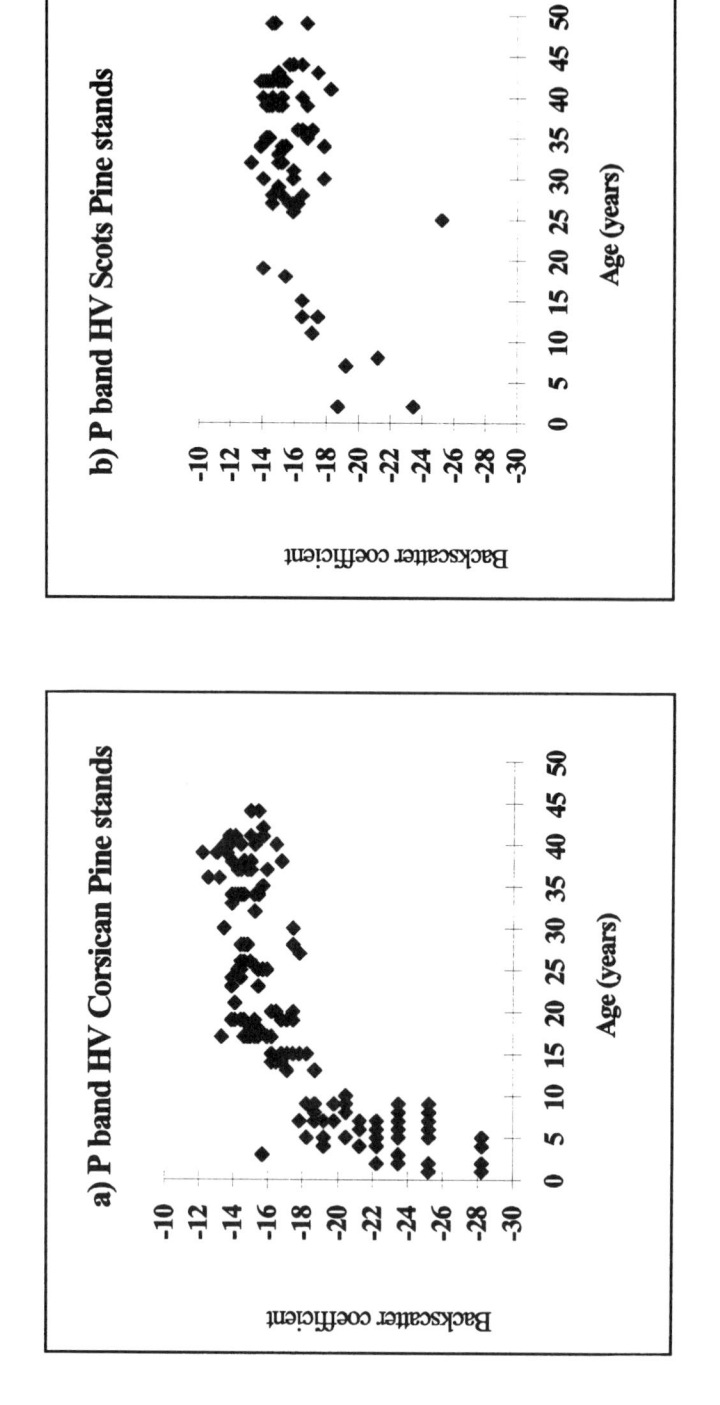

Figure 13.4 Plots of mean radar backscatter coefficient in forest stands against the age of the stand for (a) Corsican pine and (b) Scots pine stands in Thetford Forest, Norfolk

Table 13.1 *Classification accuracy*

Training class	Classification accuracy (%) for verification polygons		
	Per-pixel, 3 bands (P,L,C band HV)	Per-polygon, 3 SAR bands (P,L,C HV)	Per-polygon, 3 SAR and 3 TM bands
Corsican pine	41	73	92
Scots pine	42	81	87
Deciduous woodland	54	86	87

shows the great improvement in accuracy using this method over a per-pixel classification on radar data. The class assigned by the classification was saved as a further attribute of each polygon. It is also possible to save the classification accuracy statistics, or confidence measure, as an attribute of the polygon.

Examination of the attributes of the polygons not used in training revealed that many young Corsican pine stands (< 10 years) had been left as 'unclassified'. The lower accuracy for Corsican pine, compared with Scots pine, using only the three SAR bands may therefore be due to the relationship between age and backscatter shown in Figure 13.4a. This indicates that it might also be possible to distinguish stands of different age groups and some success has been achieved in this (Hinton, 1995).

It is also possible to improve the classification of species by combining the radar images with optical images. Table 13.1 includes results of a classification carried out on the mean image value in three SAR bands and three TM bands. The inclusion of information in the optical channels improved the classification of the young Corsican pine stands considerably. The system calculates the image statistics for pixels within the vector polygon for each input image and saves them all as polygon attributes and these attributes are then classified. It is not necessary to re-sample the various data sources to the same grid. Thus, using the data model described here, per-polygon classification is independent of the resolution of the images used. A procedure can be developed to include a check that the image resolution is appropriate to the size of the vector objects being used, for example a lower limit can be placed on the number of pixels within a polygon that can generate useful statistics.

In a fully integrated GIS it is possible to implement this classification methodology for a selected set of polygons only, either selected manually by the user or as the results of some GIS query. In an object-oriented system classification according to image statistics can be defined as a method of the polygon object class and executed as a behaviour for selected polygons. This can be of particular benefit when examining change – only those polygons where change is likely to have occurred need to be reclassified. The output is an updated attribute value, rather than a complete new coded raster the size of the input images. This saving in computation and storage requirements is an important advantage for routine, repetitive or operational use.

13.3 Vector Polygon Extraction

The method for classification described above relies on the existence of a vector data set, defining the areas over which the image statistics are gathered. In some parts of the world such vector data are hard to come by and even where vector mapping is well established it is often necessary to update such data sets. When no map data are available, it can be possible to derive a boundary vector data set directly from Earth observation images.

Several authors have described a combined process for segmentation and classification of images (e.g., Cross *et al.*, 1988: Janssen *et al.*, 1993). However, the most suitable data for identification of landscape structure may not be those best suited for the classification of the features of interest. Algorithms for the extraction of edges and boundaries from radar data are being developed (Quegan, 1995). However, such segmentation methods depend on the wavelength and resolution of the radar data, are computationally expensive and have achieved variable success. The speckle in the images makes segmentation difficult but it is just this speckle which makes per-polygon classification essential for radar images in the first place. High spatial resolution optical images may provide better structural information. Spatial units such as forest compartments or agricultural fields are clearly identifiable on such images and of sufficient size to be accurately extracted. On an operational level, such images would not be required as frequently as the images for classification because the structure of the landscape generally does not change as rapidly as the land use within each parcel.

Segmentation algorithms are generally based on one of two basic properties of pixel values: discontinuity and similarity. Using the former, abrupt changes in grey-level are identified and used to partition the image, or to detect 'edges' in the pixel value distributions. Using the latter, areas of relatively homogeneous values are thresholded or grown together using region splitting and merging algorithms. However, many such examples in the literature are concerned with the identification of the feature itself (e.g., distinguish a fence from a hedge). For per-polygon classification, it is not always necessary to know the type of boundary extracted, merely that it separates two areas of different spectral response.

The intention in the example described below is to separate the tasks of defining spatial structure from the identification (classification) of parcel contents. The method does not make any assumptions about absolute pixel values, but relies on relative variations. It is shown here that local variability can be used to identify and extract areas in an image representing relatively homogeneous cover type. In an integrated system it is then possible to verify and, if necessary, edit derived vector areas using image pixel values within the areas as a guide. Manual editing can be very time-consuming and subjective, so for any operational repetitive use an automated procedure is needed, requiring realistic computational resources.

13.3.1 Example

The variability in pixel values in a neighbourhood will be larger over an area which includes an 'edge' than over a 'non-edge', or mid-parcel region. A simple standard deviation filter has been tested to separate areas of high variability from those of low variability. Such filters are available in virtually all raster processing systems and are computationally very simple. This is just one example of an edge-detecting filter.

Various permutations on Sobel filters (e.g., Perkins, 1980; Ton et al., 1989) or region growing from seed points have also been shown to be useful.

An area of agricultural fields is visible in a SPOT panchromatic image of Norfolk (Figure 13.5; Plate III). A standard deviation filter with a 5 × 5 pixel neighbourhood was applied to the image to produce an output image. 'Edge' regions, where the standard deviation in the original image was high, were represented in the output image by large values, while homogeneous fields had small standard deviations and therefore low values in the 'edge' image. The resulting image was thresholded to separate 'edge' and 'non-edge' regions. The 'non-edge' regions, which represent areas of relatively homogeneous image values, were converted to vector polygons, using the available raster to vector conversion facilities in the system, and these are shown overlain on the image in Figure 13.5 (Plate III).

Once the edges are determined, by any filter, and built into polygon boundaries, they must be verified as representing homogeneous areas, if they are subsequently to be used for per-polygon analysis. It was possible to carry out post-processing of these derived areas, using statistics from the images (both the images used to extract the edges, and any other images available such as the ones to be used in the classification) as a guide. For example, some of the polygons shown in Figure 13.5 (Plate III) have small holes within them. In some cases an examination of the original image shows that these holes are representative of clumps of trees in the centre of the field, and are therefore features which should be retained. In some cases, however, they are an artefact of the use of neighbourhood filters. These would be caused where a small change in illumination or topography would increase the pixel value variation of the filter neighbourhood to exceed the threshold selected for 'non-edge' regions. However, the image statistics over the area of the full field would not be significantly affected by the inclusion of this small hole. In such a case the hole could validly be deleted. Similarly, the use of a 5 × 5 neighbourhood filter in this example has resulted in the edge regions being quite broad and the extracted vector areas being 'shrunk' away from the edges in the image.

A rule base has been developed for iterative editing of the derived areas, for example to 'grow' them outwards by specified distances, based on the statistics of pixels within the area from the images available. The advantage of this capability is that the resultant polygons will represent areas which are statistically homogeneous and therefore suitable for per-polygon classification. In Figure 13.5 (Plate III) some areas have been 'grown' outwards closer to the field edges visible on the image. The rules for selecting which areas to 'grow' were based solely on the statistics of the image pixels within the polygons. The polygons can retain, as attributes, information which reflects the confidence with which they have been extracted and the amount of post-processing they have undergone.

The automated procedures described above require the independent manipulation of each polygon in the image according to statistics calculated from the underlying image(s). This is reasonably simple to implement in an object-oriented system. While the results of this work have been encouraging, a number of methods for further enhancement of edges detected with filters have been suggested. An expansion/contraction technique (Perkins, 1980) could be used to expand edge regions to order to close gaps, then contracted after separate regions have been labelled. The size of expansion is increased incrementally. This iterative approach involving raster and vector processing is a possible route for the future. It may be possible to use fixed features such as known road locations as an optional addition if required. However, any method for area extraction should not be dependent on such features in order to be relevant to as wide a range of geographical areas as possible. In the longer term an interesting approach would be a link with an expert system for implementing a rule base for whether an area should be split, deleted, or grown. Multispectral methods have been also been proposed (Qiu and Goldberg, 1985) using the sum of the neighbourhood variance in each band. Particularly with future high-resolution multispectral sensors, this may become more common.

13.4 Discussion and Conclusions

The functionality required in a GIS for environmental applications is somewhat different (although overlapping) from that needed in other disciplines in which GIS are used, such as market analysis, social geography or urban planning. In particular there is much more need for raster processing facilities because environmental parameters generally are continuous with no firmly located boundaries. However, distinct political or administrative boundaries are often imposed for the monitoring and/or control of natural areas. Organizations responsible for such natural areas will have a requirement to monitor the area within their boundaries and will also look at the influence of surrounding areas (e.g., the impact of farming within a buffer zone outside Sites of Special Scientific Interest). Earth observation data are a cost-effective source of environmental information and integrated GIS are emerging to allow their analysis in conjunction with other data.

It is shown here that it is possible to perform a classification of radar image data of a forest by combining the image data with vector and tabular information. The calculation of image mean within each polygon for use in classification removes the effects of speckle in the image and the noise it introduces into the classified image. This per-polygon classification removes the need for image filtering operations which would blur the boundaries between areas. In order to carry out the processes described here a system must have the capabilities to perform multispectral image enhancement, vector editing, vector search and query, attribute database query and editing, raster query within vector boundaries, raster editing and image classification. The ability to customize the system to specific requirements is also a significant advantage and makes the system as flexible as possible.

The procedures described here would be applicable to most remotely sensed image types. The development of integrated systems in the future must consider the wide range of new sources of data becoming available. The use of high spatial and spectral resolution remotely sensed data is becoming common and facilities must be provided to derive maximum benefit from them. The temporal dimension of data must also be considered, both raster and vector. The ability to identify change between data sets and to update data sets easily is essential. Landscape structure does not usually change rapidly, although the contents of agricultural fields, for example, can change on a yearly basis. Similarly in a managed forest the stand boundaries will often remain constant while the trees are planted, mature and are felled. Therefore, for the monitoring of land use change by per-polygon classification, high-resolution imagery suitable for area extraction does not need to be acquired very often. Once a baseline polygon data set has been produced for an area, map updating systems (e.g., Brown and Fletcher, 1994) can use a variety of imagery and other data sources to keep the boundary information up to date, using some of the techniques described here.

Considering the benefits in the ability to query image values within individual vector objects it is surprising how difficult it is to perform in most GIS or image analysis systems. It is usually necessary to convert data into equivalent formats, usually into raster for subsequent overlay/masking. The ability to maintain data in their original formats means there is no need for multiple copies of the same data, thus saving storage space and there is no danger of the introduction of errors caused by the conversion process or by the existence of different versions of the same data. An object-based vector data model is considered essential for this type of analysis for anything other than very small areas. The vector data set used in the example described in Section 13.2.1 included nearly a thousand polygons. Because each object has a complete geometry it is possible to calculate a modified geometry of each object separately. This temporary geometry can be used in the procedures for editing polygons based on image statistics. In a system with a thematic data model the calculation and testing of the criteria for area editing would involve substantial format conversion, overlay operations and data volume increase.

Many possible options for the future of integrated GIS have been discussed previously (Hinton, 1996), including the addition of modelling tools and integration with expert systems. Their future is challenging, but with their continued development remotely sensed images will be increasingly used operationally in a wide range of environmental applications.

Acknowledgements

The AirSAR image of Thetford Forest was provided by NASA (JPL) as part of the MacEurope campaign. The vector data set was digitized by Huntings Technical Services from Forestry Commission maps. Dr John Baker and Dr Adrian Luckman provided valuable advice and ideas concerning SAR data processing. The SPOT Panchromatic image was donated by SPOT Image and the author gratefully acknowledges the help of M. Christophe Hutin in acquiring the image. Funding from the Department of Trade and Industry is gratefully acknowledged.

References

Archibald, P.D., 1987, GIS and remote sensing data integration, *Geocarto International*, 3, 67–73.

Baker, J.R., Mitchell, P.L., Cordey, R.A., Groom, G.B., Settle, J.J. and Stileman, M.R., 1994, Relationships between physical characteristics and polarimetric radar backscatter for Corsican Pine stands in Thetford Forest, UK, *International Journal of Remote Sensing*, **15**, 2827–2849.

Brown, R. and Fletcher, P., 1994, Satellite images and GIS: making it work, *Mapping Awareness*, **8**, 20–22.

Cross, A.M., Mason, D.C. and Dury, S.J., 1988, Segmentation of remotely sensed images by a split and merge process, *International Journal of Remote Sensing*, **9**, 1329–1345.

Ehlers, M., Edwards, G. and Bedard, Y., 1989, Integration of remote sensing with geographic information systems: a necessary evolution. *Photogrammetric Engineering and Remote Sensing*, **55**, 1619–1627.

Hinton, J.C., 1995, Integrating remote sensing and GIS for environmental applications, in M. Palmer (ed.), *Proceedings of a Seminar on Integrated GIS and High Resolution Satellite Data* (Farnborough: Defence Research Agency), Ref No: DRA/CIS/(CSC2)/5/26/8/1/PRO/1.

Hinton, J.C., 1996, GIS and remote sensing integration for environmental applications, *International Journal of Geographic Information Systems*, **10**, 877–890.

Janssen, L., Jaarsma, M. and van der Linden, E., 1990, Integrating topographic data with remote sensing for land cover classification, *Photogrammetric Engineering and Remote Sensing*, **56**, 1503–1506.

Janssen, L., Shoenmakers, R. and Verwaal, R., 1993, Integrated segmentation and classification of high resolution satellite images, in M. Molenaar, L. Janssen and H. van Leeuwen (eds), *Proceedings of an IAPR TC7 International Workshop on Multisource Data Integration in Remote Sensing for Land Inventory Applications* (Wageningen: Department of Land Surveying and Remote Sensing, Wageningen Agricultural University), 65–84.

Johansen, M.E., Tommervik, H., Guneriussen, T. and Pedersen, J.P., 1994, Using a GIS (ArcInfo) as a tool for integration of remote sensed and in-situ data in an analysis of the air pollution effects on terrestrial ecosystems in Varanger (Norway) and Nikel-Pechenga (Russia), in *Proceedings of International Geoscience and Remote Sensing Symposium* (Piscataway, NJ: IEEE), **2**, 1213–1216.

Mattikali, N.M., 1994, An integrated GIS's approach to land cover change assessment, in *Proceedings of International Geoscience and Remote Sensing Symposium* (Piscataway, NJ: IEEE), **2**, 1204–1206.

Milne, P., Milton, S. and Smith, J.L., 1993, Geographical object-oriented databases – a case study, *International Journal of Geographical Information Systems*, **7**, 39–55.

Pathan, S.K., Sastry, S.V.C., Dhinwa, P.S., Rao, M. and Majumdar, K.L., 1993, Urban growth trend analysis using GIS techniques – a case study of the Bombay metropolitan region, *International Journal of Remote Sensing*, **14**, 3169–3179.

Perkins, W.A., 1980, Area segmentation of images using edge points, *IEEE Transactions on Pattern Analysis and Machine Intelligence*, **2**(1), 8–15.

Qiu, Z-C. and Goldberg, M., 1985, A new classification scheme based upon segmentation for remote sensing, *Canadian Journal of Remote Sensing*, **11**, 59–69.

Quegan, S., 1995, Recent advances in understanding SAR imagery, in F.M. Danson and S.E. Plummer (eds), *Advances in Environmental Remote Sensing* (Chichester: Wiley), 89–104.

Ranson, K.J. and Sun, G., 1994a, Northern forest classification using temporal multifrequency and multipolarimetric SAR images, *Remote Sensing of the Environment*, **47**, 142–153.

Ranson, K.J. and Sun, G., 1994b, Mapping biomass of a northern forest using multifrequency SAR data, *IEEE Transactions on Geoscience and Remote Sensing*, **32**, 388–396.

Raper, J. and Livingstone, D., 1995, Development of a geomorphological spatial model using object-oriented design, *International Journal of Geographical Information Systems*, **9**, 359–383.

Ton, J., Jain, A.K., Enslin, W.R. and Hudson, W.D, 1989, Automatic road identification and labelling in Landsat 4 TM images, *Photogrammetria (PRS)*, **43**, 257–276.

van der Knaap, W.G.M., 1992, The vector to raster conversion: (mis)use in geographical information systems, *International Journal of Geographic Information Systems*, **6**, 159–170.

Wooding, M.G., Zmuda, A.D. and Griffiths, G.H., 1994, Crop discrimination using multitemporal ERS-1 SAR data. *Proceedings of the 2nd ERS-1 Symposium* (Noordwijk: ESA), ESA SP-361, 51–56.

Worboys, M.F., 1994. Object-oriented approaches to geo-referenced information, *International Journal of Geographic Information Systems*, **8**, 385–399.

Worboys, M.F., Hearnshaw, H.M. and Maguire, D.J., 1990, Object-oriented data modelling for spatial databases, *International Journal of Geographic Information Systems*, **4**, 369–383.

14

Per-field Classification of Land Use Using the Forthcoming Very Fine Spatial Resolution Satellite Sensors: Problems and Potential Solutions

Paul Aplin, Peter M. Atkinson and Paul J. Curran

14.1 Introduction

This chapter reports research undertaken as part of a project within the Applications Demonstration Programme, led by the British National Space Centre and the Ordnance Survey. The objective of the project was to develop an automated operational system for classifying land use with fine spatial resolution satellite sensor imagery at the local scale, that could later be extended to the national scale. Land use classes were selected according to the specifications of the National Land Use Stock System (NLUSS) and classification was performed on a per-field basis (field refers to a parcel of land including semi-natural areas, agricultural fields and urban roads, buildings and gardens) by utilizing digital vector data and geographical information systems (GIS). Per-field, as opposed to per-pixel, classification provided a vehicle within which the spatial variability and texture inherent in fine spatial resolution imagery could be utilized. In performing the classification several sources of misclassification were identified and addressed. As a result per-field classification was 8% more accurate than per-pixel classification and, further, it allowed the incorporation of texture into the classification to enhance the information presented to the user.

Advances in Remote Sensing and GIS Analysis. Edited by Peter M. Atkinson and Nicholas J. Tate.
© 1999 John Wiley & Sons Ltd.

Three issues are considered in detail in the remainder of this section: the implications of fine spatial resolution satellite sensor imagery for classifying land use, previous studies on per-field classification and the rôle of GIS in classifying land use per-field.

14.1.1 Fine Spatial Resolution Satellite Sensor Imagery

At the turn of the twenty-first century several fine spatial resolution satellite sensors will be launched and multispectral imagery with a spatial resolution of 4 m will become widely available (Aplin et al., 1997a; Atkinson et al., 1998). In particular, three US organizations are developing multispectral satellite sensors, due for launch in 1999 or 2000, with a spatial resolution as fine as, or finer than, this (Table 14.1).

The accuracy with which land use has been mapped from satellite sensor imagery at local to national scales has been limited by the relatively coarse spatial resolution of current instruments (Hill and Mégier, 1988; Townshend, 1992). For example, the Institute of Terrestrial Ecology Land Cover Map (LCM) of Great Britain was generated using Landsat Thematic Mapper (TM) imagery with a spatial resolution of 30 m and a routinely mapped parcel size of greater than 1 ha (Fuller et al., 1994b). At this spatial resolution a considerable amount of detail in the scene is obscured from the image. It is anticipated that the new sources of fine spatial resolution imagery will increase our ability to map land use at local to national scales. In particular, the minimum parcel size at which mapping takes place will be considerably smaller than that of current surveys, resulting in an increase in geometric detail and accuracy. In fact, increasing the spatial resolution of imagery used to generate the LCM has been identified as a future refinement (Fuller et al., 1994a).

Associated with an increase in spatial resolution is, commonly, an increase in the internal variability within land parcels ('noise' in the image). Thus, while the information content of the imagery increases with spatial resolution the accuracy of land use classification may *decrease* on a per-pixel basis (Townshend, 1981; Irons et al., 1985; Cushnie, 1987). Traditional automated classification techniques classify land use on the basis of the spectral distribution of the pixels within an image, whereby each pixel is associated with the most similar spectral class (defined using some distance measure in feature space) (Mather, 1987). This general method of classification, referred to as per-pixel classification because each pixel is classified individually, can produce classifications that are 'noisy' due to the high spatial frequency of the landscape. An alternative method is per-field classification, so called because fields, as opposed to pixels, are classified as independent units. In principle, per-field classification should average out the 'noise', thereby leading to an increase in both classification accuracy (Mason et al., 1988; Wyatt and Fuller, 1992; Harris and Ventura, 1995) and ease of interpretation (Fuller and Parsell, 1990; Ryherd and Woodcock, 1996) over per-pixel classified images.

Table 14.1 Future fine spatial resolution multispectral satellite sensors

Multispectral satellite sensor	IKONOS 1	QuickBird	OrbView-3
Organization	Space Imaging EOSAT	EarthWatch	ORBIMAGE
Launch date	June 1998	Late 1998	Early 1999
Spatial resolution (m)	4	4	3.3
Spectral wavebands (nm)	450–520, 520–600, 630–690, 760–900	450–520, 520–600, 630–690, 760–900	450–520, 520–600, 630–690, 760–900
Swath width (km)	11	36	8
Tilting capabilities (°)	± 45 across/along track	± 30 across/along track	± 45 across/along track
Frequency of coverage (days)	< 4	< 4	< 3

14.1.2 Previous Studies on Per-field Classification

Janssen and Molenaar (1995) suggest that the success of per-field classification depends on the relationship between (a) the spectral and spatial properties of the imagery, (b) the size and shape of the fields and (c) the land cover classes chosen. These relations suggest that replacing the coarser spatial resolution satellite sensor imagery available currently with finer spatial resolution imagery would lead to an increase in per-field classification accuracy, especially for relatively small field sizes. Also important, however, is the method used to integrate remotely sensed imagery with field boundaries.

Per-field classification, based on the integration of remotely sensed imagery and digital vector data, has been used to generate land cover and land use information for more than a decade. Catlow *et al.* (1984) overlaid per-pixel classified Landsat Multispectral Scanner System (MSS) imagery at a spatial resolution of 57 m by 79 m with vector data as a means of assessing visually, on a per-field basis, the accuracy of per-pixel classification. Subsequent studies applied per-field classification techniques to imagery with finer spatial resolutions. Harris and Ventura (1995), Janssen and Molenaar (1995) and Lobo *et al.* (1996) performed per-field classification using Landsat Thematic Mapper (TM) imagery at a spatial resolution of 30 m. Further studies used Système Probatoire d'Observation de la Terre (SPOT) High Resolution Visible (HRV) imagery (Pedley and Curran, 1991; Johnsson, 1994) and simulated SPOT HRV imagery (Mégier *et al.*, 1984) at a spatial resolution of 20 m. Mason *et al.* (1988) integrated digital vector data with Airborne TM imagery at a spatial resolution of 10 m. The accuracy of per-field classification was consistently higher than that of per-pixel classification.

There are several ways of integrating remotely sensed imagery and digital vector data to achieve per-field classification. Mason *et al.* (1988) suggested that vector data can be incorporated into the classification at three stages: before classification (pre-classifier stratification), during classification (classifier modification) and after classification (post-classifier sorting). Generally, however, recent examples of per-field classification have employed only the latter two. For example, Westmoreland and Stow (1992) and Wang *et al.* (1997) integrated cartographic data with remotely sensed imagery during classification to assess land use change on a per-field basis. This method has the benefit of computational simplicity in that per-field classification is achieved in a single stage. Alternatively, several studies have classified land cover on a per-pixel basis before integrating the classified image with cartographic data for per-field classification. The land cover class of each field is represented by a statistic, such as the mode, of the classes of all the pixels within that field (Janssen *et al.*, 1990; Mattikali *et al.*, 1995; White *et al.*, 1995). Although more computationally complex than single stage per-field classification, this method has the benefit that within-field class texture can be used to increase the accuracy and utility of the classification.

14.1.3 The Rôle of GIS

Geographical information systems (GIS) provide a means of integrating raster (e.g., remotely sensed imagery) and vector data and, thereby, performing per-field classification. There are numerous examples of this type of analysis being performed successfully (e.g., Harris and Ventura, 1995; Janssen and Molenaar, 1995). However, there are certain problems associated with the process of integration. These primarily concern the different data models of remotely sensed imagery and GIS data. Remotely sensed imagery are acquired in the raster data model as regular pixel arrays. In contrast, most thematic GIS data layers are in the vector data model as a series of points, lines and areas. The fundamental difference between these two data models limits the extent to which the two types of data can be integrated (Ehlers et al., 1989; Trotter, 1991; Fritsch, 1992; Laan, 1992). This difference can be regarded as a result of the different ways in which these two data models represent spatial information. In the vector data model data are treated as objects (cartographic data), whereby each feature has some meaning associated with it as a result of prior interpretation and identification. Raster remotely sensed imagery comprise pixels which represent the radiation associated with regularly shaped patches of the Earth's surface. As such, remotely sensed data, without further interpretation, represent a lower form of information than object-based vector data. The integration of remote sensing and GIS is limited by the inability to fully understand and implement a transformation between these two representations of information (Ehlers et al., 1989; Ehlers, 1990; Davis et al., 1991).

Ehlers et al. (1989) identify three stages of integrated remote sensing and GIS analysis: (a) separate database, cartographic and image processing systems with facilities to transfer data between them, (b) two software packages (image processing and GIS) with a shared user interface and simultaneous display, and, (c) a single software package with shared processing. Early attempts at per-field classification using GIS were limited by the relatively low level of integration offered by systems at stage one (Janssen et al., 1990; Davis et al., 1991; Ehlers et al., 1991). Improvements in technology mean that more advanced integration has become possible and several systems have achieved stage two (Hinton, 1996). However, stage two integration requires the conversion of one form of data model to another to enable joint analysis. For example, several attempts at per-field classification have involved the conversion of vector cartographic data to the raster data model (e.g., Sader et al., 1995; Carbone et al., 1996). While this process is easy to implement (ESRI, 1993), there are problems associated with data conversion. Little is known about how conversion, which can be a slow, complex process, affects the geometric accuracy of data (Hinton, 1996; Wilkinson, 1996; Congalton, 1997). Attempts have been made to quantify this. For example, Knapp (1992) performed vector to raster conversion using eight different GIS, each of which resulted in a different geometric accuracy. Alternatively, various studies have shown that different shapes and sizes of vector polygons, and different raster pixel sizes, also affect the geometric accuracy of conversion (Carver and Brunsdon, 1994; Congalton, 1997). Additionally, where errors exist in the original data, converting to another data structure may propagate these (Wilkinson, 1996). Recently, however, per-field classification has been successfully performed using a system with stage three ('full') integration (Smith et al., 1997), thereby overcoming the problem of introducing or compounding spatial errors.

In addition to their providing a means for integrating remotely sensed imagery and vector data to achieve per-field classification, GIS have additional benefits for land use surveys. The flexible data structure of GIS enables the organization and manipulation of multiple data sets (Cowen *et al.*, 1995), and the ability to update or revise these data sets as required (Janssen and Middlekoop, 1992; Ortiz *et al.*, 1997). By incorporating physical variables which affect the distribution of land use, such as those related to elevation, climate and geology, the accuracy of the land use classification can be increased (Bunce *et al.*, 1992; Carbone *et al.*, 1996). By linking to other variables, such as those related to population, transportation and land boundaries (e.g., private property, local authority, protected areas) the land use data set can be analysed in a wider social and environmental context (Kushwaha, 1996; Lobo *et al.*, 1996). Further, GIS can extrapolate local studies of land use to a wider spatial scale. For example, local land use databases maintained in GIS, such as those of local authorities, can be linked spatially to enable analysis at the regional or national scale (White *et al.*, 1995).

14.2 Study Area and Data

To execute the per-field classification a study area was selected and data were acquired from three sources: Compact Airborne Spectrographic Imager (CASI) imagery, Land-Line digital vector data and ground reference data.

14.2.1 Study Area

The study area, approximately 5 km by 8 km, on the western fringe of St Albans, Hertfordshire, UK, contained a variety of rural and urban land use types. The rural areas were largely agricultural with fields of cereals, legumes, fallow and pasture, intermixed with woodland. The urban areas comprised a mixture of residential, industrial and recreational (school playing fields, public parks, boating lake) land use.

14.2.2 CASI Imagery

Compact Airborne Spectrographic Imager (CASI) imagery was acquired to simulate fine spatial resolution satellite sensor imagery, which was not then available. This simulated imagery had a spatial resolution of 4 m in four spectral wavebands, thus matching the specifications of two forthcoming satellite sensors, IKONOS 1 and Orbview-3 (Aplin *et al.*, 1997a).

Four flightlines of imagery, running south–north, were acquired over the study area on 31 May 1996 and were supplied following roll correction. Cloud shadow covered significant portions of the study area and, therefore, the images were divided into three smaller subsets for further analysis. Of these, the first (approximately 1.5 km by 3 km) was predominantly agricultural, the second predominantly wooded and the third predominantly urban. For clarity, only the first (Figure 14.1a) will be discussed in detail in this chapter, while general reference will be made to the other two. For a fuller discussion of all three subsets see Aplin and Atkinson (1997).

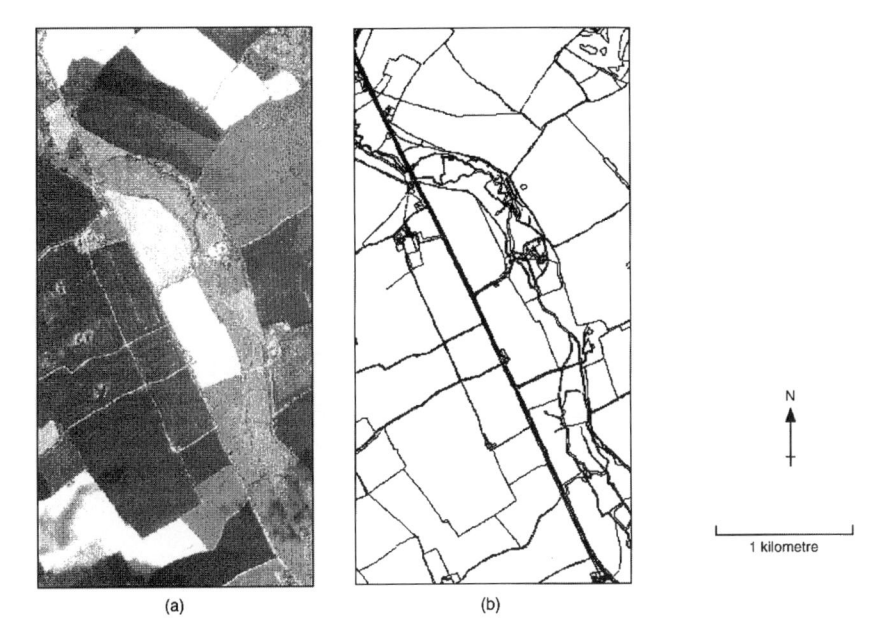

Figure 14.1 Original (a) CASI image and (b) Land-Line vector data (© Ordnance Survey, 1996) used in the classification

14.2.3 Land-Line Digital Vector Data

Land-Line digital vector data were supplied by the Ordnance Survey. These data comprised points and lines and were registered to the British National Grid (BNG) map coordinate system. Each entity (point and line) had a feature code by which it was categorized. A vector coverage was created to match exactly the area covered by the image subset (Figure 14.1b).

14.2.4 Ground Reference Data

The ground reference data required for the study were driven by the NLUSS specifications. Data were obtained on land cover classes (e.g., spring barley and wheat) and land use classes (e.g., fallow and park), and NLUSS classes were derived from these data. While a mix of land cover and land use classes were identified on the ground, these are hereafter referred to collectively as land use classes. Ground reference data on land use were acquired from three sources: a radiometric survey, communication with residents and a land use survey.

A radiometric survey was performed concurrently with the acquisition of the CASI imagery to enable atmospheric correction. Radiometric measurements of six bright and dark surfaces (winter barley, wheat, oilseed rape, running water, golf course and road) were recorded using a SPECTRON SE-590™ spectroradiometer.

Residents and landowners throughout the study area were contacted, prior to image acquisition, and asked to provide land use information on proforma maps. In the days following image acquisition, an independent land use survey was conducted to map land use in each field throughout the study area. These two sources of ground reference data were combined to generate a comprehensive land use map of the study area. This map provided a reference data set with which classes were trained and the accuracy of classification assessed.

14.3 Analysis

Following data acquisition, an analytical sequence was developed to classify land use using both per-pixel and per-field techniques (Figure 14.2). The method used for per-field classification was post-classifier sorting whereby the CASI image was classified on a per-pixel basis prior to integration with the land line coverage. There were three main stages: preprocessing the CASI image, per-pixel classification and per-field classification.

14.3.1 Pre-processing the CASI Image

Pre-processing was carried out to ensure that the CASI image was in an appropriate format for classification. This involved two steps: atmospheric correction and geometric rectification.

Atmospheric correction was performed to exemplify the procedure that could be applied to satellite sensor imagery in a national land use monitoring system. Initially, regression equations for each of the four CASI spectral wavebands were developed by relating ground-based radiometric measurements of the six surfaces with spectral values obtained from the imagery. The equations were then applied to the CASI image. This technique is dependent on the availability of ground-based radiometric data. Although, to date, such data are not available commonly for satellite sensor imagery, this availability may increase for future fine spatial resolution imagery since the suppliers of these image products intend to give customers more control over the acquisition of imagery than is currently provided with coarser spatial resolution imagery (Aplin *et al.*, 1997a). That is, customers may be able to order the acquisition of imagery at a specific time and may, therefore, acquire ground-based measurements simultaneously.

To integrate the CASI and Land-Line data it was necessary to register both data sets to a single map coordinate system. Since the Land-Line coverage was already registered to the BNG the CASI image was registered directly to the Land-Line coverage. Initially, a series of Ground Control Points (GCPs) common to both data sets were identified before transforming the CASI image to the BNG and applying a nearest neighbour re-sampling technique. A relatively large number (79) of GCPs and a high (4th) order of transformation were used in the rectification in an attempt to correct for the severe geometric distortion in the CASI image, a result of movement (yaw and pitch) of the aircraft during image acquisition, and slight geometric errors in the Land-Line coverage. However, despite these efforts, the Root Mean Square (RMS)

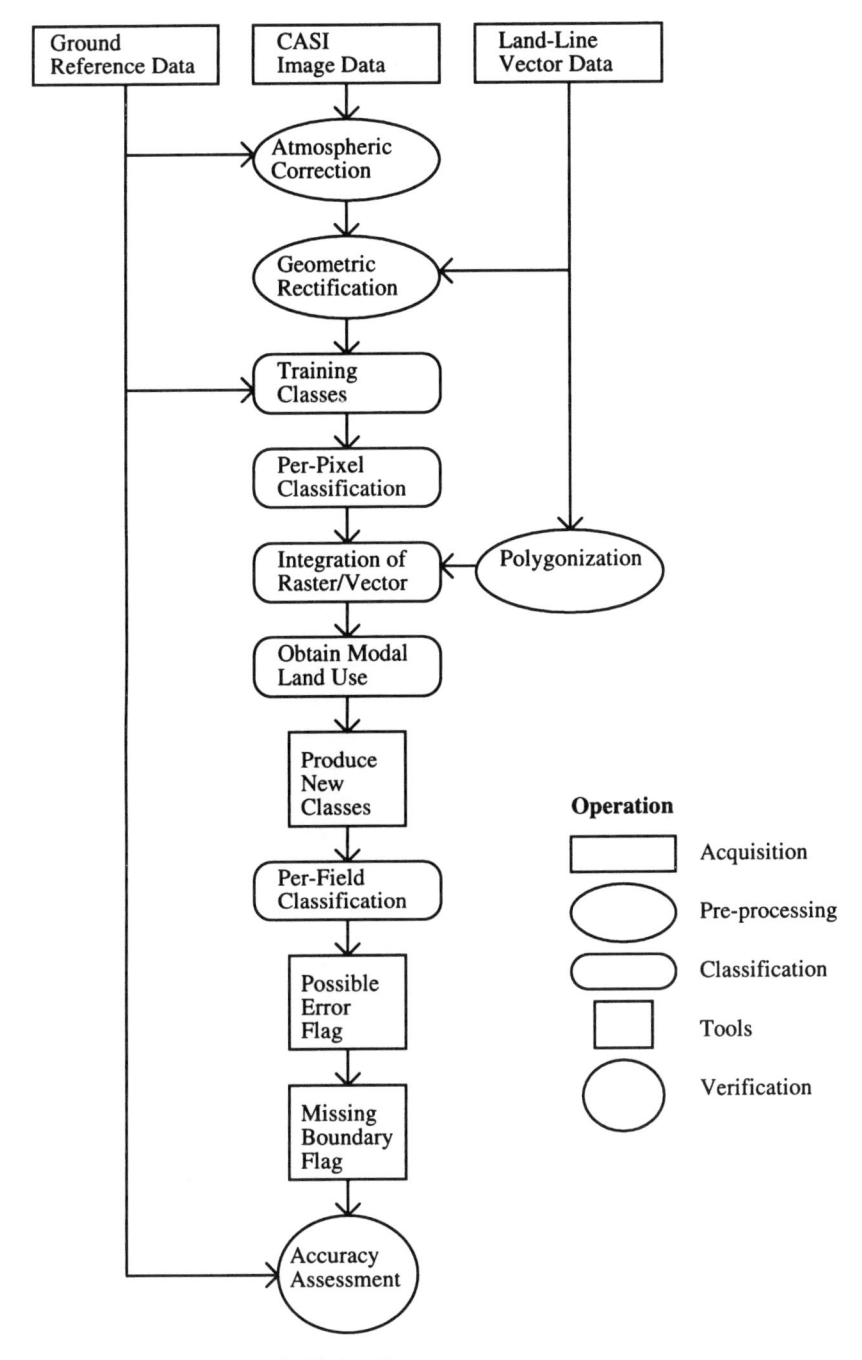

Figure 14.2 Key stages in per-field classification

error of the rectification was large (greater than 4 pixels). It is important to note, though, that for satellite sensor imagery the RMS error should be less since satellite platforms are more stable than airborne platforms and generate imagery with less geometric distortion.

14.3.2 Per-pixel Classification

The fine spatial resolution of the CASI image enabled the differentiation between similar crop types (such as wheat and barley), seasonal crops (such as spring and winter barley) and woodland categories (such as conifer and broadleaved). Thirteen land use classes based on NLUSS specifications (spring barley, winter barley, wheat, broad beans, oilseed rape, peas, bare soil, improved pasture, rough pasture, broadleaved woodland, running water, golf course and road) were selected. Remotely sensed reflectance is related to land cover and not land use (e.g., it is possible to infer from spectral values in a single pixel that the land cover is grass, but not that it is a golf course). However, in the present case, each land use class was assumed to correspond to spectrally separable land covers.

The per-pixel classifier was trained on a representative sample of each of the land use classes. Training samples were selected using the reference land use map and block training was performed, whereby blocks of pixels were selected from the centres of fields of known land cover. This method of training, which accounts for a degree of within-field variation, was preferable to point training (the selection of individual pixels) since the unit of study was the field. Fields of mixed land cover and the edges of fields were not selected for training since these would create a mixed training class.

Per-pixel classification of the CASI image used a supervised maximum likelihood classification algorithm (Figure 14.3a; Plate IVa). This classifier was selected for two main reasons. First, it is a relatively simple computational procedure available in many digital image processing systems. Second, by using the shape of the distribution of membership (as represented by the covariances) as well as the mean of the training data to identify each class, the accuracy of the classification is generally relatively high (Mather, 1987).

14.3.3 Per-field Classification

Following per-pixel classification the classified image was integrated with the Land-Line coverage and re-classified by field.

Initially, the Land-Line data were cleaned and built to create polygons. Line features which were not required in the creation of a land use coverage (such as the centre line of public roads and parish boundaries) were deleted, along with 'dangling' lines. The automatic deletion of all dangling lines, although not strictly necessary for the construction of a polygon coverage, was carried out to ensure that there were no unclosed polygons in the coverage. These occur where a gap in a polygon boundary joins two or more otherwise independent polygons. Since not all dangling lines were immediately obvious in the Land-Line data, they may have been mistaken as polygon boundaries resulting in the misinterpretation of the classified polygons. The deletion of dangling lines avoided any such misinterpretation.

Integration of the CASI image and Land-Line coverage was achieved in a geographical information system (ARC/INFO). The polygon coverage was rasterized to a 4 m pixel size, matching that of the CASI image, and the two raster data sets (CASI and Land-Line) were then combined in a single raster grid. Per-field classification was performed by calculating the modal land use class (the spatially dominant land use class) within each field (Land-Line polygon) in the raster grid and then applying this class to the entire field in the original Land-Line (non-rasterized) coverage. Figure 14.3b (Plate IVb) illustrates the per-field classified image.

14.3.4 Results

The accuracy of classification was assessed by comparing the classified images to the reference land use map. This was carried out on a *per-pixel* basis for both the per-pixel (Tables 14.2a) and the per-field (Table 14.2b) classifications whereby a random sample of 50 points was selected for each land use class on each image and cross-referenced with the land use map. A per-pixel method was used for both classifications to enable a direct comparison between the two sets of results and, in the per-field classification, to weight each field according to its size. That is, by carrying out a per-pixel accuracy assessment on a per-field classification, it is likely that more pixels will be selected from a large field than from a small field. Therefore, incorrectly classifying a large field is a greater source of misclassification than incorrectly classifying a small field. In contrast, if fields rather than pixels were selected for assessing the accuracy of a per-field classification then a direct comparison between the per-pixel and per-field classifications would not be possible and each field would have an equal weighting, regardless of size. For example, a rural field with an area of $5000 \, m^2$ would be of equal importance to an urban field with an area of $5 \, m^2$ and, therefore, the classification accuracy results would not be representative of the areal distribution of land use within the classified image.

In the following two sections, problems at the per-pixel and per-field stages of classification are identified. In most research papers such problems could be omitted, but they are presented here to clearly identify pitfalls that may be encountered by users with only moderate resources. In Section 14.3.4.3 some solutions to these problems are presented, followed in Section 14.3.4.4, by some suggestions for further processing.

14.3.4.1 *Problems Identified at the Per-pixel Stage*

The overall classification accuracy of the per-pixel classified image was 63.08%. Considerable misclassification occurred in several classes for five main reasons (Table 14.3). First, winter barley was misclassified as spring barley, because these classes were spectrally similar. Second, rough pasture was misclassified as improved pasture. This problem arose because the land *use* classes were spectrally similar. Improved and rough pasture are land use classes which are only distinguished using judgement. In fact, the land cover at the ground probably varied within each of the two land use classes to such an extent that the two classes were indistinguishable spectrally.

Table 14.2 (a) Per-pixel classification accuracies

Predicted Class	Reference Class												
	Unclassified	Spring barley	Winter barley	Wheat	Broad beans	Oilseed rape	Peas	Bare soil	Improved pasture	Rough pasture	Broadleaved woodland	Running water	Golf course	Road	Total
Unclassified	0	0	0	0	0	0	0	0	0	0	0	0	0	0	0
Spring barley	1	29	12	0	0	0	0	0	3	4	1	0	0	0	50
Winter barley	1	1	39	4	0	0	0	0	1	4	0	0	0	0	50
Wheat	1	0	0	49	0	0	0	0	0	0	0	0	0	0	50
Broad beans	14	0	0	0	13	0	2	0	0	13	3	0	0	5	50
Oilseed rape	2	0	0	0	0	48	0	0	0	0	0	0	0	0	50
Peas	1	0	0	0	0	0	49	0	0	0	0	0	0	0	50
Bare soil	2	0	0	0	0	0	0	47	0	0	0	0	0	1	50
Improved pasture	4	2	7	0	0	0	0	0	13	21	0	0	3	0	50
Rough pasture	10	0	0	0	0	0	5	0	2	30	0	0	0	3	50
Broadleaved woodland	6	0	0	13	0	0	0	0	1	18	8	4	0	0	50
Running water	12	0	0	0	0	0	0	0	0	0	6	31	0	1	50
Golf course	6	7	0	0	0	0	0	0	2	5	0	0	30	0	50
Road	21	0	0	1	0	0	1	1	0	0	1	1	0	24	50
Total	81	39	58	67	13	48	57	48	22	95	19	36	33	34	650
Producer's Accuracy (%)		74.36	67.24	73.13	100.00	100.00	85.96	97.92	59.09	31.58	42.11	86.11	90.91	70.59	
User's Accuracy (%)		58.00	78.00	98.00	26.00	96.00	98.00	94.00	26.00	60.00	16.00	62.00	60.00	48.00	

Overall Classification Accuracy (%) = 63.08

Table 14.2 (b) Per-field classification accuracies

Predicted Class	Reference Class													
	Unclassified	Spring barley	Winter barley	Wheat	Broad beans	Oilseed rape	Peas	Bare soil	Improved pasture	Rough pasture	Broadleaved woodland	Running water	Golf course	Road	Total
Unclassified	0	0	0	0	0	0	0	0	0	0	0	0	0	0	0
Spring barley	1	0	0	0	0	0	0	0	0	49	0	0	0	0	50
Winter barley	0	7	37	0	0	0	0	0	0	5	0	0	0	1	50
Wheat	1	0	0	49	0	0	0	0	0	0	0	0	0	0	50
Broad beans	11	0	0	0	39	0	0	0	0	0	0	0	0	0	50
Oilseed rape	1	0	22	0	0	27	0	0	0	0	0	0	0	0	50
Peas	2	0	0	0	0	0	48	0	0	0	0	0	0	0	50
Bare soil	0	0	0	0	0	0	0	50	0	0	0	0	0	0	50
Improved pasture	0	1	0	0	0	0	0	0	25	24	0	0	0	0	50
Rough pasture	1	0	0	0	0	0	0	0	0	49	0	0	0	0	50
Broadleaved woodland	18	0	8	0	0	0	0	0	0	14	4	5	0	1	50
Running water	50	0	0	0	0	0	0	0	0	0	0	0	0	0	50
Golf course	0	0	0	0	0	0	0	0	0	0	0	0	50	0	50
Road	42	0	0	0	0	0	0	0	0	1	0	0	0	7	50
Total	127	8	37	79	39	27	48	50	25	142	4	5	50	9	650
Producer's Accuracy (%)	0.00	100.00	62.03	100.00	100.00	100.00	100.00	100.00	34.51	100.00	0.00	100.00	77.78		
User's Accuracy (%)	0.00	74.00	98.00	78.00	54.00	96.00	50.00	98.00	8.00	0.00	50.00	14.00			

Overall Classification Accuracy (%) = 59.23

Table 14.3 Problems identified in performing both per-pixel and per-field classifications

Problem	Affected per-pixel classification	Affected per-field classification	Solution presented
1. Spectrally similar classes	Yes		
2. Land *use* classes	Yes	Yes	Yes
3. Classes with insufficient sample size	Yes	Yes	Yes
4. Classes acting as 'catch-all'	Yes	Yes	Yes
5. Cloud shadow	Yes	Yes	Yes
6. Mis-registration of raster/vector data		Yes	
7. Errors in vector data		Yes	Yes

This problem was exacerbated by problems of distinguishing the two land use classes on the ground. That is, there may have been inconsistencies in the reference land use map obtained from field survey. Hence, there was much confusion between the two classes, and in the present case this resulted in the tendency to misclassify rough pasture as improved pasture.

Third, many pixels were misclassified as broad beans, broadleaved woodland, running water or roads because these classes occupied relatively small proportions of the study area and there were not sufficient quantities of pure pixels from which to select a sample of 50. For each class in the classification a given number of pixels were misclassified due to mixed boundary pixels and so on. Classes which covered a small area, such as broad beans, for which there was only one relatively small field, had a higher proportion of misclassified pixels to pure pixels than classes which covered larger areas. As a result, a large number of the sample points of each of these four classes were referenced as unclassified.

Fourth, wheat and rough pasture were misclassified as a variety of other classes because they occupied relatively large proportions of the study area and each acted as a 'catch-all' for classes which had insufficient quantities of pure pixels.

Fifth, cloud shadow, areas of which were clearly visible in the southern portions of the original CASI image (Figure 14.1a), initially caused considerable misclassification. These areas, although visibly misclassified on the per-pixel classified image (Figure 14.3a; Plate IVa), were masked out for the purpose of the accuracy assessment and so had no affect on the final classification accuracies.

14.3.4.2 Problems Identified at the Per-field Stage

The overall classification accuracy of the per-field classified image was 59.23%, similar to that of the per-pixel classified image. There are six main reasons for the failure to provide an increase in accuracy (Table 14.3), the first four of which are similar to problems identified at the per-pixel stage: land *use* classes, classes generating an insufficient sample of pure pixels (e.g., the running water class had 'no' pure fields which resulted in all 50 sample points being referenced as unclassified), classes acting as a 'catch-all' and cloud shadow. Importantly, although areas of cloud shadow were masked out (as before) for the purpose of the accuracy assessment, where cloud shadow covered only part of a field the remainder of that field was included in the accuracy assessment. Consequently, cloud shadow had a considerable effect on accuracy. For example, since the field of rough pasture misclassified as running water in the south-east of the study area was the only field of running water in the per-field classification, all sample points were taken from this field. Similarly, since there were only relatively few pure fields of road a considerable number of sample points were selected from the large field of oilseed rape misclassified as road in the south-west of the study area (Figure 14.3b; Plate IVb).

Other causes of misclassification included mis-registration between the CASI image and the Land-Line coverage and errors in the Land-Line data. Mis-registration affected all small fields throughout the study areas and, in particular, the 'road' land use class. However, despite the large RMS error associated with geometric rectification, many small fields were classified correctly.

Errors in the Land-Line data occurred where field boundaries were missing and led to the misclassification of large portions of the study area. For example, several field boundaries were missing around four fields of wheat and three of oilseed rape in the north of the study area and this resulted in the misclassification of wheat as oilseed rape (Figure 14.3b; Plate IVb). Some of these boundaries were missing in the original Land-Line data and may have been the result of the creation of new field boundaries on the ground. Others, however, were present in the original Land-Line data (Figure 14.1a) but were removed when being polygonized because they were dangling lines.

14.3.4.3 *Solutions*

The accuracy of both per-pixel and per-field classification were increased by accounting for the causes of misclassification identified above (Table 14.3), as demonstrated in a related study (Aplin *et al.*, 1997b). Initially, misclassification was reduced by training the per-pixel classifier on land cover classes as opposed to land use classes that comprise a mixture of land covers. If the objective of a study is to extract information on land use this may be done at a later stage by using textural or syntactic information between the land cover classes.

Since the classes occupied dissimilar proportions of the study area, Aplin *et al.* (1997b) employed a more appropriate strategy for the sampling of data for training. For example, if, as in this case, certain classes (such as wheat) cover a much greater area than other classes (such as running water and roads), a more appropriate sampling strategy may be to select a different number of sample points for each class based on their areal extent. This would reduce the misclassification occurring, first, as a result of non-pure sample points being selected for the classes covering relatively small areas and, second, as a result of the class covering a large area acting as a catch-all. An alternative solution may be to select classes which occupy similar proportions of the study area. In following these practices a per-pixel classification accuracy of 78% was achieved.

Since the per-pixel classified image was used as an input to per-field classification, the accuracy of the latter is strongly dependent on that of the former. Per-field misclassification, therefore, may be reduced by reducing per-pixel misclassification.

Alternatively, such a dependence on per-pixel classification may be avoided by integrating remotely sensed imagery with vector data at a different stage in the per-field classification. Classifier modification, unlike postclassifier sorting, has no requirement for prior per-pixel classification and, therefore, will be independent of any misclassification that may arise using this technique.

Misclassification at the per-field stage may be further reduced by improving the geometric registration between the image and vector data, masking out areas of cloud shadow and identifying and repairing errors in the Land-Line data. The former was not achieved and continued to be a problem in the related study (Aplin et al., 1997b). However, as indicated above, this is likely to be less of a problem for satellite sensor imagery which has considerably less geometric distortion than airborne sensor imagery.

To avoid misclassification arising from areas of cloud shadow, these were masked out (Aplin et al., 1997b). An alternative and more complete solution to cloud and cloud shadow may be to combine imagery from different sources as inputs to per-field classification. Where cloud or cloud shadow cover certain portions of the imagery, it may be possible to combine different sources of imagery to create a cloud-free scene of the whole study area. In addition to enabling the integration of raster and vector data, GIS functionality readily enables the integration of different sources of remotely sensed imagery. Per-field classification using GIS as a means of integration is, therefore, ideally suited to this type of combined analysis. However, when combining different sources of imagery it is important that their acquisition dates are matched closely to ensure that the land cover has remained relatively constant, in terms of spectral properties, between the dates. Further, the use of multiple sources of imagery has implications for atmospheric correction, which should be performed independently on each source to ensure standardization.

In following the practices outlined above to increase the accuracy of per-field classification and using the updated per-pixel classification (the accuracy of which is stated above) as an input, Aplin et al. (1997b) achieved a per-field classification accuracy of 86%. The 8% increase in accuracy of per-field over per-pixel classification occurred largely as a consequence of removing within-field variation, a cause of misclassification at the per-pixel stage.

14.3.4.4 *Increasing Utility by Integrated Processing*

To identify errors in the Land-Line data a 'missing boundary' flag was developed, based on a within-field texture filter applied following per-field classification. This exploitation of within-field spatial and spectral information was enabled by the flexibility of the raster–vector integration process and the spatial resolution of the simulated satellite sensor imagery. For each field, the local variance was calculated, by passing a moving window over the pixels within the field, and this was compared to the proportion of modal land cover. Where both the local variance and the proportion covered by the modal land cover were small the user was informed that there may be two or more homogeneous patches of different land cover types (fields on the ground) and, therefore, missing field boundaries (Aplin et al., 1998). This 'missing boundary' flag could be supplemented in two ways to increase the accuracy of identifying missing field boundaries and, in particular, locate exactly where in each field boundaries are missing. First, dangling lines could be retained when building the polygon coverage and subsequently displayed to indicate the presence of unclosed field boundaries. As suggested in Section 14.3.3, the danger of doing this is that, where they are difficult to identify, dangling lines may be omitted from this process, resulting in misinterpretation. Second, a within-field edge detection algorithm could be used to identify borders between homogeneous patches of different land use types represented by spatial clusters of pixels.

Classification errors at the per-field stage were further reduced by developing an additional tool to identify fields with a high likelihood of misclassification. Within-field spatial and spectral information were exploited again to develop a 'possible error' flag. This flag involved examination of the proportion of each field occupied by the modal land cover class. For example, where the modal land cover occupied greater than 80% of a field it was unlikely to be misclassified. Where the modal land cover occupied less than 40% of a field it was more likely to be misclassified and manual checking was performed.

Proportional measures of land cover were also used to alter the properties of land cover classes at the per-field stage to make them more representative of the land cover on the ground. The classes in the per-pixel classification were manipulated to produce and assign new useful classes in the per-field classification. This was done by using the proportion of *all* land cover classes within each field (rather than only the modal land cover) to assign land cover per field. For example, a mixed woodland class was created from the input per-pixel classes of conifer and broadleaved woodland (Aplin *et al.*, 1997b). Where either of these classes was the modal land use for a given field and both classes occupied more than a given proportion of the field, mixed woodland was assigned to the field.

These three useful tools are only a small selection of what it is possible to achieve with integrated remotely sensed and vector data. Such processing is likely to be central to future integrated systems designed to handle fine spatial resolution satellite sensor imagery.

14.4 Conclusions

Per-pixel classification of fine spatial resolution simulated satellite sensor imagery resulted in misclassification, not only as a consequence of internal variability within fields, but also due to the use of land use as opposed to land cover classes, a poor sampling strategy for accuracy assessment and the presence of areas of cloud shadow. Per-field classification was able to overcome the problem of internal variability within fields in some instances, but in others, where much misclassification occurred at the per-pixel stage, this resulted in the selection of an incorrect class as the modal land use. In such fields, misclassification at the per-pixel stage was compounded at the per-field stage where classification error equalled 100%. That is, the entire field was misclassified. Misclassification also occurred at the per-field stage as a result of poor geometric registration and errors in the original Land-Line data. Overall, the accuracy of per-pixel classification was slightly greater than that of per-field classification. With the advent of a new era of fine spatial resolution satellite sensor imagery, users will need to be aware of the potential for such errors in classification, and of the relatively low classification accuracies which may result.

All identified sources of misclassification (with the exception of spectrally similar classes and poor geometric registration) were accounted for by selecting a more appropriate classification scheme and sampling strategy, and by removing errors in the vector data. Where these practices were followed, per-field classification resulted in an accuracy of 86%, an increase of 8% over that of per-pixel classification. Errors in geometric registration will be less of a problem for satellite sensor imagery which contains less geometric distortion than airborne sensor imagery.

The flexibility of the integration process and the fine spatial resolution of the simulated satellite sensor imagery enabled the development of three tools to increase the utility of per-field classification. A 'missing boundary' flag was developed based on a per-field local variance, a 'possible error' flag was implemented by manipulating the proportion of modal land cover in each field, and new useful classes were produced by manipulating the proportion of all land cover classes in each field.

It is believed that the forthcoming multispectral fine spatial resolution satellite sensor imagery will have a significant impact on the way we map land cover and land use. In particular, these sources of imagery may lead to an increase in the geometric detail and accuracy with which land cover can be mapped over that of current coarser spatial resolution imagery. Per-field classification, through the integration of remotely sensed imagery and vector data, is an attractive means of generating accurate land cover information from this type of imagery. Notably, this method of classification makes use of the within-field spatial variability and texture inherent in fine spatial resolution imagery. GIS have the functionality to enable this type of analysis and can provide additional benefits for land cover surveys including the integration of related data sets and the ability to update or revise land cover information as required.

Acknowledgements

This chapter reports on work undertaken as part of a project entitled Landuse And National Digital Mapping from Advanced Satellite Sensors (LANDMASS) within the Applications Demonstration Programme, led by the British National Space Centre and the Ordnance Survey. The results presented here were presented earlier in a LANDMASS Completion Report (Aplin and Atkinson, 1997).

The authors are grateful to Atlantic Reconnaissance for acquiring the CASI imagery, the Natural Environment Research Council Equipment Pool for Field Spectroscopy for supplying the spectroradiometer, the residents and landowners at the study area who provided land use information, the team of fieldworkers from the Department of Geography, University of Southampton, who mapped the land use of the study area, the GeoData Institute, University of Southampton, for assistance with computing, and the two anonymous referees for their useful and constructive comments.

References

Adinarayana, J. and Rama Krishna, N., 1996, Integration of multi-sensed images for improved landuse classification of a hilly watershed using geographical information systems, *International Journal of Remote Sensing*, **17**, 1679–1688.

Al-Garni, A.M., 1996, A system with predictive least-squares mathematical models for monitoring wildlife conservation sites using GIS and remotely-sensed data, *International Journal of Remote Sensing*, **17**, 2479–2503.

Aplin, P. and Atkinson, P.M., 1997, *Work Package 6 Completion Report, Application Demonstration Program – Landuse And National Digital Mapping from Advanced Satellite Sensors (LANDMASS)* (Southampton: Ordnance Survey).

Aplin, P., Atkinson, P.M. and Curran, P.J., 1997a, Fine spatial resolution satellite sensors for the next decade, *International Journal of Remote Sensing*, **18**, 3873–3881.

Aplin, P., Atkinson, P.M. and Curran, P.J., 1997b, Using the spectral properties of fine spatial resolution satellite sensor imagery for national land cover and land use mapping, in G. Guyot and T. Phulpin, (eds), *Physical Measurements and Signatures in Remote Sensing* (Rotterdam: Balkema), 661–668.

Aplin, P., Atkinson, P.M. and Curran, P.J., 1998, Identifying missing field boundaries to increase the accuracy of per-field classification of fine spatial resolution satellite sensor imagery, *Proceedings of the 27th International Symposium on Remote Sensing of Environment: Information for Sustainability*, (Oslo: Norwegian Space Centre), 399–402.

Atkinson, P.M., Aplin, P. and Curran, P.J., 1998, A clean sweep of the skies, *Remote Sensing Supplement*, 12–15, in *GIS Europe*, **7**(5), and *Mapping Awareness*, **12**(4).

Bunce, R.G.H., Barr, C.J. and Fuller, R.M., 1992, Integration of methods for detecting land use change, with special reference to Countryside Survey 1990, in M.C. Whitby (ed.), *Land Use Change: The Causes and Consequences*, Institute of Terrestrial Ecology symposium number 27 (London: HMSO), 69–78.

Carbone, G.J., Narumalani, S. and King, M., 1996, Application of remote sensing and GIS technologies with physiological crop models, *Photogrammetric Engineering and Remote Sensing*, **62**, 171–179.

Carver, S.J. and Brundson, C.F., 1994, Vector to raster conversion error and feature complexity: an empirical study using simulated data, *International Journal of Geographical Information Systems*, **8**, 261–270.

Catlow, D.R., Parsell, R.J. and Wyatt, B.K., 1984, The integrated use of digital cartographic data and remotely sensed imagery, *Earth-Orientation Applications in Space Technology*, **4**, 255–260.

Congalton, R.G., 1997, Exploring and evaluating the consequences of vector-to-raster and raster-to-vector conversion, *Photogrammetric Engineering and Remote Sensing*, **63**, 425–434.

Cowen, D.J., Jensen, J.R., Bresnahan, P.J., Ehler, G.B., Graves, D., Huang, X., Wiesner, C. and Mackey, Jr, H.E., 1995, The design and implementation of an integrated geographic information system for environmental applications, *Photogrammetric Engineering and Remote Sensing*, **61**, 1393–1404.

Cushnie, J.L., 1987, The interactive effect of spatial resolution and degree of internal variability within land-cover types on classification accuracies, *International Journal of Remote Sensing*, **8**, 15–29.

Davis, F.W., Quattrochi, D.A., Ridd, M.K., Lam, N.S-N., Walsh, S.J., Michaelsen, J.C., Franklin, J., Stow, D.A., Johannsen, C.J. and Johnston, C.A., 1991, Environmental analysis using integrated GIS and remotely sensed data: some research needs and priorities, *Photogrammetric Engineering and Remote Sensing*, **57**, 689–697.

Ehlers, M., 1990, Remote sensing and geographic information systems: towards integrated spatial information processing, *IEEE Transactions on Geoscience and Remote Sensing*, **28**, 763–766.

Ehlers, M., Edwards, G. and Bedard, Y., 1989, Integration of remote sensing with geographic information systems: a necessary evolution, *Photogrammetric Engineering and Remote Sensing*, **55**, 1619–1627.

Ehlers, M., Greenlee, D., Smith, T. and Star, J., 1991, Integration of remote sensing and GIS: data and data access, *Photogrammetric Engineering and Remote Sensing*, **57**, 669–675.

ESRI, 1993, *ARC Command References J-Z* (Redlands, CA: Environmental Systems Research Institute).

Fritsch, D., 1992, Analysis of remote sensing data in geographical information systems, *EARSeL Advances in Remote Sensing*, **1**(3), 60–65.

Fuller, R.M. and Parsell, R.J., 1990, Classification of TM imagery in the study of land use in lowland Britain: practical considerations for operational use, *International Journal of Remote Sensing*, **11**, 1901–1917.

Fuller, R.M., Groom, G.B. and Jones, A.R., 1994a, The land cover map of Great Britain: an automated classification of Landsat Thematic Mapper data, *Photogrammetric Engineering and Remote Sensing*, **60**, 553–562.

Fuller, R.M., Sheail, J. and Barr, C.J., 1994b, The land of Britain, 1930–1990: a comparative study of field mapping and remote sensing techniques, *The Geographical Journal*, **160**, 173–184.

Harris, P.M. and Ventura, S.J., 1995, The integration of geographic data with remotely sensed imagery to improve classification in an urban area, *Photogrammetric Engineering and Remote Sensing*, **61**, 993–998.

Hill, J. and Mégier, J., 1988, Regional land cover and agricultural area statistics and mapping in The Department Ardeche, France, by use of Thematic Mapper data, *International Journal of Remote Sensing*, **9**, 1573–1595.

Hinton, J.C., 1996, GIS and remote sensing integration for environmental applications, *International Journal of Geographical Information Systems*, **10**, 877–890.

Irons, J.R., Markham, B.L., Nelson, R.F., Toll, D.L., Williams, D.L., Latty, R.S. and Stauffer, M.L., 1985, The effects of spatial resolution on the classification of Thematic Mapper data, *International Journal of Remote Sensing*, **6**, 1385–1403.

Janssen, L.L.F. and Middlekoop, H., 1992, Knowledge-based crop classification of a Landsat Thematic Mapper image, *International Journal of Remote Sensing*, **13**, 2827–2837.

Janssen, L.L.F. and Molenaar, M., 1995, Terrain objects, their dynamics and their monitoring by the integration of GIS and remote sensing, *IEEE Transactions on Geoscience and Remote Sensing*, **33**, 749–758.

Janssen, L.L.F., Jaarsma, M.N. and Van der Linden, T.M., 1990, Integrating topographic data with remote sensing for land-cover classification, *Photogrammetric Engineering and Remote Sensing*, **56**, 1503–1506.

Johnsson, K., 1994, Segment-based land-use classification from SPOT satellite data, *Photogrammetric Engineering and Remote Sensing*, **60**, 47–53.

Knaap, W.G.M. van der, 1992, The vector to raster conversion: (mis)use in geographical information systems, *International Journal of Geographical Information Systems*, **6**, 159–170.

Kushwaha, S.P.S., Subramanian, S.K., Chennaiah, G.C., Ramana Murthy, J., Kameswara Rao, S.V.C., Perumal, A. and Behera, G., 1996, Interfacing remote sensing and GIS methods for sustainable rural development, *International Journal of Remote Sensing*, **17**, 3055–3069.

Laan, F.B. van der, 1992, Integration of remote sensing in a raster and vector GIS environment, *EARSeL Advances in Remote Sensing*, **1**(3), 71–80.

Lobo, A., Chic, O. and Casterad, A., 1996, Classification of Mediterranean crops with multisensor data: per-pixel versus per-object statistics and image segmentation, *International Journal of Remote Sensing*, **17**, 2385–2400.

Mason, D.C., Corr, D.G., Cross, A., Hogg, D.C., Lawrence, D.H., Petrou, M. and Tailor, A.M., 1988, The use of digital map data in the segmentation and classification of remotely-sensed images, *International Journal of Remote Sensing*, **2**, 195–215.

Mather, P.M., 1987, *Computer Processing of Remotely-Sensed Images: An Introduction* (Chichester: Wiley).

Mattikalli, N.M., Devereux, B.J. and Richards, K.S., 1995, Integration of remotely sensed satellite images with a geographical information system, *Computers and Geosciences*, **21**, 947–956.

Mégier, J., Mehl, W. and Rupelt, R., 1984, Per-field classification and application to SPOT simulated, SAR and combined SAR-MSS data, *Proceedings of the 18th International Symposium on Remote Sensing of Environment* (Ann Arbor, MI: University of Michigan), 1011–1018.

Ortiz, M.J., Formaggio, A.R. and Epiphanio, J.C.N., 1997, Classification of croplands through integration of remote sensing, GIS and historical database, *International Journal of Remote Sensing*, **18**, 95–105.

Pedley, M.I. and Curran, P.J., 1991, Per-field classification: an example using SPOT HRV imagery, *International Journal of Remote Sensing*, **12**, 2181–2192.

Ryherd, S. and Woodcock, C., 1996, Combining spectral and texture data in the segmentation of remotely sensed images, *Photogrammetric Engineering and Remote Sensing*, **62**, 181–194.

Sader, S.A., Douglas, A. and Liou, W-S., 1995, Accuracy of Landsat-TM and GIS rule-based methods for forest wetland classification in Maine, *Remote Sensing of Environment*, **53**, 133–144.

Smith, G.M., Fuller, R.M., Amable, G., Costa, C. and Devereux, B.J., 1997, CLEVER mapping: an implementation of a per-parcel classification procedure within an integrated GIS environment, *Proceedings of the 23rd Annual Conference of the Remote Sensing Society, Observations and Interactions* (Nottingham: Remote Sensing Society), 21–26.

Townshend, J.R.G., 1981, The spatial resolving power of earth resources satellites, *Progress in Physical Geography*, **5**, 32–55.

Townshend, J.R.G., 1992, Land cover, *International Journal of Remote Sensing*, **13**, 1319–1328.

Trotter, C.M., 1991, Remotely-sensed data as an information source for geographical information systems in natural resource management: a review, *International Journal of Remote Sensing*, **5**, 225–239.

Wang, R.S.M., Roberts, S.A. and Efford, N.D., 1997, An object-based approach to integrate Landsat TM data within a GIS context for detecting land use changes at urban–rural fringe areas, *Proceedings of the 23rd Annual Conference of the Remote Sensing Society: Observations and Interactions* (Nottingham: Remote Sensing Society), 179–184.

Westmoreland, S. and Stow, D.A., 1992, Category identification of changed land-use polygons in an integrated image processing/geographic information system, *Photogrammetric Engineering and Remote Sensing*, **58**, 1593–1599.

White, J.D., Kroh, G.C. and Pinder III, J.E., 1995, Forest mapping at Lassen Volcanic National Park, California, using Landsat TM data and a geographical information system, *Photogrammetric Engineering and Remote Sensing*, **61**, 299–305.

Wilkinson, G.G., 1996, A review of current issues in the integration of GIS and remote sensing data, *International Journal of Remote Sensing*, **10**, 85–101.

Wyatt, B.K. and Fuller, R.M., 1992, European applications of space-borne earth observation for land cover mapping, *Proceedings of the Central Symposium of the 'International Space Year' Conference* (Munich; European Space Agency), 655–659.

15

Modelling Soil Erosion at Global and Regional Scales Using Remote Sensing and GIS Techniques

Nick A. Drake, Xiaoyang Zhang, Eva Berkhout, Rogario Bonifacio, David I.F. Grimes, John Wainwright and Mark Mulligan

15.1 Introduction, Aims and Rationale

Soil erosion is recognized as a major problem arising from agricultural intensification, land degradation and possibly global climatic change. However, the extent of the problem is hard to quantify as field measurements of erosion are rare, time-consuming, and are usually only acquired over restricted temporal and spatial scales. To date global mapping of erosion has involved looking at sediment yield or questionnaire surveys. Walling and Webb (1983) generated global maps of sediment yield data; however, problems associated with coupling and the variable sediment dynamics of the fluvial system mean that these data are not directly comparable to hill slope erosion. Middleton and Thomas (1997) used questionnaire surveys to gain an insight into erosion severity but this approach only provides qualitative information and suffers from bias on the part of the people being questioned.

One way to provide a quantitative and consistent approach is to model erosion over large areas. Our ability to do this is restricted by problems of parameterizing models because data requirements are large, including information on vegetation, soil, topography and climate. Remote sensing and GIS hold great promise in this regard as they allow distributed models to be applied over wide areas at a number of different scales. Such methods have the advantage that rather than providing static maps, as the methods outlined above do, erosion model predictions can be updated on a monthly or even daily basis. However, the modelling of soil erosion at regional and global scales may be problematic because soil erosion models have generally been built and calibrated only at the field or catchment scale and may not be directly applicable at the regional and global scale. The objective of this study is to investigate the advantages and disadvantages of modelling soil erosion at global and regional scales using remote sensing and GIS. The ultimate aim is to develop an operational soil erosion modelling and monitoring system, that can be used to quantify the erosion problem over large areas and identify areas where management is needed.

Probably the simplest way to identify areas in need of management is to identify regions where erosion is high. However, this strategy is limited because, in areas of high erosion, damage may already be severe. A strategy more suited to sustainable use of soil resources is to implement a long-term monitoring programme and identify areas where erosion is accelerating. These areas can then be targeted for the application of soil conservation measures before soil erosion has become a problem. In this chapter, a soil erosion model is implemented to show how it can provide information that would allow implementation of both these strategies, albeit at different scales. Using global scale data it is currently only possible to implement the first strategy as the temporal resolution of the data is often only long-term monthly averages (10–70 years) that simply provide long-term average erosion rates. The regional scale data allow continuous monitoring on a daily basis and thus should tolerate implementation of the second strategy. At this scale it could potentially be done in real time for little cost.

15.2 Background

GIS provides a logical medium in which to implement soil erosion models as it allows the models to be applied in a distributed manner, ancillary data can be readily combined with remotely sensed imagery, and GIS functions can be used to calculate some of the necessary variables such as slope.

The mapping and monitoring capabilities of remote sensing provide useful inputs into soil erosion studies. Price (1993) has shown that there is strong correlation between radiance and total soil loss. This relationship exists because many of the factors that exert an influence on erosion also have an influence on the reflectance of the surface or the atmosphere (e.g., topography, vegetation cover, particle size, surface roughness, soil and rock type, soil organic matter content, rainfall, soil moisture content and porosity) and thus can potentially be estimated by remote sensing.

Some of these parameters can be mapped using standard techniques whose advantages and limitations are well known, such as production of DEMs using photogrammetric methods, classification for soil and vegetation mapping, NDVI for estimating the amount of green vegetation and cold cloud duration for rainfall estimation. However, estimation of the other parameters outlined above is more problematic for at least four reasons. First, their effect on the spectral response of surfaces may be small (e.g., porosity and surface roughness at visible and infrared wavelengths). Second, the effect of different parameters may be similar and thus hard to estimate unambiguously (e.g., moisture and organic matter both reduce the reflectance of soils in the visible and infrared). Third, the wavelengths at which some of these parameters exert the largest influence on the spectral response are not currently available operationally. Finally, techniques that allow estimation of some parameters do not yet exist (e.g., soil organic matter content). These problematic variables currently need to be measured in the field or estimated using indirect methods. Nevertheless, advances in technology and radiative transfer modelling are beginning to provide techniques that may help in the retrieval of some of these more elusive parameters. For example, Johnson et al. (1992) have produced a model to determine mineral abundance and grain size from reflectance data and it may be possible to obtain porosity using bidirectional reflectance distribution function (BRDF) measurements.

Numerous studies have shown the potential of remote sensing in soil erosion mapping. At the most basic level, estimates of erosion risk can be derived by classification of pixels according to the percentage of bare soil (Zhou and Folving, 1994). Graetz et al. (1986) improved on this scheme by estimating vegetation cover from Landsat MSS data and combining this with slope generated from a DEM to produce an index of erosion. Ideally, estimates of erosion should consider soil, climate, vegetation and topography, with these factors combined in the form of a model. Empirical soil erosion models utilizing this information have been implemented using remote sensing. Classification of TM imagery has been used to estimate the Crop Management factor in the Universal Soil Loss Equation (USLE) (Jurgens and Flander, 1993; Price, 1993). While useful, this technique suffers from the fundamental limitations of the USLE in terms of both scale and temporal frequency of the model predictions (Wischmeir, 1976). A comparison of methods for producing maps of vegetation-related variables for soil erosion studies using TM imagery was carried out by de Jong (1994), who found that NDVI was the most useful. This author also used remote sensing to map soils. Spectral matching techniques were applied to imaging spectroscopy data and the resultant soil map combined with NDVI, a DEM and climate data in a GIS to implement the SEMMED erosion model.

To date, all soil erosion models implemented using remote sensing and GIS have used high spatial resolution imagery. However, availability of GIS and remote sensing data that allow implementation of erosion models currently occurs at three general scales, each with limitations on the parameters that can be derived, their temporal frequency and cost (Table 15.1). There is a large amount of cheap data available at the global scale (1 degree to 10 minutes) for inputs into global climate models, but the spatial resolution and to some extent the temporal resolution is limited. At the regional scale (1–8 km) imagery at high temporal resolution is readily available to estimate parameters such as rainfall and vegetation cover. It is probable that erosion predictions at these two scales will indicate an environmental predisposition to erosion, rather than actual erosion rates, as many man-made features affecting erosion rates in agricultural areas (e.g., terracing and drainage channels) will be at the sub-grid scale. At the local scale (10 to 100 m) the high spatial resolution may allow definition of these features and thus more accurate modelling. Furthermore, the generally high spectral resolution at this scale provides a wealth of information on surface properties. However, there are limitations to erosion modelling at this scale: the temporal resolution is

Table 15.1 Summary information on the different scales at which data are available for soil erosion modelling

Scale	Spatial resolution	Temporal resolution	Example of type of data	Cost
Global	1° to 10'	Yearly to monthly	Rainfall, NDVI	Free to a few $100
Regional	1–8 km	Hourly, daily and decads	Meteosat, AVHRR, Rainfall Radar	Free to a few $100
Local	10 to 80 m	16–18 days	Landsat, SPOT ERS-1	$200–5000

Table 15.2 Types of soil erosion

Water erosion
Sheetwash erosion
   Rainsplash detaches soil particles that are removed by overland flow
Rill erosion
   Overland flow concentrated into rivulets detaches and removes soil particles
Gully erosion
   Development of rills into larger features by similar processes. May also develop by:
      Stream/gully bank collapse
      Undercutting causes failure and channel flow removes the detached material
Solution
Materials are dissolved and transported away in solution
Wind erosion
Wind detaches and transports soil particles

lower, precluding estimation of important parameters such as rainfall, and data are expensive especially if numerous images need to be purchased for monitoring purposes.

15.3 The Soil Erosion Model

Soil erosion is a general term for a complex set of processes (Table 15.2) and there are no united models that consider all its aspects. Remote sensing lends itself to water erosion more than wind erosion because there are no currently available techniques of estimating wind speed at the land surface whereas there are numerous techniques for estimating rainfall. The different types of water erosion outlined in Table 15.2 vary in terms of ease of recognition in the field and thus the ease with which their detrimental effects can be assessed. Rill erosion, for example, is easily recognized. Sheetwash erosion varies; it may be slow and insidious or rapid and spectacular. Insidious sheet erosion is probably the most dangerous form of water erosion as it may not be observable and so no action can be taken to counter it.

    This chapter outlines the implementation of the following sheetwash erosion model of Thornes (1985, 1989)

$$E = k\, OF^2\, s^{1.67} e^{-0.07v} \tag{1}$$

where E is erosion in mm/day or mm/month depending on the time step of the model, k

is a soil erodibility coefficient calculated from soil grain size, OF is overland flow (mm per time step) derived from sub-models of varying complexity, s is the slope in degrees and v the vegetation cover (%). The model was chosen because it can be readily implemented in a GIS and all the parameters of the model that vary over short-term time scales can potentially be estimated from remote sensing.

15.4 Global Scale Soil Erosion Modelling

Overland flow was calculated using a C++ program employing data from the *Global Ecosystems Database* (GED) CD-ROM (Kineman and Ohrenschall, 1992) and point data interpolated from the *World Weather Guide* (Pearce and Smith, 1993) using IDRISI. All other soil erosion equation parameters were calculated and the model implemented with the IDRISI GIS using data from the GED. The model was implemented on a lat./long. co-ordinate system for a region covering Europe, Asia, Arabia and northern Africa on a 10-minute grid as this is the most common resolution in the GED. This provides a maximum pixel size of 341.9 km² at the equator and minimum of 162.2 km² in the north of the study area.

15.4.1 Vegetation Cover

The GED contains maximum monthly global normalized difference vegetation index (NDVI) data from 1985 to 1990. These data were used to calculate the average monthly vegetation cover for this period. The raw NDVI data contained numerous blank areas where there was cloud cover for the whole month. These areas were filled with the average values for the preceding and following months.

Numerous authors have found a relationship between either total or green vegetation cover and NDVI. An important question to be answered for global scale studies is whether there is a consistent relationship that transcends different ecosystems. Table

Table 15.3 The relationship between NDVI and % vegetation cover. The data of Blackburn and Milton (1995) were acquired using a Spectron, we recalculated their NDVI to simulate AVHRR bandpasses

R^2	Number of samples	Sensor type	Ecosystem type and land use	Source
0.911	26	AVHRR	Deciduous forest	Blackburn and Milton (1995)
0.812	22	AVHRR	Semi-natural Mediterranean, Semi-arid	Kennedy (1989)
0.236	21	ATM	Coniferous forest	Danson (1989)
0.410	33	TM	Semi-natural Mediterranean and agricultural	de Jong (1994)
0.910	36	TM	Sugar beet	Malthus et al. (1993)

15.3 summarizes the various studies. It is clear that the correlations vary considerably. Malthus *et al.* (1993) and Blackburn and Milton (1995) produce the highest correlations. The former data set produces a surprisingly high correlation considering the data were acquired over different soil backgrounds and some of the sites suffered from Beet Yellows Virus. Blackburn and Milton (1995) looked at a single site over time and thus eliminated the effect of variable soil background but included senescent vegetation at the end of the growing season. The data of Kennedy (1989) show more scatter which she attributes to soil background effects. The correlation of de Jong (1994) is low probably because his estimates are of total cover (e.g., including branches and other non-photosynthetic plant material), whereas those of Kennedy (1989) only include green leaves and those of Malthus *et al.* (1993) and Blackburn and Milton (1995) are dominated by them. The low correlation of Danson (1989) may be due to the limited range of his cover estimates.

The cases where the NDVI and v values were published were collated (Figure 15.1) and the following regression equation calculated to convert NDVI to v:

$$v = 93.07466 \text{ NDVI} + 8.79815 \tag{2}$$

The combined correlation of these studies is surprisingly high ($R^2 = 0.66$) though there is much scatter particularly at low values of NDVI (which is presumably due to the effect of different soil backgrounds). Some of the scatter at intermediate values of NDVI could be explained by different relationships for different ecosystems (e.g., the data from Danson (1989) and Kennedy (1989)). There are, however, other explanations. First, the different sensors used may produce different results, second, each study used different methods to estimate cover, and, finally, the cover estimates were produced by different observers and there is known to be a lack of consistency in this regard (Holm *et al.*, 1984).

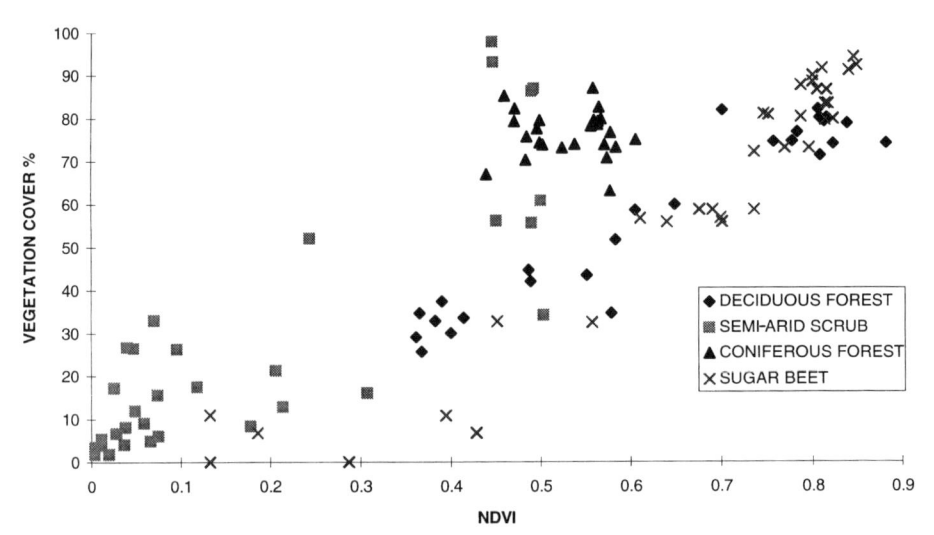

Figure 15.1 *Scattergram showing the relationship between percentage vegetation cover and NDVI for those studies in Table 15.2 that measured percentage green vegetation cover or whose measurements were dominated by it*

NDVI is currently the only globally available remote sensing estimate of vegetation cover. Clearly it has limitations at low values where variable soil backgrounds have notable effects, and in situations where there are large amounts of non-green material in the canopy. However, the only alternative to using remote sensing is to model the vegetation growth and this is probably fraught with more problems than NDVI, the main one being modelling the numerous anthropogenic impacts on the vegetation. These impacts are observed directly by remote sensing.

15.4.2 Soil Erodibility

The coefficient k is controlled primarily by soil texture and soil organic matter content and to a lesser extent by structure and permeability. Though global soil texture information is available within the GED there is no information on organic matter content. To circumvent this problem k was calculated using Boolean algebra employing the relationship between soil texture and k outlined by Olsen (1981). This soil texture triangle assumes soils have a 2% organic matter content, fine granular structure, and moderate permeability. Though these assumptions clearly do not apply to all the soils studied here they are used as a first-order approximation while methods for globally estimating soil organic matter content, soil structure and permeability are being investigated.

15.4.3 Slope

There are numerous global DEMs available at different scales from which slope can be calculated using a GIS (e.g., 10 minute, 5 minute and 30 arc seconds). The problem with calculating slope from these global DEMs is that it is progressively underestimated at coarser scales. We calculated slope from the largest scale DEM (the EROS Data Center 30 arc second DEM at http://edcwww.cr.usgs.gov/landdaac/) and then averaged the slope to 10 minutes. Although this procedure produced steeper slopes than simply calculating slope from 10 minute data, slope is still, however, underestimated.

15.4.4 Overland Flow

To implement the soil erosion model overland flow needs to be modelled on a monthly basis. There are numerous overland flow models that could be selected. The main criteria that have been used to select a model for global overland flow are simplicity, realism, and availability of data at the global scale. The SCS model has been selected as it requires few parameters, and is both realistic and robust.

The empirical SCS model (Soil Conservation Service, 1972) was developed by studying overland flow in many small experimental watersheds. The basis of the model is that the ratio of overland flow (OF) to effective storm rainfall ($Pe = P - Ia$, where P is rainfall and Ia is initial abstraction) is equal to Fa/S where Fa is the water retained in the watershed and S the potential retention

$$OFp = \frac{(r_i - Ia)^2}{(r_i + 0.8S)} \qquad (3)$$

where OFp is the overland flow in a rainfall event, $Ia = 0.2S$, S is the potential retention $= (25400/CN) - 254$ (mm), and CN is the runoff curve number that can be estimated from published tables (e.g., Rawls et al., 1993) and is dependent on factors such as land cover, soil type and soil texture. Since overland flow will occur when r_i exceeds Ia (r_i being the rainfall per rain day), the following equation is applied when this occurs

$$r_i = Ia + i\Delta r \qquad (4)$$

$$\Delta r = \frac{(r_{max} - Ia)}{n} \qquad (5)$$

where there are n different classes of rainfall intensity, $i = 1,...,n$, and the maximum rainfall per rain day (r_{max}) is assumed to be 500 mm.

This model has been adjusted to run on an average monthly basis where $OF(i)$ is calculated in the following manner:

$$OF(i) = \sum_{i=1}^{n} \Delta r OFpJ \qquad (6)$$

OFp is overland flow per n, and J is the rain day frequency density function which is assumed to be

$$J = \frac{Jo}{ro} e^{-r_i/r_0} \qquad (7)$$

where Jo is the total number of rain days per month, and ro is the mean rainfall per rain day (mm).

The data used to calculate the parameters in the modified SCS model are illustrated in Figure 15.2. J was calculated from the average monthly rainfall data available on the GED and the number of rain days per month interpolated from the *World Weather Guide*. The runoff curve numbers were calculated using the following parameters from the GED.

15.4.4.1 Soil Texture

Calculated on the basis of soil texture of each horizon.

15.4.4.2 Soil Type

Four soil types are classified according to the soil textural triangle:

Group A: loam, sand, loamy sand, silt.
Group B: sandy clay loam, silty clay loam, sandy loam and silt loam.

Modelling Soil Erosion at Global and Regional Scales Using Remote Sensing and GIS 249

```
                    ┌─────────────────┐
                    │  OVERLAND FLOW  │
                    └─────────────────┘
                       ↑         ↑
            ┌──────────┘         └──────────┐
   ┌────────────────┐              ┌────────────────┐
   │    Rainfall    │              │ Curve number CN│
   └────────────────┘              └────────────────┘
     ┌──────────────────┐            ┌──────────────────┐
     │Monthly precipitation│         │Soil texture in each│
     └──────────────────┘            │horizon           │
        ┌──────────────────┐         └──────────────────┘
        │Monthly rain days │           ┌──────────────────┐
        └──────────────────┘           │Soil depth of horizons│
                                       └──────────────────┘
                                          ┌──────────────┐
                                          │Land use      │
                                          └──────────────┘
                                            ┌──────────────────┐
                                            │Monthly AVHRR-    │
                                            │NDVI              │
                                            └──────────────────┘
                                              ┌──────────────────┐
                                              │Cultivation       │
                                              │intensity         │
                                              └──────────────────┘
```

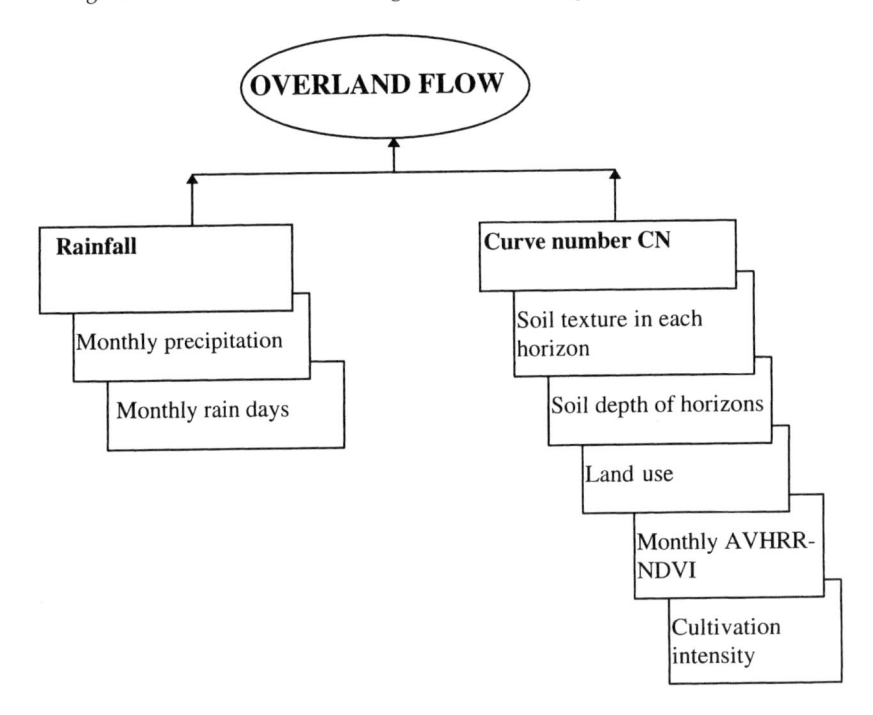

Figure 15.2 *Data used for modelling overland flow with modified SCS model*

Group C: sandy clay, silty clay and clay loam.
Group D: clay.

15.4.4.3 Land Cover

Seven classes of land cover are defined according to Matthew's Vegetation Types (Kineman and Ohrenschall, 1992). These are cultivated land, pasture or range land, meadow, wood or forest land, dirt, brush–weed–grass mixture with brush the major element and wood–grass combination.

15.4.4.4 Hydrologic Condition

This consists of poor, fair and good condition and is estimated on the basis of vegetation cover derived from monthly AVHRR-NDVI data as described above.

15.4.4.5 Conservation Treatment

Good conservation treatment and poor conservation treatment are divided in the light of Matthew's Cultivation Intensity (Kineman and Ohrenschall, 1992). This parameter is only used for cultivated land.

15.4.5 Global Results

The result of the global scale soil erosion modelling is shown in Figure 15.3. In qualitative terms the distribution of erosion seems reasonable in most areas though the rates of erosion may be somewhat low. Quantitatively validating global scale models is notoriously difficult. It is particularly problematic in the case of erosion models as field studies of erosion are never implemented at this scale and are not conducted over long enough time periods to provide average annual values. The use of data from other sources such as catchment sediment yields and lake sedimentation rates are complicated by the need to understand the sediment delivery ratio (Walling, 1988).

Not surprisingly, the highest rates are found in mountainous areas. Here most of this erosion occurs at the end of the dry season when low vegetation cover coincides with a high rainfall intensity. The highest erosion is 950 mm/year in the Himalayas which though unverified sounds reasonable. There is a low magnitude (usually less than 0.2 mm/year) but notable halo of erosion around all deserts regardless of topography because the intermediate vegetation cover and rainfall provide an environment susceptible to soil erosion (Langbein and Schumm, 1958).

There are notable differences in erosion within regions that experience monsoon rainfall. India, for example, experiences higher erosion than the Sahel in most areas. This is probably due to the fact that in many parts of India the monsoon breaks quite suddenly providing intense rainfall on parched ground with low cover, while in the Sahel rainfall gradually builds up over a period of three months or so allowing vegetation cover to increase before the intense rainfall starts.

There are a few obvious errors in Figure 15.3 in northern latitudes. Scotland and eastern Norway, which are not noted for soil erosion problems, provide higher erosion estimates than south-east Spain, which is known to suffer from this phenomenon. These high erosion rates at high latitudes are caused by low predicted vegetation cover in the winter months when the region receives most of its rainfall. The low NDVI values that cause the low cover estimates are due to the presence of partial cloud cover in the AVHRR pixels throughout the winter months. This limitation of NDVI means that the model provides spuriously high erosion rates anywhere north of about 52° latitude, and thus cannot be used in these regions.

The model results are useful for quantifying the erosion problem, and provide a baseline from which to assess the effect of climate change on erosion. In addition they could also be used to manage erosion by identifying areas that are erosion prone and then implementing control measures. A limitation of this management strategy is that regions that suffer high erosion rates may already be severely degraded. A more sustainable strategy would be to determine areas where erosion is accelerating and implement control measures before it reaches a dangerous level. This requires continuous monitoring of erosion.

15.5 Regional Scale Soil Erosion Modelling and Monitoring: The Walia Catchment, Mali

To estimate erosion continuously we have implemented the soil erosion model in the

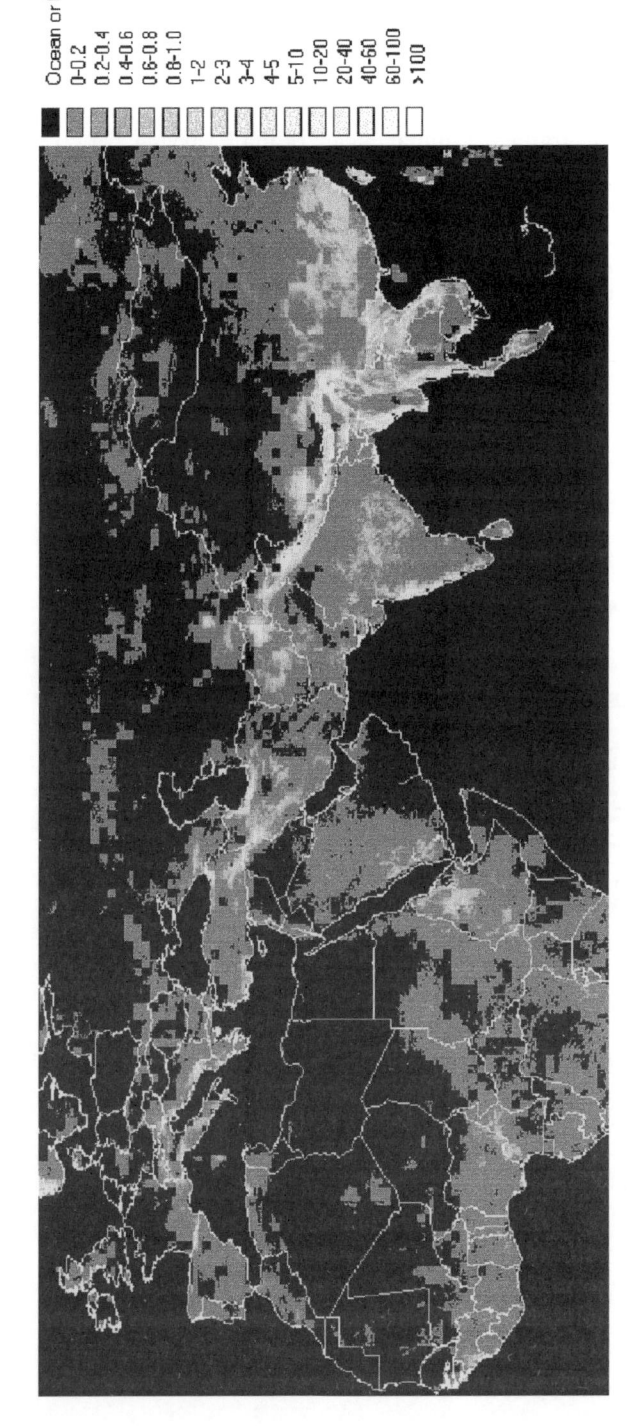

Figure 15.3 *Soil erosion modelled at the global scale (mm/year)*

252 *Advances in Remote Sensing and GIS Analysis*

Walia catchment, Mali, at the 4 km scale using daily data that are largely derived from remote sensing. In this section we have tested the feasibility of implementing an erosion monitoring system during the 1987 wet season. This was a rather dry year but one where all necessary data were available.

The location of the catchment is shown in Figure 15.4. It is characterized by a strong south–north rainfall gradient (annual mean rainfall varying between 700 mm and 400 mm). The rainy season, associated with the passage of the Inter Tropical Convergence Zone, lasts from June to September and the catchment has a marked intra- and inter-annual variation in vegetation cover and spatial extent as is typical of the Sahelian zone. The surface area is 85 000 km^2 and there are a total of 19 raingauges within and around the catchment.

15.5.1 Estimating Overland Flow and Vegetation Cover

Previous work (Grimes *et al.*, 1993, 1995) demonstrated the potential of using rainfall

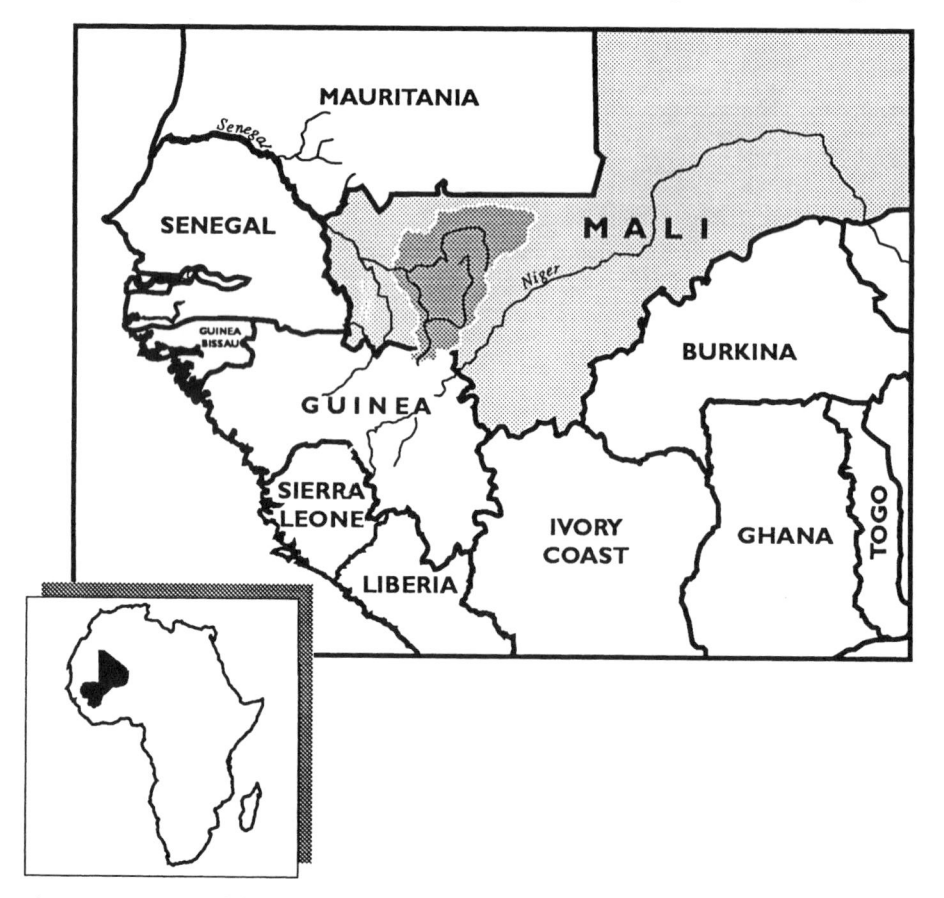

Figure 15.4 Map of the Walia catchment. The catchment is shown as the darker shaded area in southern Mali

estimates, derived from the Thermal Infra Red imagery of Meteosat, for daily overland flow forecasts in large catchments such as the Nile and Senegal, and this methodology is adopted here. Overland flow was estimated using the Pitman (1976) model, a simple runoff model designed for implementation is semi-arid environments. Because the model has been adapted to explicitly deal with evapotranspiration and soil moisture it may better simulate the large variations in soil moisture storage that occur during the transition between wet and dry seasons in this environment. The way a model partitions water during this period where moisture deficits are high and vegetation cover is low may significantly affect modelled erosion rates.

Daily mean catchment rainfall was estimated from Meteosat TIR images using a cloud-top threshold temperature T_t which distinguishes between rainy and non-rainy pixels as determined via a comparison with raingauge data. Cold Cloud Duration (CCD) images are then prepared, showing the number of hours with brightness temperature below T_t for each pixel. The CCD value is assumed to be linearly related to rainfall amount, P where

$$P = a_0 + a_1 CCD \qquad (8)$$

The coefficients a_0 and a_1 are determined by regressions of pixel CCD values against contemporaneous raingauge data for non-zero CCD pixels only. Mean catchment rainfall is then calculated by applying Equation 8 to the area mean CCD for the part of the catchment with CCD > 0. A separate calibration for each month is estimated using that month's raingauge data (Table 15.4).

The Walia catchment flow is modelled as being dominated by overland flow with contributions from soil water in excess of the soil storage capacity. The Pitman overland flow model (Pitman, 1976) is a simple bucket model that assumes a symmetrical triangular statistical distribution function for the infiltration rate within the catchment. The model allows for interception and has two subsurface water stores corresponding to soil moisture and groundwater. The maximum and minimum infiltration rates (Z_{MAX} and Z_{MIN}) are adjusted for the soil moisture deficit SWD that was assumed to be 0 at the start of the monitoring period

$$SWD = S_m / S_{max} \qquad (9)$$

where S_m is the soil moisture and S_{max} the maximum soil moisture storage.

Overland flow (OF) is calculated as a function of Z_{MAX} and Z_{MIN} where:

$$C = 4/2^{(2\ SWD)} \qquad (10)$$

$$Z_3 = C\ Z_{MAX} \qquad (11)$$

Table 15.4 Calibration parameters and correlation coefficients for the Walia catchment

	Jul. 87	Jul. 88	Aug. 87	Aug. 88	Sep. 87	Sep. 88
T_t (°C)	−40	−40	−40	−40	−40	−40
a_0 (mm)	−0.1	1.4	1.9	4.6	−0.2	3.6
a_1 (mm/h)	0.9	1.3	0.9	0.8	1.0	0.6
P	0.8	0.9	0.7	0.8	0.8	0.4

$$Z_1 = C\, Z_{MIN} \tag{12}$$
$$Z_2 = (Z_3 + Z_1)/2 \tag{13}$$

OF is 0 for $P \leq Z_1$, however, for $Z_1 \leq P \leq Z_2$:

$$OF = \frac{2(P - Z_1)^3}{3(Z_3 - Z_1)^2} \tag{14}$$

for $Z_2 \leq P \leq Z_3$:

$$OF = P - Z_2 + \frac{2(Z_3 - P)^3}{3(Z_3 - Z_1)^2} \tag{15}$$

and for $P \geq Z_3$:

$$OF = P - Z_2 \tag{16}$$

An evaporation algorithm has been embedded in this model to update the soil moisture store on a daily basis. Each pixel is assumed to be divided into a vegetated and a non-vegetated fraction, the vegetation cover being calculated from ARTEMIS decadal maximum NDVI composites. Evaporation from the soil and the vegetated fraction is treated differently. Soil evaporation E_S is given by

$$E_S = (1 - v)\, \beta_S\, E_{SPot} \qquad E_{SPot} = (2/3)\text{PET} \tag{17}$$
$$\beta_S = 1 \qquad\qquad\qquad \text{if Rainfall} > \text{PET} \tag{18}$$
$$\beta_S = 1 - e^{(-2S/S_{max})} \qquad \text{if Rainfall} < \text{PET} \tag{19}$$

where PET is potential evapotranspiration derived from maps of long-term averages provided by the FAO and the β factors account for effects of soil water depletion on evapotranspiration (Serafini and Sud, 1987). Vegetation transpiration is given by

$$E_v = v\, \beta_v\, E_{VPot} \qquad E_{VPot} = \text{PET-}I \tag{20}$$
$$\beta_V = 1 \qquad\qquad\qquad \text{if Rainfall} > \text{PET} \tag{21}$$
$$\beta_V = 1 - e^{(-6.8S/S_{max})} \qquad \text{if Rainfall} < \text{PET} \tag{22}$$

where I is interception and is assumed to vary linearly with v.

The actual and modelled hydrograph at the catchment outlet is shown in Figure 15.5. There is reasonable agreement between the estimated and actual overland flow. Periods of over- and underestimation of river flow are noticeable but overall the results are very encouraging particularly considering that the flow represents only 1% of the seasonal rainfall. These results suggest that operational modelling of overland flow for a semi-arid catchment in real time is feasible from satellite and climatological data. If there was access to satellite receivers the results could be obtained within an hour or so. If the model implementation relied on the NOAA Satellite Active Archive (SAA) anyone with Internet access could obtain results the same day; however, at the time of writing a search for the relevant Mali data suggests that only 4 km spatial resolution AVHRR data are available.

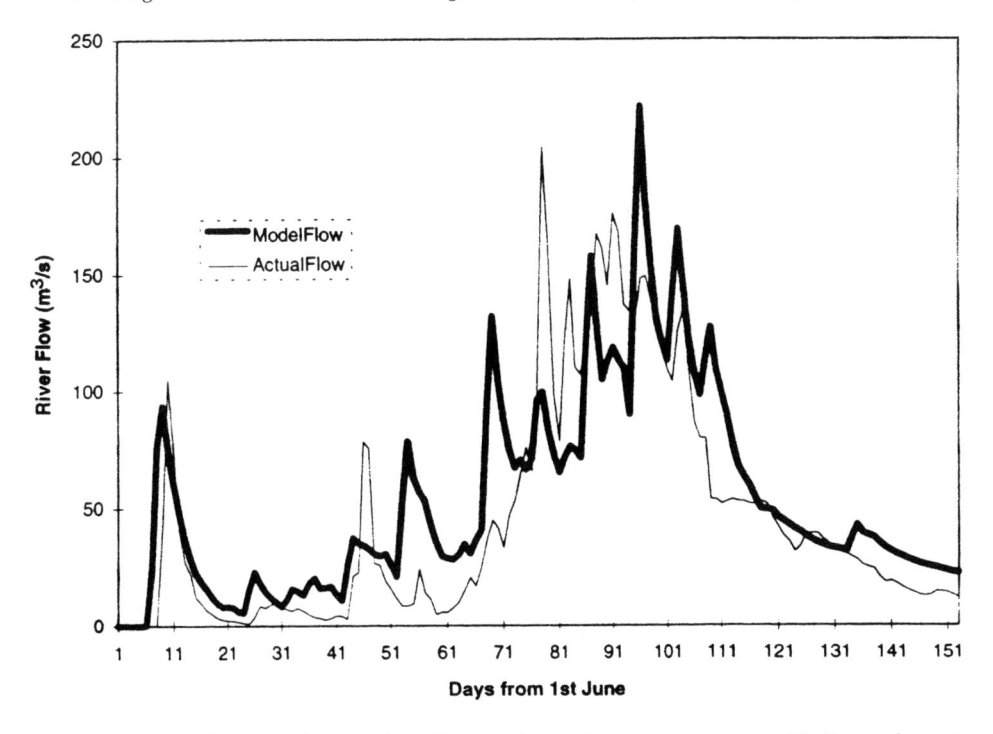

Figure 15.5 Modelled and actual river flow for the 1987 wet season in the Walia catchment using the Pitman model

15.5.2 Estimation of k and s

These parameters in the soil erosion model were estimated in a similar manner to that of the global modelling. Slope was estimated from the EROS data centre 1 km DEM in a similar manner to that of the global study.

Coefficient k was estimated from the FAO soil map of the world. At this scale the data are very generalized but provided the only readily available information. The six soil units present in the catchment (Luvic Arenosol, Ferric Luviosol, Eutric Regosol, Petroferic Lithosol, Eutric Cambisol, Eutric Nitosol) were digitized and k calculated from the texture of the published type sections in the same manner as the global study by assuming a 2% organic matter content. Though organic matter data were provided in the type sections they were not used as they often appeared to be high for semi-arid soils (e.g., the Eutric Cambisol is 8.5%) because the type sections were sometimes derived from more humid regions. No texture information was provided for the Petroferic Lithosol probably because it is lightly weathered bedrock regolith and thus most likely consists largely of stones. Because of this it was assigned a k of 0.

15.5.3 Regional Modelling and Monitoring Results

Analysis of these results is at a preliminary stage. As most erosion occurs only after

Table 15.5 Regional scale soil erosion model parameters and results

Date	Mean rainfall (mm/day)	Mean runoff (mm/day)	Mean vegetation cover (%)	Maximum erosion (mm/day)
7 Jun. 87	18.2	0.11	19.5	0.00016
8 Jun. 87	18.8	0.15	19.5	0.00030
25 Jun. 87	12.7	0.04	30.2	0.00001
13 Jul. 87	14.7	0.07	39.2	0.00002
23 Jul. 87	17.2	0.16	40.5	0.00008
5 Aug. 87	11.7	0.04	42.9	0.00001
7 Aug. 87	18.7	0.26	42.9	0.00019
16 Aug. 87	15.4	0.10	43.1	0.00011
29 Aug. 87	13.5	0.12	46.2	0.00010
3 Sep. 87	18.3	0.40	52.7	0.00069
10 Sep. 87	15.9	0.21	52.7	0.00000
12 Oct. 87	9.9	0.03	35.6	0.00042

significant amounts of rainfall and runoff, to date only the 12 most intense rainfall events have been modelled (Table 15.5). The highest erosion rate occurred on 3 September 1987. This was the day with the highest amount of runoff (a mean of 0.4 mm/day with a maximum of 1.51 mm/day) and one of the most intense rainfall events (18.3 mm/day). The maximum erosion rate is 0.00068 mm/day but the majority of areas that are subjected to erosion experienced between 0.0004 and 0.0005 mm/day. Some areas that experience high runoff produce no erosion because they have been assigned a k of 0, thus the average for the whole catchment is very low. Summing the 12 erosion events produces erosion rates of between 0.00009 and 0.0015 mm/year; if we assume that we have modelled all erosive rainfall events this compares favourably with the global model results of 0.0001 to 0.0003 mm/year when one considers that the global scale data need to be averaged to that of the regional scale.

The model shows a marked variability in the ability of similar sized rainfall events to produce erosion. Rainfall events of similar magnitude to the one on 3 September 1987 at the start of the wet season (e.g., 18.8 mm/day on 8 June 1987) produce less overland flow and less erosion even though there is a lower vegetation cover because more water is consumed in infiltration to replenish the soil moisture store. On the other hand moderately high rainfall events at the height of the wet season (e.g., 15.9 mm/day on 10 September 1987) produce no erosion even though they generate notable overland flow because at this time the vegetation cover is at its maximum. Thus the low erosion rates in Walia may be due to three factors of the Sahelian environment: (a) the rapid response of vegetation to rainfall in the Sahel (Tucker et al., 1991); (b) the gradual build up of rainfall intensity as the wet season progresses and (c) the lack of intense rainfall events in what is a relatively dry year. To determine whether the low rainfall is an important factor we are currently running the model for the 1988 wet season when there was more rainfall than usual.

15.6 Discussion and Conclusions

The low erosion rates in Walia may be due to the nature of the environment; however, we are probably underestimating erosion rates at the global scale also. These low values may be due to three factors: (a) problems with the model, (b) our implementation of the model (e.g., the model scale), and (c) errors in the calculation of the model parameters. These potential problems are discussed in turn below.

Though simple, the model has been found to be effective in previous implementations (Thornes, 1989; Wainwright, 1994). When calculating model parameters simplifications and assumptions were made and the quandary may lie here. In order to isolate which parameters in the model are most likely to be a problem a sensitivity analysis of the model for the average parameter values of the Walia catchment was conducted and the following order of sensitivity was found $k < s < OF < v$. Coefficient k is the least sensitive parameter and though we calculated it in a simplistic manner the values we obtained do not seem low (apart from the areas of 0 k in the regional study). Thus k cannot explain low erosion rates in the global study or the low maximum rates in the regional study. Slope is a relatively insensitive parameter but does suffer problems related to scale discussed below.

Overland flow is a relatively sensitive variable, and certainly the one it is hardest to estimate. The good fit to the Senegal River hydrograph suggests that we are not underestimating it at the regional scale. Though the model is designed to run on average daily data, by using average daily rainfall, and not considering a rainfall intensity distribution we may be masking the short periods of intense rainfall that generate brief periods of Hortonian overland flow. This potential problem is currently being investigated. At the global scale there has been a limited validation of these data by comparing runoff coefficients and there is reason to believe that overland flow is underestimated by the SCS model and this may to a certain extent explain the low erosion rates at the global scale.

Vegetation cover is the most sensitive parameter especially at low values. Though there are numerous problems associated with using NDVI to estimate v, the most important one is that NDVI is a better predictor of green vegetation cover than total cover and will thus tend to underestimate v. This would produce high erosion rates rather than low ones.

The problem of low predicted erosion may lie with applying a model developed and tested at the local scale. Thus the scaling of the model parameters needs to be considered. Little work has been conducted on the scaling of overland flow; however, there is no reason to believe that there will be differences at local, regional and global scales. Coefficient k is probably overestimated at the global scale as the soil map does not consider regions where there is no soil (e.g., rock outcrops) that will have a low k. Slope is clearly scale dependent. Using slope averaged over large areas we are underestimating it and this could explain the low erosion rates. Vegetation cover is also highly scale dependent. When one considers a 5 km^2 area of 50% vegetation cover, we have no information on the distribution of vegetation within the pixel and assume that it is evenly distributed throughout the area. However, there are a large number of spatial distributions vegetation can adopt, other than a uniform one, that can give the same average cover within a pixel but may cause different erosion rates. Vegetation could be distributed as a gradient across the pixel, and in the most extreme case the gradient could be from 0 to 100%. The cover could, on the other hand, be distributed as discrete patches such as fields where half the pixel could have say 100% cover and the other half could be bare. Finally, a combination of two or more of the three possibilities could be occurring.

This problem of vegetation cover scaling has been considered for global climate models (Koster and Suarez, 1992); however, it does not appear to have been considered for soil erosion models at the scales discussed here presumably because few people have attempted to implement erosion models at this scale. The question that needs to be answered to quantify it is 'how do these various possible spatial distributions of sub-pixel cover affect erosion?'. To answer this question the soil erosion model was implemented using synthetic data where all parameters were kept constant ($k = 0.2$, $OF = 25$ mm/day, $s = 0.15$ m/m, $v = 50\%$). The only thing that was varied was the spatial distribution of the vegetation and its spatial resolution. To accomplish this synthetic images of vegetation covering a 5120 m² area were created at a spatial resolution of 10 m. In the first image the vegetation was distributed as a north–south gradient from 0 to 100%, in the second the vegetation was distributed evenly throughout the region, and in the third it was distributed as square patches of 100 m² with an equal number of patches in the pattern of a chess-board, with either 100% or 0% v. These images were then decreased in resolution by sequentially averaging groups of four adjacent pixels providing images with a spatial resolution of 10, 40, 160, 640, 2560 and 5120 m² that span the change between local and regional scales.

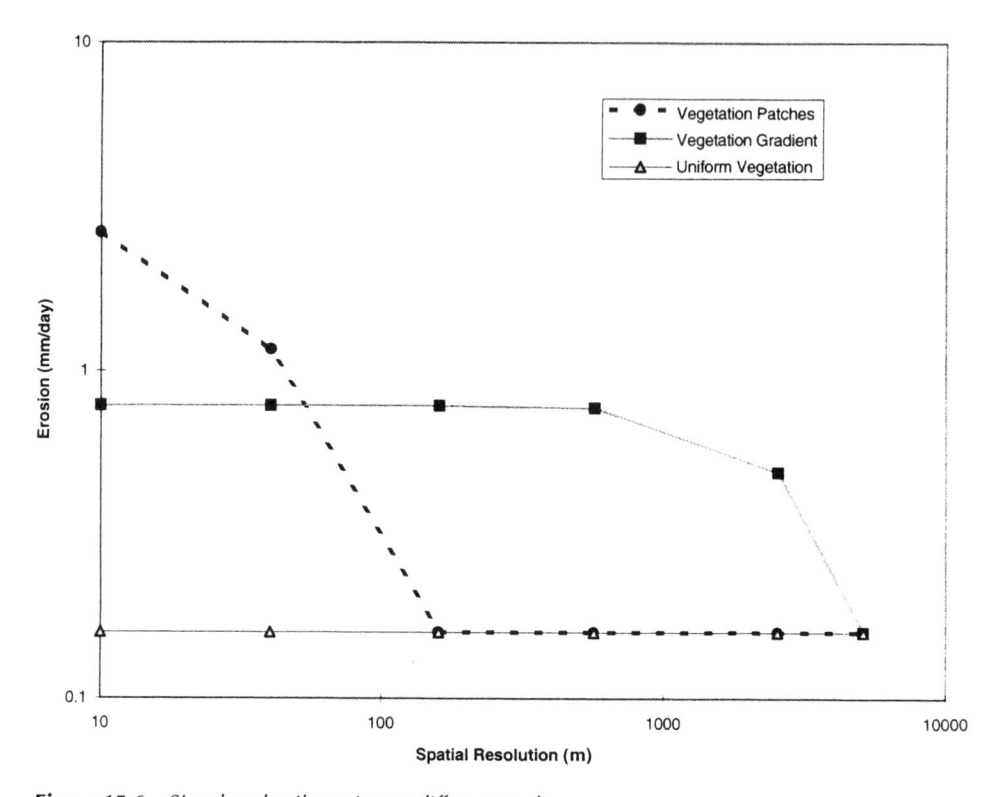

Figure 15.6 *Simulated soil erosion at different scales*

The resultant erosion is shown in Figure 15.6. It is clear that the amount of erosion varies considerably between the different spatial distributions and scales. By assuming that vegetation is uniformly distributed within the pixel we always get the minimum possible erosion rate (in this case 0.16 mm/day), and these estimates do not vary with spatial resolution. When the vegetation cover is arranged as evenly distributed 100 m² patches, erosion is much higher at smaller spatial resolutions (2.63 mm/day at 10 m) and rapidly falls off as the averaging process causes the vegetation cover to spill out of the field boundaries damping down erosion in bare areas. At a spatial resolution of 160 m all pixels are equal mixtures (50%) and erosion has decreased to 0.16 mm/day. When the spatial distribution of the vegetation is a gradient, erosion changes little with decreasing spatial resolution until the erosion nears about a quarter the size of the gradient. This occurs because like is being averaged with like. Following this there is a rapid tailing off of erosion to the level of evenly distributed vegetation cover.

It is evident from this simulation that at regional and global scales erosion predictions for areas of discontinuous vegetation cover can be reduced by an order of magnitude from those at the local scales simply by averaging out the small-scale variations in vegetation cover. The scaling of vegetation cover can explain, at least in part, the underestimation of erosion at regional and global scales. This underestimation will occur in agricultural areas where vegetation tends to be distributed as patches or in areas where there are small-scale gradients in vegetation cover. Though this scaling phenomenon is a limitation of global and regional scale modelling, it may not preclude implementation of sustainable management strategies as we are more interested in the long-term trends than their magnitude.

It would be useful, however, to overcome this scaling problem and to this end it will be necessary to obtain an estimate of the sub-pixel vegetation distribution. It is possible that this could be done by using sensors that provide a measure of the BRDF as this should vary with sub-pixel vegetation structure. An alternative for the regional scale modelling is to periodically acquire a local scale satellite image to gain an insight into the sub-pixel vegetation distribution; however, the Walia catchment would require three TM images to do this so it is an expensive option.

One problem in identifying temporal trends in erosion is that it is episodic and thus long-term monitoring may be needed. If the monitoring is carried out on a daily basis large amounts of data will have to be processed to run the model. However, as the majority of erosion only occurs in a few intense rainfall events it is questionable whether all rainfall events need to be modelled for erosion and overland flow. CCD can be used to identify intense rainfall events; however, to model overland flow for short periods during the wet season we also needed an estimate of initial moisture conditions. It is possible that this could be supplied by remote sensing using the method outlined by Gillies and Carlson (1995).

Due to the fact that there are limitations to each scale at which soil erosion models can be applied using remote sensing and GIS an operational monitoring system could use all scales. The global scale modelling could be used to identify areas prone to erosion. These could be monitored using regional scale modelling and areas that are identified as having accelerating erosion could be subjected to local scale modelling using images acquired just before large erosion events identified by the regional scale modelling to gain a more accurate estimate of erosion. Global scale modelling could also be used to determine whether the increased erosion is due to anthropogenic activity (e.g., agricultural intensification) or natural causes (e.g., an intense rainfall event at a time of year when vegetation cover is low).

References

Blackburn, G.A. and Milton, E.J., 1995, Seasonal variations in the spectral reflectance of deciduous tree canopies, *International Journal of Remote Sensing*, **16**, 709–721.
Danson, F.M., 1989, Factors affecting the remotely-sensed response of coniferous forest canopies, PhD thesis, University of Sheffield.
de Jong, S.M., 1994, Applications of reflective remote sensing for land degradation studies in Mediterranean environment, *Netherlands Geographical Studies*, **177**, 1–237.
Elvidge, C.D., 1990, Visible and infrared reflectance characteristics of dry plant materials, *International Journal of Remote Sensing*, **12**, 1775–1795.
Gillies, R.R. and Carlson, T.N., 1995, Thermal remote-sensing of surface soil-water content with partial vegetation cover for incorporation into climate-models, *Journal of Applied Meteorology*, **34**, 745–756.
Graetz, R.D., Pech, R.P., Gentle, M.R. and O'Callaghan, J.F., 1986, The application of Landsat image data to rangeland assessment and monitoring: the development and demonstration of a land image-based resource information system (LIBRIS), *Journal of Arid Environments*, **10**, 53–80.
Grimes, D.I.F., Milford, J.R. and Dugdale, G., 1993, The use of satellite rainfall estimates in hydrological modelling, *Proceedings of the First International Conference of the African Meteorological Society*, Nairobi, Kenya, February 1993, 206–274.
Grimes, D., Bonifacio, R., Dugdale, G. and Diop, M., 1995, Flow forecasting of a semi-arid catchment with NOAA/AVHRR and METEOSAT data, *Proceedings of the 1995 Meteorological Satellite Data Users Conference*, Winchester, UK, 4–8 September 1995, 551–557.
Holm, A.R., Curry, P.J. and Wallace, J.F., 1984, Observer differences in transect counts, cover estimates and plant size measurements on range monitoring sites in an arid shrubland, *Australian Rangeland Journal*, **6**, 98–102.
Johnson, P.E., Smith, M.O. and Adams, J.B., 1992, Simple algorithms for remote determination of mineral abundances and particle sizes from reflectance spectra, *Journal of Geophysical Research*, **97**, 2649–2657.
Jurgens, C. and Flander, M., 1993, Soil-erosion assessment and simulation by means of SGEOS and ancillary data, *International Journal of Remote Sensing*, **14**, 2847–2855.
Kennedy, P.J., 1989, Monitoring the phenology of Tunisian grazing lands, *International Journal of Remote Sensing*, **10**, 835–845.
Kineman, J.J. and Ohrenschall, M.A., 1992, *Global Ecosystems Database, Version 1.0 (on CD-ROM)* (Boulder, CO: NOAA).
Koster, R.D. and Suarez, M.J., 1992, Modelling the land surface boundary in climate models as a composite of independent vegetation stands, *Journal of Geophysical Research*, **97**, 2697–2715.
Langbein, W.B. and Schumm, S.A., 1958, Yield of sediment in relation to mean annual precipitation, *Transactions American Geophysical Union*, **39**, 1076–1084.
Malthus, T.J., Andrieu, B., Danson, M.F., Jaggard, K.W. and Steven, M.D., 1993, Candidate high spectral resolution infrared indices for crop cover, *Remote Sensing of Environment*, **46**, 204–212.
Middleton, N. and Thomas D., 1997, *World Atlas of Desertification*, (Chichester: Wiley).
Olsen, G.W., 1981, *Soils and the Environment. A Guide to Soil Surveys and Their Application* (London: Chapman & Hall).
Pearce, E.A. and Smith, C.G., 1993, *World Weather Guide* (Oxford: Helicon).
Pickup, G., Chewings, V.H. and Nelson, D.J., 1993, Estimating changes in vegetation cover over

time in arid rangelands using Landsat MSS data, *International Journal of Remote Sensing*, **43**, 243–263.

Pitman, W.V., 1976, A mathematical model for generating river flows from meteorological data in South Africa, Report Number 276, Hydrological Research Unit, Dept of Civil Engineering, University of Witwatersrand.

Price, K.P., 1993, Detection of soil erosion within Pinyon-Juniper woodlands using Thematic Mapper (TM) Data, *Remote Sensing of Environment*, **45**, 233–248.

Rawls, W.J., Ahuja, L.R., Brakensiek, D.L. and Shirmohammadi, A., 1993, Infiltration and soil water movement, in D.R. Maidment (ed.), *Handbook of Hydrology* (New York: McGraw-Hill), 5.1–5.51.

Serafini, Y.V. and Sud, Y.C., 1987, The time scale of the soil hydrology using a simple water budget model, *Journal of Climatology*, **7**, 585–591.

Settle, J.J. and Drake, N.A., 1993, Linear mixing and the estimation of ground cover proportions, *International Journal of Remote Sensing*, **14**, 1159–1177.

Soil Conservation Service, 1972, *SCS National Engineering Handbook Section 4*, Hydrology, (USDA).

Thornes, J.B., 1985, The ecology of erosion, *Geography*, **70**, 222–234.

Thornes, J.B., 1989, Erosional equilibria under grazing, in J. Bintliff, D. Davidson and E. Grant (eds), *Conceptual Issues in Environmental Archaeology* (Edinburgh: University Press), 193–210.

Tucker, M.E., Dregne, H.E. and Newcomb, W.W., 1991, Expansion and contraction of the Sahara Desert from 1980 to 1990, *Science*, **253**, 299–301.

Wainwright, J., 1994, Anthropogenic factors in the degradation of semi-arid regions: a prehistoric case study in southern France, in A.C. Millington and K. Pye (eds), *Environmental Change in Drylands* (Chichester: Wiley), 285–304.

Walling, D.E., 1988, Erosion and sediment yield research, some recent perspectives, *Journal of Hydrology*, **100**, 113–141.

Walling, D.E. and Webb, B.W., 1983, Patterns of sediment yield, in K.J. Gregory, (ed.), *Background to Palaeohydrology* (Chichester: Wiley), 69–100.

Wischmeir, W.H., 1976, The use and misuse of the USLE, *Journal of Soil and Water Conservation*, **31**, 5–9.

Zhou, X. and Folving, S., 1994, Application of spectral mixture modelling to the regional assessment of land degradation: a case-study from Basilicata, Italy, *Land Degradation and Rehabilitation*, **5**, 215–222.

16
Extracting Information from Remotely Sensed and GIS Data

Peter M. Atkinson and Nicholas J. Tate

16.1 Introduction

The objective of this final chapter is to provide a framework within which remote sensing and GIS analysis can be viewed. This framework includes concepts such as data and information, reality and model, raster and vector data structures (also referred to as data models) and field and object-based models. The framework, thus, encompasses some concepts crucial to remote sensing and GIS analysis and, further, it touches upon some important philosophical questions. The chapter could have been presented at the start, rather than the end of the book. Here, the chapter is used as an opportunity to synthesize the preceding chapters which cover some disparate themes.

16.2 Models of Data and Information

The one theme which is common to all the chapters of this book is a focus on spatial data: data representing explicitly some property varying in N-dimensional Euclidean space. It is for spatial data that we wish to construct a conceptual framework. To provide a starting point we need to distinguish clearly between reality and data (Figure 16.1). Data are a representation, abstraction or model of reality (Gatrell, 1991; Burrough and McDonnell, 1998). Data are not reality itself. This statement, while appearing satisfactory, does come with a proviso. Since we might consider measurement made directly with our senses, such as sight, to produce data in our brains then we must concede that our thoughts are not reality either. There is a very good reason for

---

Advances in Remote Sensing and GIS Analysis. Edited by Peter M. Atkinson and Nicholas J. Tate.
© 1999 John Wiley & Sons Ltd.

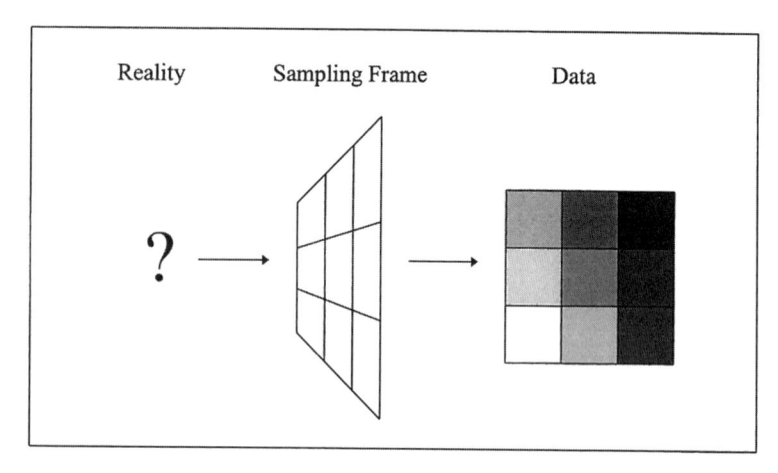

Figure 16.1 Data modelled as a function of the underlying reality and the sampling framework

saying this which relates to the sampling framework (i.e., the sample size, sampling scheme and sampling density). The sampling framework provides a neat separator between reality and data since all data are a function of the underlying reality (whatever that might be) and the sampling framework (Woodcock and Strahler, 1987; Jupp et al., 1988, 1989; Atkinson, 1993) (Figure 16.1). In this sense, data obtained directly by our senses are no different to data obtained by remote sensing. A second important consequence of the distinction between reality and data is that operations performed on data, such as the data analysis tools of remote sensing and GIS, relate to those data and not to the reality they represent.

An important choice often made by GIS users is at a conceptual level between object and field models, and, at a more practical level, between models of data in the form of raster and vector data structures. Furthermore, it may sometimes be necessary to convert from one data structure to another, as in the chapters in the final section of this book. The dichotomy provided by these representational models, in particular, raises some important philosophical questions relating back to the distinction between reality and data and, more specifically, to the way that our brains work. Most people would agree that the vector-based GIS provides maps which are more appealing visually than the equivalent maps represented in the raster-based GIS (see, for example, Aplin et al., Chapter 14, this volume). It is possible to understand this preference by thinking through the functioning of the human visual–sensory system.

The human eye provides at the retina an image which is observed by an irregular spatial array of rods and cones. The analogy in the remote sensing case is that an image is provided in optical wavelengths of the electromagnetic spectrum, which is observed by a regular spatial array of sensors. Of course, in remote sensing an image is constructed line-by-line as the sensor platform passes over the surface of the Earth, but that is not important. Both scenarios provide data in a form akin to the raster data structure. The important point is that much primary data acquisition, whether by remote sensing or the human eye, involves the raster representation, which is as close to reality as we can get.

Data in the raster data structure can be conceptualized in terms of the field model. We can define a two-dimensional field as any single valued function of location in two-dimensional space (Mark, 1999). In this model the observed values of the variable of interest (data) can be treated as realizations of a spatial stochastic process (generating model), and this model is useful for making inferences about the property which the data represent.

Handling data in the raster data structure can be expensive in terms of computer time (Mather, 1987). For example, it is common for a remotely sensed image (single waveband) to contain 1000 by 1000 pixels, that is, 1 000 000 discrete values. It would be surprising, therefore, if the human brain were to store and analyse data in this way. Rather, the human brain works by classifying the data it receives into (usually functional) objects such as cars, buses, roads, houses, and so on. Treating reality in this object-based way reduces the complexity of the real world enormously making the analysis and storage of data much more efficient (Burrough and McDonnell, 1998). However, the important point is that these functional objects do not exist as such in the real world: they are mental constructs and relate only to human perception and conception. Although it is possible to encode raster images as a series of objects placed either in a mosaic or on a background, this is not common in remote sensing. Perhaps the reason for not adopting the object-based view is that in remote sensing the focus of attention is usually the natural world, not the human world of objects. In this context, description in terms of fields is more appropriate.

The vector data structure lends itself to the description of entities which we term 'objects'. A notable exception to this is provided by census data where 'objects' are chosen rather arbitrarily as census wards, enumeration districts and so on. For census data such 'objects' have little physical meaning. There exist many techniques for the analysis of objects both in the raster and vector data models (e.g., Serra, 1982) and many of these comprise much of the functionality of modern GIS. In the vector data model, series of points taken together form features such as houses (nodes), railway lines, rivers (arcs) and agricultural parcels, forest stands and gardens (polygons). Here the features are represented by recording their boundaries and it is clear that to observe, one must have *a priori* knowledge of the features being recorded. Thus, the vector data structure, most often coupled with the object-based model, represents a different kind of measurement to that of the raster data structure (most often coupled with the field model) and is one which is further from reality (the natural world) and closer to the object-based processing of our human brains. It is not surprising then that the raster data structure is used in remote sensing and the vector data structure is used commonly for management purposes, for example, of public utilities. Further, it should be clear that the vector data structure is appealing because generally it relates information on objects. We will explore this theme further and define the terms data and information below.

A complication arises in that both the raster and vector data structures provide insufficient representation of geographical boundaries. As Mark (1999) notes, the raster GIS does not represent object boundaries at all, whereas the vector GIS is restricted to the repesentation of crisp boundaries only. Clearly many geographical boundaries are not as distinct as the vector representation would suggest. The often transitory boundary between different soil types is a familiar example. The development of fuzzy set theory and fuzzy membership functions allows the representation of objects with indistinct boundaries, which can be utilized in remote sensing and GIS in both a conceptual context and a practical classification context as evidenced in the chapters in this book by Mather (Chapter 2) and Foody (Chapter 3).

16.3 Data and Information

The terms data and information are often used interchangeably, and often incorrectly. Here simple definitions of the two terms are given to provide a distinction between them and to provide a clear framework within which to understand spatial data handling in general.

We have already encountered data described as a model of reality (Figure 16.1). Of particular interest here are spatial data where the locations as well as the attributes (values) of the data are important. If data are in the form of actual numbers (for digital data) or implied numbers (for analogue data), then what is information? The answer is that information exists in the differences, or more generally the relations, between data. This is an important distinction and one which requires substantiation.

Consider a single value, say 0.53. According to the above definition this single value contains strictly zero information and we can all agree that it means little on its own. If we were to tell you that the value represents the percentage of the population of the UK who drink water you would recognize the value as too small relative to an *a priori* framework. The information would exist in the relation between the datum and the *a priori* knowledge. If on the other hand we provided a second value of 0.67 you would recognize the first value as small *relative* to the second.

In spatial data analysis one of the richest sources of information is that which can be extracted between image and ground variables (for remote sensing) or between data layers (for GIS) (Davis *et al.*, 1991). In remote sensing it is common to train and validate statistical estimators such as regression, and classifiers using ground data (measurements of the property of interest obtained at the ground at the locations of a few representative pixels) (Schowengerdt, 1997). In the case of regression, the ground variable is regressed on the image pixels to obtain a model of the relations between the remotely sensed image and the property of interest. It is this model which encapsulates the information required to predict the property of interest everywhere from the remotely sensed image. Chapters 2 and 3 of this book involve a similar approach based on a fuzzy model. Similarly, much of the power of GIS lies in the ability to analyse and manipulate multiple data layers through, for example, the overlay operation.

For multiple variables we often develop explicit models of the relations between them. However, for a single variable distributed spatially the information is contained implicitly in the relations between the values. The reason that we do not need to develop explicit models of the information in spatial data is that this is exactly what the human brain is good at: our brains do the work automatically for us, observing and perceiving relations in spatial data usually within an object-based framework. Of course, it may be desirable to model spatial relations explicitly, for example to quantify such relations for use in subsequent analysis, and this is the aim of the majority of the chapters of this book.

It may be instructive in relation to the preceding discussion to consider how information is conveyed in an image in the raster data model. Clearly, for a reasonably sized image there are many more possible relations between data than there are data. To determine how much information exists in a single image it is necessary to limit the search to the relations between immediate neighbours, which for a raster image, are the pixels immediately above, below, to the right and left (Atkinson, 1995). This effectively avoids double counting. Given such a quantitative definition of information it is possible to determine the amount of redundancy in an image; basically variation which is not informative. Such redundancy arises because pixels close together tend to be more alike than pixels further apart, a concept known as spatial dependence which, as in this book, is commonly represented with the variogram. Thus, an image with no spatial dependence would contain much information, whereas an image with much spatial dependence would contain less information. Paradoxically, the latter case is often regarded as more informative and this is for two reasons. First, an absence of spatial dependence is rare and usually implies noise or error which is generally of little or no interest. Second, humans, as explained above, usually interpret the world as being comprised of objects which necessarily lead to spatial dependence in the raster data model. It is partly because the vector data model is so informative in terms of the objects represented, with no unwanted variation, that it is so appealing to the human brain.

16.4 Conclusions

We have looked briefly at the distinctions between reality and data, between the raster and vector data structures and between the field and object-based models. A further distinction was drawn between data and information and the importance of this distinction was stressed. Some of the assertions made will no doubt be questionable in that they may not fit everyone's perception, or in some cases may conflict with an already accepted model in a different discipline. However, it is important that such a framework is developed and applied, for example to help formulate research goals.

From the discussion of data and information, it is clear that much of the interesting research in spatial data analysis lies in exploring, visualizing and modelling the *relations* between data, including the relations implicit in spatial data. Mapping as a technical objective is also important for many organizations producing environmental and socio-economic data and increasingly sophisticated means for producing such maps form a large part of what we do. However, one cannot help feeling that the real breakthroughs in understanding will come from the modelling of multivariate and spatial relations.

Acknowledgement

Much of the thinking for this chapter was undertaken while the lead author was a post-doctoral researcher at the University of Bristol funded by the Leverhulme Trust.

References

Atkinson, P.M., 1993, The effect of spatial resolution on the experimental variogram of airborne MSS imagery, *International Journal of Remote Sensing*, **14**, 1005–1011.

Atkinson, P.M., 1995, A method for describing quantitatively the information, redundancy and error in digital spatial data, in P. Fisher (ed.), *Innovations in GIS III* (London: Taylor & Francis), 85–96.

Burrough, P.A. and McDonnell, R.A., 1998, *Principles of Geographical Information Systems. Spatial Information Systems and Geostatistics* (New York: Oxford University Press).

Davis, F.W., Quottrochi, D.A., Ridd, M.K., Lam, N.S-M., Walsh, S.J., Michaelson, J.C., Franklin, D.A., Stow, D.A., Johannsen, C.J. and Johnston, C.A., 1991, Environmental analysis using integrated GIS and remotely sensed data: some research needs and priorities, *Photogrammetric Engineering and Remote Sensing*, **57**, 689–697.

Gatrell, A.C., 1991, Concepts of space and geographical data, in D.J. Maguire, M.F. Goodchild and D. Rhind (eds), *Geographical Information Systems: Principles and Applications*, Vol. 1 (London: Longman), 119–134.

Jupp, D.L.B., Strahler, A.H. and Woodcock, C.E., 1988, Autocorrelation and regularization in digital images I. Basic theory, *IEEE Transactions on Geoscience and Remote Sensing*, **26**, 463–473.

Jupp, D.L.B., Strahler, A.H. and Woodcock, C.E., 1989, Autocorrelation and regularization in digital images II. Simple image models, *IEEE Transactions on Geoscience and Remote Sensing*, **27**, 247–258.

Mark, D.M., 1999, Spatial representation: a cognitive view, in P. Longley, M.F. Goodchild, D.J. Maguire and D. Rhind (eds), *Geographical Information Systems: Principles, Techniques, Management and Applications* (New York: Wiley), 81–89.

Mather, P.M., 1987, *Computer Processing of Remotely Sensed Images* (Chichester: Wiley).

Openshaw, S., 1977, The modifiable areal unit problem, *Concepts and Techniques in Modern Geography, Catmog 38* (Norwich: Geo-Abstracts).

Schowengerdt, R.A., 1997, *Techniques for Image Processing and Classification in Remote Sensing* (New York: Academic Press).

Serra, J., 1982, *Image Analysis and Mathematical Morphology* (London: Academic Press).

Woodcock, C.E. and Strahler, A.H., 1987, The factor of scale in remote sensing, *Remote Sensing of Environment*, **21**, 311–322.

Index

1991 Census 196
accuracy (*also see* accuracy assessment) 9, 17
accuracy assessment (*also see* accuracy) 30, 157
Advanced Spaceborne Thermal Emission and Reflectance Radiometer (ASTER) 148
air flow 98
air temperature, minimum 97–114
airborne imagery 222
Airborne Visible/Infrared Imaging Spectrometer (AVIRIS) 148–57
Arc-Info GIS 99
artificial neural network (ANN) 9–11, 18, 32, 188
  accuracy 23–27
  activation function 21, 43
  architecture 10, 20, 26
  back propagation 28, 43
  binary diamond 24
  data input and scaling 25
  data layers 21, 25–6
  epochs 43
  equations 22
  error 23
  error rate 22
  error surface 22
  feed forward 18–19, 43–6
  generalization 20–6
  Hopfield 18, 41
  Kohonen 18
  learning parameters 25
  learning rate 22
  momentum term 22, 43
  multi-layer perceptron (MLP) 18
  net input 21
  network error 22
  number of training iterations 25
  processing 22
  processing units 18, 24–5
  radial basis function 24
  rule of thumb 20
  sigmoid activation function 21–3
  training 23–6, 42
  weights 25
  weights 22
atmospheric correction 148, 226
attribute graph 176
autocorrelation (*also see* spatial correlation) 101, 110–14

Bartholomew's data 99, 110
Bernoulli distribution 122
beta distribution 121–4
bidirectional reflectance distribution function (BRDF) 243
bimodal distribution 129
British National Grid (BNG) 225
British National Space Centre (BNSC) 219

Census of Population 189
Central Postcode Directory 189
Chambers Twentieth Century Dictionary 8
choropleth map 189
class
  discrimination 10
  identification 10
  memberships 7, 9

classification 8, 17
  accuracy 10–12, 20–4, 30–1, 43, 190, 211, 226–33
  appropriateness 25
  based on variogram 138
  Bayesian 135, 144
  completely crisp 30–1, 45–7
  co-occurrence matrix 135, 172
  fully fuzzy 31–2
  fuzziness 18, 30–1
    continuum of 30
  fuzzy 10–11, 28–32
  hard 7–15, 27–31
  maximum likelihood 9–11, 19–28, 135, 197, 211, 226
  memberships 26–32
  minimum distance to mean 135
  multispectral image 168
  of clouds 42
  of land use 219
  of soil 8
  per-parcel 209, 220–33
  per-pixel 173, 208, 220
  region-based 173
  soft 7–12, 30–2
  spectral/textural 138
  subjectivity of 25
  supervised 18–19, 26, 42
  testing 29–32, 140–2
  textural 135–44, 188
  training 29–32, 140–2
  training set 20–6
  unsupervised 18
  non parametric 19, 42
  parametric 19, 42
classified imagery, urban 190
cloud
  cores 48
  motion 39–43
  objects 41–5
  pixels 40
  shadow 232
  structures 48
  systems 40
  tracers 40
  vectors 40–3
clouds 39–44, 49–50
CNSD 82
coherent structures 47
cokriging (also see kriging) 77–91, 149, 157–61
  block 81
cold cloud duration (CCD) 253
Compact Airborne Spectrographic Imager (CASI) 224–6

confusion matrix (also see contingency matrix) 140–3, 230–1
conservation treatment 249
contingency matrix (also see confusion matrix) 10
coregionalization 91, 158
correlation coefficient 80, 85–90
correlogram 149
cover fraction 120–30
cross-correlation 84, 149–50
cross-correlogram 150
cross-semivariance 84
cross-validation 84
cross-variogram (also see variogram) 84, 90, 137–9, 43

dangling line 235
dasymetric mapping 198
data, nature of 263–7
data conversion
  raster-to-vector 208, 215
  vector-to-raster 208
data cube 157
data quality 186
data transfer 208
DEM, slope of 247
density profile 185, 203
Department of the Environment (DoE) 190, 202
digital elevation model (DEM) 77–83, 97, 242
dimensionality of data 20
dispersion variance 124–9
distributed model 242
DMSP Special Sensor Microwave Imager (SSM/I) 76

edge detection 42, 214
elevation 77–85, 108
enumeration district (ED) 189, 192, 196–9
Environment Agency 75
EROS Data Center 247, 254
error
  matrix 10
  propagation 17
ERS-1 synthetic aperture radar (SAR) 76
ERS-2 synthetic aperture radar (SAR) 76
evapotranspiration 254
extrapolation 19

factor analysis 14
field of view (FOV) 120–5

fitness for purpose   186
Forestry Commission Stock Map   209
fractal analysis   186–201
fractal geometry   187–201
Freeman chain-codes   173
frequency distribution   9
frontal system   78
fuzzy
  connectedness   48
  membership probabilities   11–12, 28, 44–5
  rules   45
  segmentation   45
  threshold   47
fuzzy *c*-means   28
fuzzy logic   24

generalization   9, 189
genetic programming   24
geology   147
  alunite   150
  buddingtonite   150
  calcite   159
  dolomite   159
  illite   154
  kaolinite   150
geometric correction   18
geophysical field   119
geostatistics   77–91, 100, 122
GERIS   159–63
Global Ecosystems Database (GED)   245
graph matching   179
graph searching   180
graph theory   176–9
Greenwich Mean Time (GMT)   97
ground control point (GCP)   226
ground data   30–2

histogram, co-occurrence   49
Hughes phenomenon   19
hydrograph   254
hydrologic condition   249
hyperspectral data   20

IGIS   208
IKONOS-1   224
image processing   9, 48–50
imaging spectrometry   147–8
indicator function   121–4
information, nature of   263–7
integrated raster/vector processing   207
interpolation   78–80

inverse distance weighting (IDW)   77–91

joint distribution   161
joint probability   154

knowledge-based techniques   181
kriging   77–91, 113, 149
  block   81
  block indicator   153
  equations   86
  indicator   150–4
  ordinary   103, 157
  simple   164
kriging variance, block indicator   153
kurtosis   129

Lagrange multiplier   83
land cover   24–31, 167–9, 226, 249
Land Cover Map (LCM) of Great Britain   220
land use   167–9, 226–32
  non-residential   197
  residential   197
  urban   188
Land-Line data   225
Landsat multispectral scanner system (MSS)   180
Landsat Thematic Mapper (TM)   13, 76, 119–26, 141, 187, 190, 213, 220–2, 243
  generalized   101
  ordinary   102, 115
  weighted   102
least squares fit   80
LEWIS   148
locally adaptive   46

machine vision   48
maximum cross-correlation   39
maximum NDVI composites   254
Medium Resolution Imaging Spectrometer (MERIS)   148
Meteorological Office   42, 75–7, 97
meteorological station   75
meteorology   39
  forecasting   39
Meteosat imagery   43–6
microwave sensor   76
MIDAS   196
missing boundary flag   234
mixed pixels   11, 27–32
mixture model   8, 13

mixture model (*cont.*)
  end member spectra 13
  end members 12
modifiable areal unit 187
multisource data analysis 18
multispectral data 19, 137

National Land Use Stock System (NLUSS) 219, 225
neurons 10
NOAA Advanced Very High Resolution Radiometer (AVHRR) 76, 119
non-stationarity 91
normal distribution 20
normalized difference vegetation index (NDVI) 119, 161, 243

object-based database 208
object-based model 263, 265
optimal estimation 82
Orbview-3 224
ordination 8
Ordnance Survey (OS) 180, 189, 219, 225
overland flow 247–57

photogrammetry 242
Pitman model 253
pixel 7–13, 27
Pocket Oxford Dictionary 8
polygon, shrunken 210
population surface 192, 196
possible error flag 235
postal geography 196
principal components analysis 14
  eigenvalues 15
  eigenvectors 15
probability mass function 122
proportional cover 120

quadtree 11

radar 209
  backscatter coefficient 210
  noise 210
  polarization 209
  speckle 209
radiometry 226
random field 263–5
random function 82, 101
random variable 82, 100–1
raster data 222, 265

region growing 42
Region Search Map 173
regionalized variable 82, 120–4
regression 80–91, 99–115
  residuals 101, 113–14
regularization 121–30
root mean square (RMS) error 47, 77–90, 103–8, 228, 233
RS-GIS 186, 207
RS-GIS integration 207
rule-based systems 181

sampling, systematic 14
sampling framework 264
SAR 209–13
  HV polarization 211
  P polarization 211
scale 12–15, 40, 242
  spatial 40
  temporal 12, 40
scale dependence 185–7
second derivative 50
second-order information 168
segmentation 41–6, 214
semantic network 178
semantic relations 169
semivariance (*also see* variogram) 82
SEMMED erosion model 243
sensitivity analysis 257
signal-to-noise ratio 50
simulation, sequential 161–3
snakes 42
snow
  area extent 75
  clearance 97
  cover 75–88, 126–30
  depth 76, 77, 78, 80
  dynamics 80
  fall 78
  pack 77–80
  water equivalent 75
Snow Survey of Great Britain 75–8
soil erodibility 247
soil erosion 241
  global 241
  global scale model 245, 250
  mapping 243
  model 242–44
  model validation 250
  monitoring system 252
soil moisture 256
soil texture 248
soil type 248
spatial analysis 185

Index   273

spatial correlation (*also see* autocorrelation)   99, 112–13
  of residuals   100
spatial form of classified imagery   189
spatial resolution   27, 76, 120–32
  fine   219–20
spatial structure   168–71
spatial topology   173
spectra   147–50
Spectral Angle Mapper   13
spectral matching   149
spectral unmixing   157, 187
spectrometer   148–9
SPOT High Resolution Visible (HRV)   10, 180–90, 222
SPOT panchromatic   215
standard deviation filter   214–15
stationarity   91
stochastic simulation
structural graphs   18
structural inference
sub-pixel   28, 157
support   76–7, 80–
syntactic pattern r(
  encoding   172
  geometric repre
  structural infor

taxonomy   8
texture analysis
thematic maps
Thermal Infra
TIGER files
topography
trend   103
t-test   150

UK Censu
UKBORI
uncertain
Universal Soil L...   ,E)   243
urban concentric zones   19.
urban fractal model   201

urban modelling   191
urban morphology   185–92
urban net density   191
urbanity   186
US National Laboratory   42

varimax   14
variogram (*also see* cross-variogram)   85–90, 99, 122, 135
  absolute   137
  average   123
  bounded model   82
  directional   136
  experimental   102, 108
  exponential model   102, 112, 120, 127
  model   101–15, 123
  nested model   123–7
  nugget variance   101–15, 128–30, 157
  omnidirectional   137
  punctual   125–8
  range   87, 101–8, 123, 157
  robust   103
  sill   101–15, 123
  structured component   83
  unbounded model   82
vector data   222, 265
vector polygon   209
vector polygon extraction   213
vegetation cover   259
vegetation index   119

waveband   140
weather
  anticyclonic   105–15
  microclimate   110
  models   39
  prediction   40
  World Weather Guide   245
weather logs   103
white noise   128

XRAG   177–8